D0843417

Henry F. Diaz   Roger S. Pulwarty   (Eds.)

Hurricanes

Springer
Berlin
Heidelberg
New York
Barcelona
Budapest
Hong Kong
London
Milan
Paris
Santa Clara
Singapore
Tokyo

Henry F. Diaz   Roger S. Pulwarty   (Eds.)

Hurricanes

Climate and Socioeconomic Impacts

With 77 Figures and 30 Tables

Dr. Henry F. Diaz
NOAA/ERL/CDC
325 Broadway
Boulder, CO 80303, USA

Dr. Roger S. Pulwarty
Cooperative Institute for Research in Environmental Sciences
Campus Box 449
University of Colorado
Boulder, CO 80309, USA

ISBN 3-540-62078-8 Springer-Verlag Berlin Heidelberg New York

Library of Congress Cataloging-in-Publication Data
Hurricanes : climate and socioeconomic impacts / [edited by Henry F. Diaz, Roger S. Pulwarty].
Includes bibliographical refereces and index.
ISBN 3-540-62078-8
1. Hurricanes—Social aspects—United States. 2. Hurricanes—Social aspects—Caribbean Area. 3. Hurricanes—Economic aspects—United States. 4. Hurricanes—Economic aspects—Caribbean Area. 5. Climatic changes—Atlantic Ocean Region. I. Diaz, Henry F. II. Pulwarty, Roger S., 1960- .
QC945.H89 1997
363.34'922'091821—dc21

Cover design: Erich Kirchner, Heidelberg
Typesetting: Camera ready by editor
SPIN: 10526008 32/3136 - 5 4 3 2 1 0 – Printed on acid-free paper

Acknowledgments

The editors wish to convey our appreciation to all of the contributing authors for their dedication and effort to this book project. We also thank the reviewers for generously giving of their time. We hope they agree, as we do, that their efforts have been rewarded.

We would like to give special thanks to Craig Anderson whose help was critical in enabling the manuscripts to be transformed into camera-ready form. We also thank Ms. Kathleen Salzberg for her assistance in the editorial process.

The book is an outgrowth of a conference held at the U.S. National Hurricane Center, located, at the time, in Coral Gables, Florida. Funding assistance has been provided by the National Oceanic and Atmospheric Administration (NOAA), and we wish to thank Robert Stockman for his help in enabling this support.

The picture of Hurricane Gilbert on the front cover, is from a NOAA satellite photo taken on 13 September, 1988, at the time of its greatest intensity. The digital data was obtained courtesy of the Louisiana State University, Earth Scan Laboratory in Baton Rouge, Louisiana.

Foreword

Hurricanes have historically caused much more damage to property and loss of life in the United State than have earthquakes. In spite of this past record, the earthquake "community" in the United States has done a much better job than has the hurricane "community" in assessing long-term risks and creating programs to address those risks. However, both of these "communities" still need considerable improvements in these areas of risk assessment and in the development and implementation of effective mitigation programs. Without realistic assessments of risk and the development of cost-effective mitigation strategies, the economic health of the United States and many other countries could be at great risk.

There have been many projections of climate changes and their potential impacts on the number, strengths, and locations of future tropical cyclones (hurricanes) made by reputable scientists as well as by those who have little scientific basis for such projections. Some of the predictions of cataclysmic changes in the not distant future have attracted considerable attention. At the same time, counterarguments have been presented which indicate little change for the foreseeable future. Such a divergence in scientific opinions has resulted in a similar divergence of opinions by powerful political leaders. The result has been a lack of any comprehensive program to deal with the long-term hurricane "threat" to the nation.

This book reports on a workshop titled *Atlantic Hurricane Variability on Decadal Time Scales: Nature, Causes and Socioeconomic Impacts* which was held at the National Hurricane Center on February 9-10, 1995. The potential climatic change issues were discussed along with socioeconomic issues, including such things as private insurance losses and availability for the future. This workshop brought experts with differing views on the range and "predictability" of potential climatic changes and their impacts on tropical cyclone activity together with those who say that the simple natural variability of hurricane activity is greater than that likely from any foreseeable climatic changes. This book provides an excellent summary of that workshop, and in my opinion, for one of the first times, puts the hurricane "risk" in its proper perspective. That is, regardless of whether or not there is some long-term climatic change that is taking place that will affect hurricane frequencies, strengths and "strike" locations, a simple return to frequencies and "strike" locations of hurricanes of the decades of the 1940s and 1950s will cause massive economic losses that will have large impacts on the economy of the nation. All of these studies show the urgent need for a comprehensive program for hurricanes, similar to the National Earthquake Hazards Reduction Program (NEHRP) for earthquakes. Such a program needs a strong research component to improve 1- to 3-day forecasts and associated warnings as well as improvements in seasonal and decadal predictions. A major component of this proposed "National Hurricane Hazards Reduction Program" must also emulate the NEHRP with strong programs for proper land use and building construction in high hurricane hazard areas.

The multidisciplinary material presented in this book should prove interesting and informative to all who have an interest in hurricanes and their impacts. I believe that the reader will find that this material makes a strong case for a need for action now to deal with the hurricane problem. Hopefully, this workshop and the associated book can become a catalyst for addressing hurricane-related problems that have been ignored far too long.

Robert C. Sheets
Lake Placid, Florida

Preface

This book is an outgrowth of a workshop titled *Atlantic Hurricane Variability on Decadal Time Scales: Nature, Causes and Socioeconomic Impacts* which was held at the National Hurricane Center in Coral Gables, Florida, on February 9-10, 1995.[1] Participants included acknowledged experts in the fields of meteorology and climate, and researchers from the hurricane-related hazards community. Several disciplines were represented, among them ecology, engineering, geography, meteorology, political science, and sociology. They included hurricane forecasters, university researchers, and representatives from federal, state, and local agencies, and the insurance industry.

Two major questions and two major objectives are addressed in this volume. The key questions are: 1) can we expect increased hurricane activity, in the coming decades, to levels more typical of earlier decades? and 2) do regional and national level responses to the hurricane threat support the assumptions concerning event frequency and population density that have become the basis for strategies of preparedness and preventative planning? The key objectives are (1) to assess some of the social, cultural, and economic currents, in light of previous hurricane landfall patterns and future landfall scenarios that constrain or enable the acceptance of stricter mitigation guidelines, and (2) to consider whether planners are utilizing the available scientific information to optimal benefit.

The book is divided into four sections. Section A deals with the principal climatic controls that determine the basic characteristics of tropical storms. It covers such topics as the fundamental mechanisms that control the genesis, geographical domains, and ultimate decay of these cyclone systems. Variability of hurricane frequency and landfall patterns on decadal time scales, including possible variations due to anthropogenic warming, are also considered. Section B addresses issues of the impact of hurricanes and their consequences in the context of climatic variability. Section C highlights the policy implications of long-term changes in hurricane landfall frequency in light of recent demographic and economic changes, including the role of development and mitigation strategies. Lastly, in Section D, the impacts on the Property Casualty Insurance industry of major events in the United States in the past few years are reviewed, and some of the options that are available for minimizing such losses in the future are identified. This is done by linking the nature of variability and forecasts to "principles of insurability."

In Section A (*Climatological Perspectives*), the focus is on documenting what we know about the large-scale climatic controls modulating Atlantic hurricane activity, while considering what may happen in the future given both natural and anthropogenic changes in climate. The chapters by Diaz and Pulwarty, and by Gray, Sheaffer, and Landsea reflect on long-term changes in observed large-scale

---

[1]The National Hurricane Center has since moved to a new address, and is now located on the Florida International University campus at 11691 S.W. 17th St., Miami, FL 33165-2149.

atmospheric patterns that modulate the risk of hurricane damage along the U.S. eastern seaboard. For instance, from the mid-1940s until the late-1960s a variety of atmospheric circulation characteristics were distinctly different from patterns prevalent since the 1970s. This section also summarizes previous work on the character of the teleconnections - large-scale interactions between widely separated regions of the globe - between tropical Pacific Ocean sea surface temperature, West African monsoon rainfall, North Atlantic Ocean circulation anomalies, etc. The association between El Niño event years and the character of tropical cyclone activity in the Atlantic provides a physical basis for long-range prediction of seasonal tropical storm activity. The question of future changes in Atlantic hurricane frequency and intensity resulting from anthropogenic forcing of the global climate system is considered by K. Emanuel, who delineates the fundamental conditions influencing the maximum strength of these great storms. Bengtsson, Botzet, and Esch evaluate the possible influences of global warming, due to a doubling of carbon dioxide, on hurricane frequency and intensity using a general circulation model of the coupled ocean-atmosphere system.

In Section B (*Impacts of Hurricane Variability*), Rappaport and Fernández-Partagás summarize the loss-of-life statistics associated with Atlantic tropical cyclones since the discovery of the Americas. The number of fatalities are stratified by region, the distribution by decade and century, and total losses are estimated. Doyle and Girod model the fate of mangrove forests as influenced by hurricanes over the past century. They discuss the future of the tropical mangrove systems in South Florida under a global-warming scenario of enhanced frequency of severe hurricane landfall in that region. A recurrence interval of major storms of around 30 years over the last century is the major factor controlling mangrove ecosystem dynamics in South Florida. Model results of climate-change scenarios indicate that future mangrove forests are likely to be diminished in stature with a higher proportion of red mangrove. Finally, Rodríguez documents a wide range of hurricane impacts from storms that have affected Puerto Rico in the recorded history of the island. Special emphasis is placed on contemporary developments, such as increasing coastal and national populations, increases in urbanized areas, and in the numbers of older people, which could cause major problems for disaster response planners the next time a major hurricane threatens or strikes the island.

Section C (*Vulnerability and Policy Issues)* opens with a study by Pielke and Pielke, who use trend data on population and property at risk for 168 counties from Texas to Maine, to show that present hurricane research has had limited influence on policies and decisions in the broader community. They argue that the question of use and value of scientific information on hurricane variability is a difficult analytical question involving the integrated assessment of the process of decision making, including national agendas for research.

The next chapter is by Pulwarty and Riebsame who argue that dampening feedbacks or equilibrium between social development and hazards, usually assumed by scientists, may not actually exist and that development is actually an open-ended process that can lead to progressive hazardousness in the light of long-term variability of hurricane incidence. They argue that framing the problem under an "engineering" approach has short-circuited more comprehensive approaches and usually ignores important trends in development that point, not only to increasing vulnerability, but also to decreased resilience in particular communities. A framework for the definition and assessment of vulnerability and risk, based on previous studies, is offered.

Watson and Vermeiren, through the Caribbean Disaster Mitigation Project of the Organization of American States and U.S. Agency for International Development, have developed a Geographic Information System or GIS-based model which incorporates the variability of storm generation and movement, changes due to development (damage to reefs, etc.) and natural changes (land subsidence, etc.) into a model for hurricane hazard assessment. Their study addresses the use of this model for planning coastal projects and insurance considerations.

Felts and Smith report on a study of the perceptions of policy makers and decision makers regarding the potential threat posed by future changes in climate. It is noteworthy that a substantial number of decision makers are beginning to form opinions (one way or the other) about the impact that global climate change is likely to have on them, personally and professionally, and that for many, this information derives from popular sources of information, such as newspapers and magazine articles.

Finally, a chapter by Jamieson and Drury describes the role of the U.S. Federal Emergency Management Agency in orchestrating a national effort to reduce the potential of both human and property losses from hurricanes (and from other natural hazards as well) along the eastern seaboard of the United States.

The last two chapters (Section D) consider the following questions. Do improved predictive capabilities actually warrant increases in insurance premium rates and would this be supported by regulatory agencies? The industry may be caught between state regulations requiring that they offer insurance at a rate that is affordable for the consumer and at the same time their need to charge enough to remain solvent after a disaster. The chapters by Roth and by Clark indicate that while short-term forecasts have limited applications in the insurance industry the possibility for some indication of changes on at least a 10-year timescale has very strong implications. Property damage estimates for future storms are extremely large for vulnerable metropolitan areas such as the Galveston-Houston area ($80 billion), New Orleans ($52 billion), southeast Florida ($104 billion), Virginia-Maryland ($68 billion), and New England ($104 billion). With these figures it is not difficult to envision a national economic catastrophe for the future. Successive $7 billion events, such as from Hurricane Hugo in 1988, would do severe damage to the property-casualty insurance industry in the U.S. and abroad. Communities experiencing a 20+ year interval between hurricane landfall events are considerably less inclined to modify or actively enforce building codes, zoning laws, and other controls, and are thus more susceptible to damage.

Central to a successful community response is the recognition of the risk factor involved and its sound integration in the planning perspective. As Gilbert White points out (see Chapter 1 by Diaz and Pulwarty) the goal of making wise use of hazardous areas now and for the indefinite future calls for an earnest examination of the information and policy affecting development decisions and the ways in which they will shape a sustainable society. We hope that this book will prove to be a substantial contribution to that effort.

H. Diaz
R. Pulwarty
Boulder, Colorado

Table of Contents

Section A

CLIMATOLOGICAL PERSPECTIVES

1 Decadal Climate Variability, Atlantic Hurricanes, and Societal Impacts: An Overview

Henry F. Diaz[1] and Roger S. Pulwarty[2]

[1]NOAA/ERL/CDC
325 Broadway, Boulder, CO 80303, U.S.A.

[2]Cooperative Institute for Research in Environmental Sciences
Campus Box 449
University of Colorado, Boulder, CO 80309, U.S.A.

Abstract

The level of societal risk to hurricane impact is a function of the frequency, strength, and duration of landfalling hurricanes, and of the degree of preparedness and types of mitigation strategies available to and employed by different segments of society. The goals of this book are twofold. First, we hope to bring together into one volume the state-of-the-art knowledge regarding hurricane variability on different time scales (from seasonal to decadal), the status of the current hurricane prediction capability (principally in the United States and Europe), and to consider some of the science-based implications of future climatic variability as it might influence the frequency and intensity of these great storms. A second objective is to explore a wide range of socioeconomic issues related to historical and recent impacts of hurricane activity in the Atlantic, to assess how these vulnerabilities may arise in different parts of the affected region (the Caribbean, Gulf of Mexico and U.S. East Coast), and to highlight the possible role of insurance in mitigation strategies. It is possible to increase societal vulnerability to a given event with and without changes in hurricane frequency. We hope to raise awareness of the potential impacts that global climate change may have on the climatology of tropical storm systems and to identify what those changes may mean in the context of socioeconomic trends and planning in the region.

1.1 Introduction

The destructive power of hurricanes has been amply demonstrated in recent years, as major storms have pounded Caribbean island nations and the United States mainland. The damage wrought by Hurricane Andrew, which struck South Florida on 24 August 1992, ranks as the most costly storm disaster in United States history, with long-term estimates of losses exceeding $30 billion. It is estimated that Hurricane Hugo, which struck the environs of Charleston, South Carolina, in 1988, resulted in a greater loss of timber than was due to the Yellowstone fires and the Mount St. Helens eruption combined. The exposure of human communities to natural phenomena has been increasing as a result of population growth and movements into areas of risk (Burton and Kates 1964; Riebsame et al. 1986; Glantz 1994; Pielke and Pielke, Chap. 8, this volume); the result has been an increase in economic losses in recent decades. It remains crucial to identify the relative contributions of natural, social, economic, and political factors to these losses. Glantz and Price (1994) point out that regional and national institutions that develop strategic plans for disaster preparedness often do not take into consideration the possible implications of climate change, or at least climate variability on greater than year-to-year timescales. A major focus of this book is to evaluate some of the changes in hurricane frequency that have occurred for about the past century, and to place these natural changes in the context of important socioeconomic trends in the United States and the Greater Caribbean Basin. In particular, we wish to highlight the changing level of risks related to the tremendous growth in coastal population throughout this region. We also consider some possible consequences to the countries in the region that may result from changes in global climate as it might relate to changes in Atlantic hurricane climatology, whether such changes arise from natural climate variability, or from anthropogenic forcing of the climate system.

In the following sections, we first consider in general terms some major aspects of the large-scale climate of the North Atlantic sector that lead to hurricane frequency changes of interannual to decadal time scales (we refer the reader to Chapter 2 by Gray, Sheaffer, and Landsea for a more in-depth analysis of large-scale climatic features that influence Atlantic hurricane variability). This is followed by a brief overview of some important aspects of the interaction between humans and natural hazards that are critical variables in the degree of risk exposure in the region to tropical storms. We conclude by emphasizing the need to take advantage of the benefits of recent scientific advances in hurricane forecasting, and to incorporate their use within the development planning process.

1.2 Hurricane Climatology

Quite apart from any modification of the climatology of Atlantic hurricanes from anthropogenic effects, there is abundant evidence that variations in hurricane activity have occurred on decadal (and longer) timescales. By "decadal timescale," we denote temporal variations lasting on the order of 10 to 50 years. This variability is important in its own right and can either exacerbate or ameliorate any potential anthropogenic effects in the future.

Several investigators have documented the likelihood of the existence of low-frequency changes in the occurrence, intensity, and principal tracks of Atlantic hurricanes during the past century (Shapiro 1982a, 1982b) and over a longer period (Walsh and Reading 1991). Chapter 2 by W. Gray and collaborators, and references therein, provides a comprehensive overview of these historically recorded changes in Atlantic hurricane activity in the post World War II period. Figures 1.1 and 1.2 illustrate the significant differences that have occurred in the frequency with which these storms have affected different segments of the Atlantic and Gulf of Mexico coasts of the United States. In the first two decades of this century, hurricane landfall occurred rather frequently along the Gulf of Mexico and the Caribbean Sea (Fig. 1.1). During the next two decades (Fig. 1.2), the pattern shifted to the Atlantic coast of the U.S., with hardly any storm system affecting the Gulf region of Texas and Louisiana (Hebert and Taylor 1979a, 1979b). The 1950-1980 period was one of relatively infrequent landfalls by intense hurricanes with the 1970s experiencing the lowest incidence of hurricane activity in the century (Fig. 1.3)

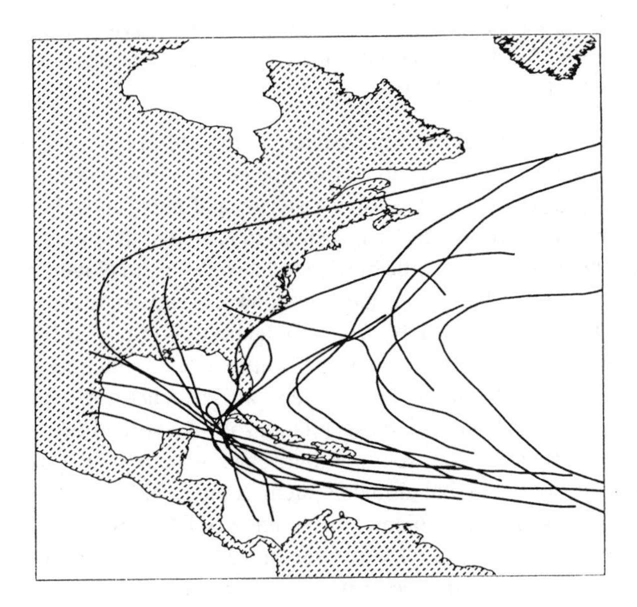

Fig. 1.1. Tracks of moderate (Category 3) and stronger hurricanes for the decade of the 1900s (1900-1910). The figure illustrates the more frequent occurrence of hurricanes in the Gulf of Mexico and the Carribean during some decades of the 20th century.

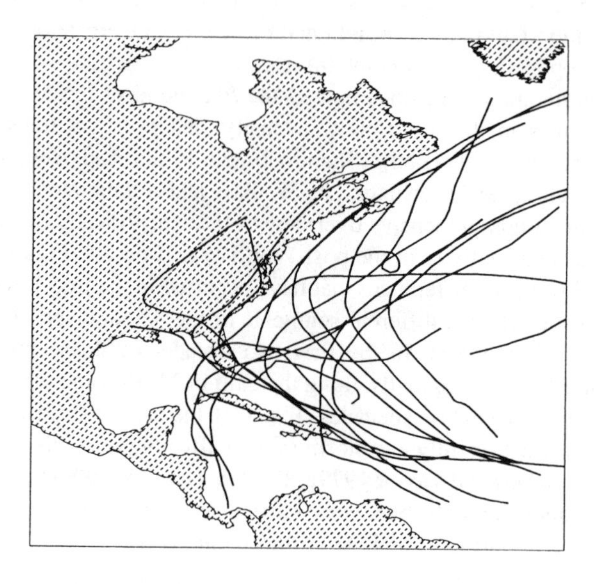

Fig. 1.2. Tracks of moderate (category 3) and stronger hurricanes for the decade of the 1920s (1920-1930). The figure illustrates the shift of hurricane tracks toward the mid-Atlantic coast, compared to 1900-1910.

Gray and Landsea (1992) and Landsea and Gray (1992) have documented some large-scale climatic relationships that affect the environment within which Atlantic tropical disturbances develop into topical storms and hurricanes. For instance, the intensity and numbers of tropical disturbances in Africa in the months immediately preceding the height of the Atlantic hurricane season (May to November), are found to modulate the risk of hurricane damage along the eastern seaboard. These and other studies (Hastenrath 1990a, b, 1995; Goldenberg and Shapiro 1993; Gray et al. 1992, 1993, 1994) have also documented the influence of a number of other large-scale features of the climate system which can influence the level of Atlantic hurricane activity, and some of these relationships have and are being used to forecast the character of the summer hurricane season in the Atlantic. Such important features of the global climate system as the El Niño/Southern Oscillation (ENSO) and the tropical quasi-biennial oscillation (QBO) have also been shown to have an association (e.g., Gray 1984; Gray and Sheaffer 1991) with the level of tropical storm and hurricane activity in the tropical Atlantic (including the Caribbean region and the Gulf of Mexico). Gray et al. (Chap. 2, this volume) consider some aspects of low-frequency (decadal scale) variations in tropical Atlantic hurricane activity in the context of the high natural variability of tropical cyclone activity quite apart from any changes that may result from human-mediated changes in global climate (see Maul 1993).

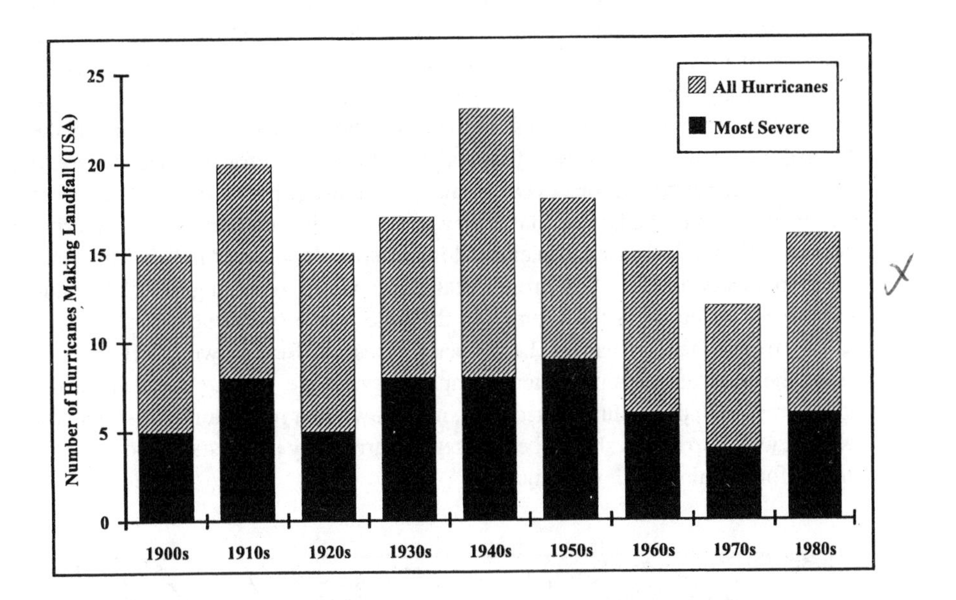

Fig. 1.3. Number of hurricanes making landfall in the conterminous United States. Lower shade bars denote landfall frequency for the most intense (category 3-5) storms, after Karl et al. (1995)

Figure 1.4 shows the sea level pressure (SLP) difference, for the months of May-October, of years with the highest tropical storm totals (totals exceeding one standard deviation from the record mean), minus the years with the lowest totals (totals below one standard deviation from the mean) in the North Atlantic region. It is observed that during years with high hurricane activity, SLP in the tropical Atlantic, and especially the Caribbean and Gulf of Mexico region, is below normal. The shaded box in the lower left area of Fig. 1.4 was chosen to construct and index of both the seasonal sea level pressure anomaly there, and the variation in the number of tropical storms. Figure 1.5 illustrates the changes in Atlantic hurricane frequency since 1930, in comparison with this SLP index, which, as depicted in Fig. 1.4, represents variations in the large-scale atmospheric circulation over the full North Atlantic Ocean. The graphs, one showing the year-to-year variations in SLP in the western Caribbean, and the other showing variations in the low frequency or decadal scale features, display a strong inverse relationship between the number of named storms and the seasonal character of the large-scale atmospheric pressure field.

The time frame of the observed multidecadal variability in hurricane characteristics such as frequency and intensity, and shifts in the tracks is comparable to the time intervals associated with economic development plans and management strategies in the wider Caribbean region, including the southern United States. Each year, on average, one or more hurricanes strike the shores of the Caribbean, Gulf coast, or eastern United States coastline. Clearly, an increase in the frequency and/or intensity of these storms could have tremendous impacts on planning horizons, debt restructuring, environmental management, tourism, and the hazard perception of countries in the affected regions. Estimates indicate that if the activity in the 1970s had been similar to the two decades prior, then losses for that decade would have risen to about $3 billion a year in the U.S. alone (Sheets 1994). The recent hurricane disasters in the United States may or may not portend that we are entering a renewed period of increased tropical cyclone activity. It is, nevertheless, important that the socioeconomic and environmental implications of changes in hurricane activity that may occur in the future - in the time frame of the next two or three decades - from whatever causes, be properly assessed. Below, we briefly review some of the most important hurricane impacts issues we feel are relevant to the topics addressed in this volume. Primarily, it is shown that reduction of disaster risk associated with hurricanes should be an integral part of any local, state, and regional strategies for sustainable development.

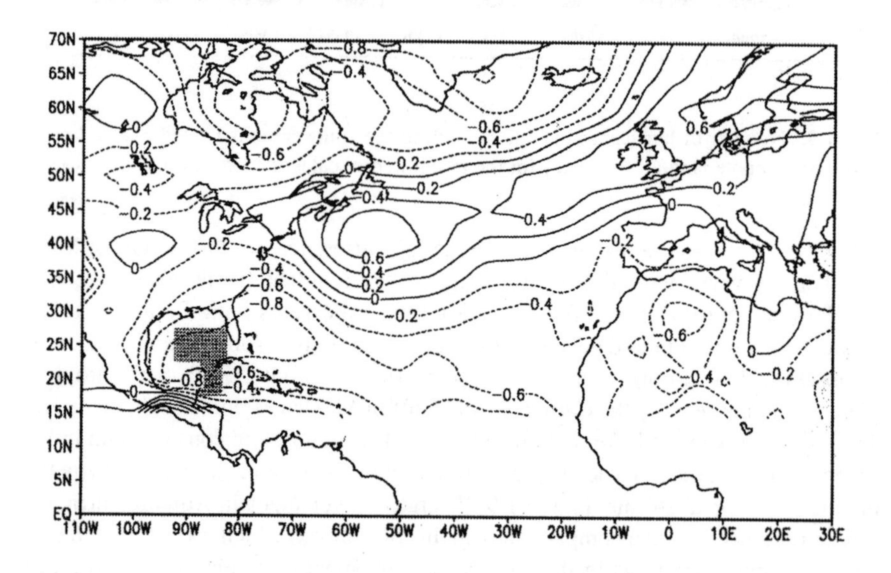

Fig. 1.4. Map showing May-October sea level pressure differences between years with high and low tropical storm activity, contour values in millibars. The shaded area in lower left was used to construct indices in Fig. 1.5.

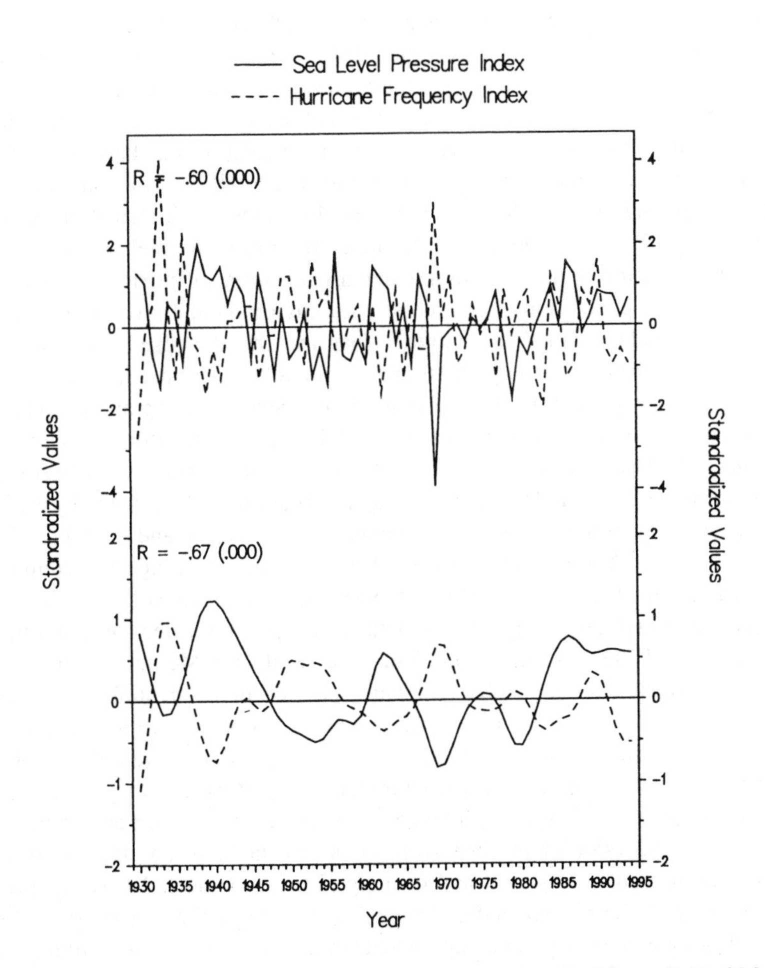

Fig. 1.5. Index of sea level pressure variations in the western Caribbean-Gulf of Mexico region in comparison with changes in the level of tropical storm activity for 1930-1994. Top graph shows the year-to-year variations, while the lower graph shows both series after filtering out the high frequency components. Both correlations (-0.60 and -0.67, respectively, for the actual and filtered series) are significant at the 1% level.

1.3 Societal Aspects

Because of the great potential for hurricane-caused destruction of property and loss of life, the science of hurricane forecasting has always been intimately connected with natural disaster preparedness, and hurricane forecasters have a long history of working closely with emergency management authorities usually on an event by event basis. Such activities have included working on simulated emergency evacuations based on storm surge simulations, conducting numerous workshops and training exercises, and ensuring efficient exchange of communication of potential risks and warnings. This close working relationship between hurricane forecasters and emergency management authorities at all levels is maintained on a routine basis and ensures that the best available scientific information related to hurricane threats is provided to the relevant authorities on a timely basis.

Nevertheless, the trend towards greater urbanization in hurricane-prone areas, growth of aging populations, the necessity for environmental protection, and other changes in landuse practices suggests that the threats posed by conceivable variations in hurricane activity should be continually reappraised. Assessments of vulnerability may provide useful tools for relating development planning to inherent socioeconomic disadvantages in the context of longer term natural variability and evolving risk (Mitchell 1990). We note, however, that methods for collecting data on social vulnerability and for merging that information with data on physical risks and vulnerability of the physical environment are at an early and evolving stage of development (see Mather and Sdasyuk 1991). Indeed, it was this shortcoming that provided the impetus for the meeting of experts on which this volume is based.

Population has dramatically increased during the past three decades in portions of the U.S. Gulf of Mexico and Atlantic coasts, as well as in Mexico and most of the Caribbean island nations, with attendant increases in residential and industrial vulnerability to hurricanes (Sheets 1985). There are about 44 million people living in the hurricane-prone coastal counties from Texas to Maine with continued rapid growth in the sunbelt states. Infrastructure has not kept pace with growth. In most coastal regions, it now takes much longer to evacuate a threatened area than it did a decade ago. The value of insured property at risk in these coastal counties (not counting flood insurance) is now approaching $2 trillion (Sheets 1994). Appropriate guidelines on development and redevelopment in high risk areas appear to be required along with the implementation and enforcement of stronger building codes in areas where they are applicable.

A major concern is that the plausible range of insured losses in several regions, and the likely impact of such losses on affected insurers and other financial institutions, has not been accurately estimated. In a short span of time, the cost of Hurricane Andrew exceeded the property insurance premiums paid over the 1970-1992 period. Increased premiums result in canceled or reduced coverage, further creating problems for financial investors. It is estimated that a major hurricane striking both Miami and Fort Lauderdale would result in claims of $60 billion, a significant percentage of the $160 billion that the insurance industry reserves for catastrophes (see Clark, chap. 14, this volume). Since Hurricane Andrew, 11 insurance companies have become insolvent and some 40 companies are either pulling out or greatly curtailing insuring property in Florida. Had Hurricane Andrew's track over South Florida been displaced only 20 miles to the north, two different studies show that losses would have exceeded $40 billion in southeast

Florida alone. A continuation of the same track would have put New Orleans under 18 to 20 feet of water. The meteorological conditions that would create the difference in the two courses of movement are essentially undetectable with our present observing systems (Sheets 1994).

Some studies of natural disasters have emphasized the significant vulnerability of poorer countries. Small nations are likely to suffer disasters of larger national scale than large countries. For instance, after a long decade of economic decline and stagnation, the Jamaican economy had started to revive with growth rates of 3.6 and 5.6% for 1986/87 and 1987/88, respectively (Barker and Miller 1990). The projected growth rate for 1988/89 was 4.5%. The actual growth rate turned out to be about 0.6% in the aftermath of Hurricane Gilbert. No sector of the economy escaped disruption, with agriculture, and in particular subsistence gardens, faring the worst (Barker and Miller 1990). Severe losses were reported in the tourist industries of Mexico and Jamaica as a result of damage to beaches, wildlife, and coral reefs. In addition to the first-order social and economic impacts, hurricanes can effect the balance of trade and foreign indebtedness for years after their occurrence (Bender 1989). Thus, the unanticipated occurrence of a sequence of weaker events, or of successive strong events may put small and locally dependent economies beyond short-term recovery and rehabilitation. Assessments of societal vulnerability and resilience inextricably link the impact of natural variations to trends in resource use and development, the dynamics of regional political economy, and the range of practical choices available (see Pulwarty and Riebsame, Chap. 9, this volume). Major structural developments are the luxuries of a developed society (e.g. fresh water storage, seawalls, strapped roofing, etc.), while in developing areas the direct effects of wind and rain may be have greater livelihood consequences. However, damages in developed areas have increased dramatically with increased coastal development even as storm-related mortality has decreased.

The means through which federal agencies, communities and industries at risk integrate the needed information into their disaster preparedness and development plans is not always uniform or compatible. Important in this assessment is the perception of the relative importance of hurricane risk and its acceptance by government agencies, the media, and the private sector. The desired response to hurricane watches and warnings is heavily dependent on how this information is carried through by the news media (Baker 1980; Sheets 1985).

1.4 Concluding Remarks

Four major issues and questions are addressed in this volume. The first looks at the question, Can we expect increased hurricane activity in the coming decades to levels that were more prevalent during earlier periods? A second issue, deals with regional and national level responses to the hurricane threat, and whether they support the assumptions concerning event frequency and population density that have become the basis for strategies of preparedness and preventative planning. A third objective was to discuss some of the social, cultural, and economic trends, in light of previous hurricane landfall patterns and future landfall scenarios, that would promote the acceptance of stricter development guidelines, and, finally, to consider whether planners are utilizing the available scientific information to optimal benefit.

The result of the meeting and the papers that followed, which make up this volume was to identify some important hurricane hazards issues that demand the attention of policymakers from the federal to the local level. First, it is important to understand that changes in the level of Atlantic hurricane activity spanning several years to decades have occurred in the observational record. Such shifts, likely associated with global changes in circulation patterns of the coupled ocean-atmosphere system, may result in the future in enhanced levels of hurricane-related hazards to coastal populations and the potential for increased property losses compared to similar periods in the past.

Improvements in scientific knowledge about hurricanes, technology, and communication advances has resulted in a large decline in the number of casualties resulting from these powerful storms. By the same token, population increases and the lure of the coast as places to live and play have placed ever greater numbers of people and property at risk from these storms. The changes in the potential level of hurricane damage are great enough within the context of natural variations in climate, let alone possible modification of global climate arising from increased concentrations of greenhouse gases in the atmosphere that could alter the climatological background in which these storms develop. Therefore, it appears that improved communication and dialogue about these issues at different levels of society are necessary if we are to reduce exposure to this hazard.

References

Baker, E.J. (ed.), 1980. Coping with hurricane evacuation difficulties. Hurricanes and coastal storms. Report No. 33, Florida Sea Grant College, Gainesville, 13-18.

Barker, D. and Miller, D., 1990: Hurricane Gilbert: anthropomorphising a natural disaster. *Area* **22**, 107-116.

Bender, S., 1989. Disaster Prevention and Mitigation in Latin America and the Caribbean: A look to the 90's. Environment Department and Human Resources Development Division of World Bank. Washington, D.C.

Burton, I. and Kates, R.W., 1964. The Floodplain and the Seashore. *Geographical Review*, **54,** 366-385.

Glantz, M.H. (ed.), 1994. *The Role of Regional Organizations in the Context of Climate Change*. NATO ASI Series. Springer-Verlag, Berlin, 208 pp.

Glantz, M.H. and Price, M.F., 1994. Summary of Discussion Sessions. In: Glantz, M (ed.), *The Role of Regional Organizations in the Context of Climate Change*, Springer-Verlag, Berlin, 33-56.

Goldenberg, S.B. and Shapiro, L.J., 1993. Relationships between tropical climate and interannual variability of North Atlantic tropical cyclones. *20th Conference on Hurricanes and Tropical Meteorology*, San Antonio, TX, American Meteorological Society, 102-105.

Gray, W.M., 1984. Atlantic seasonal hurricane frequency. Part I: El Niño and 30 mb quasi-biennial oscillation influences. *Monthly Weather Review*, **112**, 1649-1668.

Gray, W.M. and Landsea, C.W., 1992: African rainfall as a precursor of hurricane-related destruction on the U.S. east coast. *Bulletin of the American Meteorological Society*, **73**, 1352-1364.

Gray, W.M. and Sheaffer, J.D., 1991: El Niño and QBO influences on tropical cyclone activity. In: Glantz, M.H., Katz, R.W., and Nicholls, N. (eds.), *Teleconnections Linking Worldwide Climate Anomalies*. Cambridge University Press, Cambridge 257-284.

Gray, W.M, Landsea, C.W. Mielke, P.W., and Berry, K.J., 1992. Predicting Atlantic seasonal hurricane activity 6-11 months in advance. *Weather and Forecasting*, **7**, 440-455.

Gray, W.M, Landsea, C.W. Mielke, P.W., and Berry, K.J., 1993. Predicting Atlantic basin seasonal tropical cyclone activity by 1 August. *Weather Forecasting*, **8**, 73-86.

Gray, W.M., Landsea, C.W., Mielke, P.W. Jr., and Berry, K.J., 1994. Predicting Atlantic basin seasonal tropical cyclone activity by 1 June. *Weather and Forecasting*, **9**, 103-115.

Gray, W.M. and Sheaffer, J.D., and Landsea, C.W., 1996. Climate trends associated with multidecadal variability of Atlantic hurricane activity (Chapter 2, this volume).

Hastenrath, S., 1990. Tropical climate prediction: A progress report, 1985-90. *Bulletin of the American Meteorological Society*, **71**, 819-825.

Hastenrath, S., 1990. Decadal-scale changes of the circulation in the tropical Atlantic sector associated with Sahel drought. *International Journal of Climatology*, **10**, 459-472.

Hastenrath, S., 1995. Recent advances in tropical climate prediction. *Journal of Climate*, **8**, 1519-1532.

Hebert, P.J. and Taylor, G., 1979a. Everything you always wanted to know about hurricanes. Part I. *Weatherwise*, **32**, 60-67.

Hebert, P.J. and Taylor, G., 1979b. Everything you always wanted to know about hurricanes. Part II. *Weatherwise*, **32**, 100-107.

Landsea, C.W. and Gray, W.M., 1992, The strong association between western Sahel monsoon rainfall and intense Atlantic hurricanes. *Journal of Climate*, **5**, 435-453.

Mather, J. and Sdasyuk, G., (eds.), 1991. *Global Change: Geographical Approaches*. University of Arizona Press, Tucson, 289 pp.

Karl, T.R., Knight, R.W., Easterling, D.R., and Quayle, R.G., 1995. Trends in U.S. climate during the Twentieth Century. *Consequences*, **1**, 3-12.

Maul, G.A., (ed.), 1993. *Climatic Change in the Intra-Americas Sea*. United Nations Environment Programme, 389 pp.

Mitchell, J., 1990. Human dimensions of environmental hazards: Complexity, disparity, and the search for guidance. In: Kirby, A. (ed), *Nothing to Fear: An Examination of Risks and Hazards in American Society*. University of Arizona Press, Tucson, 301 pp.

Riebsame, W.E., Diaz, H.F., Moses, T., and Price, M., 1986. The social burden of weather and climate hazards. *Bulletin of the American Meteorological Society*, **67**, 1378-1388

Shapiro, L.J., 1982a. Hurricane climatic fluctuations. Part I: Patterns and cycles. *Monthly Weather Review*, **110**, 1007-1013.

Shapiro, L.J., 1982b: Hurricane climatic fluctuations. Part II: Relation to large-scale circulation. *Monthly Weather Review*, 110: 1014-1023.

Sheets, R., 1985. The National Weather Service hurricane probability program. *Bulletin of the American Meteorological Society*, **66**, 4-13.

Sheets, R., 1994. The Natural Disaster Protection Act of 1993 Statement to the U.S. House of Representatives Committee on Public Works and Transportation Subcommittee on Water Resources and Environment, 13 pp.

Walsh, R. and Reading, A. 1991. Historical changes in tropical cyclone frequency within the Caribbean since 1500. *Würzburger Geographische Arbeiten*, **80**, 199-240.

White, G.F., 1994. A perspective on reducing losses from natural hazards. *Bulletin of the American Meteorological Society*, **75**, 1237-1240.

2 Climate Trends Associated with Multidecadal Variability of Atlantic Hurricane Activity

William M. Gray,[1] John D. Sheaffer,[1] and Christopher W. Landsea[2]

[1]Department of Atmospheric Science
Colorado State University
Fort Collins, CO 80523, U.S.A.

[2]Hurricane Research Center
NOAA/AOML
4301 Rickenbacker Causeway
Miami, FL 33149, U.S.A.

Abstract

Anomalous long-term variations of ocean heat transport offer an attractive and intuitively creditable explanation for many long-term climate trends. Multidecadal variations of intense Atlantic hurricane activity are but one manifestation of an extensive array of regional and global climate trends which appear to be linked to variations of heat transport by the Atlantic thermohaline circulation. In addition to influencing Atlantic tropical cyclones and the closely associated West African monsoon, Atlantic thermohaline variability appears to be linked to global SST variations and related trends occurring throughout the global climate system. Consequently, understanding decadal trends in hurricane activity may be critically dependent on understanding the somewhat broader issue of decadal variations of the major ocean circulations.

The net transport of warm surface layer water to high latitudes by the so-called "Atlantic conveyor belt" (i.e., thermohaline) circulation is sensitive to surface-layer salinity anomalies in the "deep water" formation areas of the North Atlantic. A major decrease of surface-layer salinity appeared over portions of these areas during the late 1960s which reduced ocean water density and slowed the surface water sinking process associated with deep water formation. This trend, in turn, lead to diminished northward heat transport by the ocean hence, to cooling of ocean surface temperatures in much of the North Atlantic and warming of SSTs in much of the South Atlantic. These regional Atlantic SST anomalies initiated the atmospheric circulation anomalies associated with the long running Sahel drought and the associated decrease of intense Atlantic hurricane activity in recent decades.

At approximately the same time, the ocean surface also cooled in much of the North Pacific while strong SST warming occurred in much of the Southern

Hemisphere Atlantic, Indian, and Pacific Ocean areas. This global distribution of altered ocean surface temperatures has been directly linked to altered patterns of Atlantic and West African surface pressure and monsoon circulations. These global climate changes also include the energetics of ENSO and related variables in the tropical Pacific and Indian Oceans and numerous "teleconnected" interactions between the tropical Pacific, the North Pacific, North America and Europe.

At present there are few long-term observational data for making reliable direct estimates of trends in the net Atlantic Ocean conveyor transport. Consequently, no such information is available for detecting and further anticipating forthcoming decadal trends in hurricane activity. Needed research is suggested which includes surveys and the synthesis of additional trend data for the specification of plausible and physically consistent global interactions linking the Atlantic conveyor circulation and other decadal trend associations in the global climate system. In this way, some of these global data may yield factors which are potentially useful for forecasting the onset and termination of new decadal trends of hurricane activity. This prospect is examined in data for the most recent 50-year period and in prior realizations of similar concurrent climate trends in earlier historical data.

2.1 Background

The development of intense hurricanes occurs only with a very favorable set of atmospheric and oceanic conditions. As the tropical Atlantic region is typically only moderately accommodating to hurricane development, rather modest climatological deviations in this region can strongly alter the amount of intense hurricane activity (Gray 1979; Gray et al. 1993). Because seasonal trends in some key ocean-atmosphere conditions in the tropical Atlantic are predictable months in advance, we enjoy a surprising degree of seasonal predictability for Atlantic hurricane activity (see Gray 1984a, 1984b; Gray et al. 1991, 1993, 1994). The discussion in this chapter centers on a comparative review and synthesis of recent multidecadal trends of Atlantic hurricane activity and the multiple, concurrent, and apparently related regional and global climate trends during the past 50-100 years. As most large scale decadal and longer-term climate trends are in some way linked to variations of the oceanic thermohaline circulations, we argue that these observed decadal variations of hurricane activity and most of the concurrent trends in global climate are likely all linked to multidecadal scale variations of the Atlantic thermohaline circulation.

Comparative multidecadal distributions of hurricane tracks are shown in Fig. 2.1 for the tropical West Atlantic-Caribbean area during two recent 24-year periods. The results in Fig. 2.1 provide a basic perspective on the nature of recent decadal trends in the incidence of hurricanes. It is unlikely that the extent of the climatic processes responsible for the obvious and persistent differences shown in Fig. 2.1

are restricted to the tropical Atlantic. Hence, we seek to interpret these changes in terms of additional regional and global climate trends which are approximately coherent with recent observed multidecadal changes of hurricane activity. To this end, we present a summary of some of the more interesting recent multidecadal climate trends which suggest the presence of global mode of climate variability which is related to recent variations of hurricane activity and modulated by long-term variations of the global ocean circulations. A partial depiction of some of these widely spaced and diverse climate trends, both in the Atlantic region and in more remote locations as well, is shown in Fig. 2.2. These global climate trends include anomalous excursions in both Pacific and Atlantic basins and in teleconnections extending between tropical areas and into the middle and high latitudes. Although detailed data for many of these scattered climate indices exist for only the last 50 years or so, those with a hundred or more years of data also indicate prior realization of multidecadal climate covariability with Atlantic hurricane activity. A more complete discussion of these global climate trends is given below in part four of this chapter.

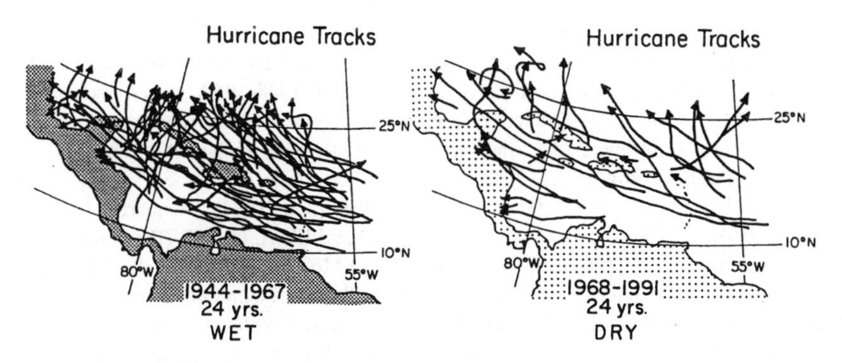

Fig. 2.1. Comparison of composite tracks of hurricanes in the western tropical Atlantic-Caribbean region during 1944 to 1967 (24 years) versus the more recent 1968 to 1991 period (also 24 years). From Landsea et al. 1994.

Strong northward transport of heat occurs in the ocean surface layer throughout the Atlantic Ocean basin. This transport circulation is associated with a net evaporative loss of water from the ocean surface along the north-south extent of the Atlantic. This net evaporation leads to the creation of relatively salty surface water conditions in the North Atlantic. As salinity dominates temperature in determining the density of sea water, chilling of this comparatively saline water in open areas of the far North Atlantic leads to the formation of very dense water which can sink to great depths in the ocean; this process is termed "deep water formation." Extensive sinking of surface water in the North Atlantic maintains a net northward moving compensation circulation of comparatively warm and salty surface layer water. This northward circulation is often termed the "Atlantic conveyor belt" (Broecker 1991). The basic features of this long recognized circulation process are illustrated in the two maps shown in Fig. 2.3, taken from early work by Stommel (1957). The circles plotted on each map in Fig. 2.3 represent terminus points for the transport streamlines for upper (U) and lower (L) level circulation features.

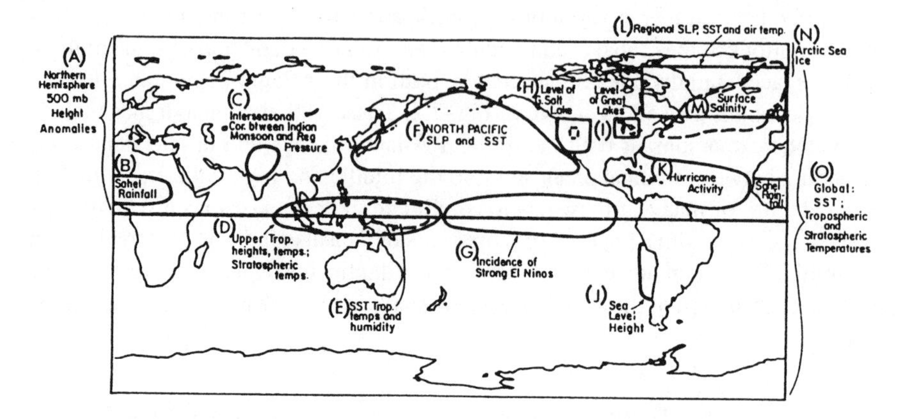

Fig. 2.2. Summary showing some of the widely spaced and diverse climate indices which experienced a distinct change of conditions between 1945-1969 versus more recent conditions during 1970-1994. The letters labeling each element in the figure keys it to one or more references as follows: (A) Shabbar et al. (1990); (B) Nicholson (1989), Landsea (1991); (C) Parthasarathy et al. (1991); (D) Reid and Gage (1993), Sheaffer and Gray (1995); (E) Gaffen et al. (1991), Graham (1990); (F) Barnett (1984), Trenberth (1990); (G) Enfield (1989); (H) Kay and Diaz (1985), McKenzie and Eberli (1987); (I) Manchard et al. (1988); (J) Nunez et al. (1992); (K) Gray (1990); (L) Moses et al. (1987), Kushnir (1994), Hurrell (1995); (M) Lazier (1980), Serreze et al. (1992), Dickson and Brown (1994); (N) Walsh and Chapman (1990), Aagaard (1995), Mysak et al. (1990); (O) Hsiung and Newell (1986), Barnett (1984), Folland et al. (1991), Street-Perrott and Perrott (1990), Angell (1988).

Reduced salinity or "freshening" dramatically lowers ocean surface water density which can, in turn, inhibit deep convective overturning in the ocean for extended periods. Significant freshening of the ocean surface can occur in various ways including anomalous melting of pack ice, precipitation, and advection (see Walsh and Chapman, 1990). Various data suggest that repeated multidecadal episodes of weakening net northward heat transport to the far North Atlantic may be linked to freshenings of the upper surface layer and decreased deep water formation in the high-latitude Atlantic region (Bjerknes 1964; Broecker 1991; Bryan and Stoffer 1991; Weaver et al. 1994). A freshening event of this sort, termed the "Great Salinity Anomaly" (or GSA), appeared during the mid-1960s (Lazier 1980) and has persisted for several decades, in rough concurrence with the observed recent 25-year change in intense hurricane activity. The GSA event has been qualitatively linked to the concurrent and apparently ongoing cooling of Northern Hemisphere Sea Surface Temperatures (SST) and warming of SSTs in much of the Southern

Hemisphere oceans in recent decades (Street-Perrott and Perrott, 1990). Climate records indicate that a similar North Atlantic freshening event may have occurred during the early part of this century (Dickson et al. 1990; Kushnir 1994; Hurrell 1995). Paleoclimate data also suggest similar but comparatively more massive freshenings have occurred on multicentury scales in the North Atlantic and may be implicated with millennium-scale climate trends during the glacial-interglacial transitions (see Gordon 1986; Broecker 1991).

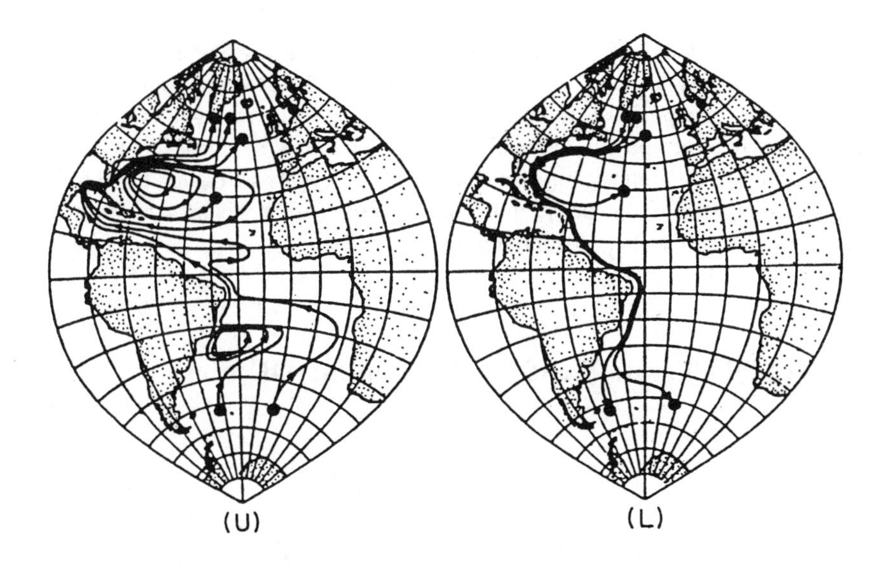

Fig. 2.3. Conceptual illustrations of the upper surface layer (U; left panel) and lower deep water (L; right panel) ocean transport streamlines in the Atlantic Basin. The large dots show the terminus points of the primary circulations (from Stommel, 1957).

The precise rate of net northward heat transport to high latitudes by the Atlantic "conveyor" process varies with latitude and is not accurately known for any location. However, consensus estimates for ocean conveyor transport across 40°N range from 15 to 30% of the total northward heat transport by the atmosphere at the latitude for the entire Northern Hemisphere (see OUCT, 1989). Therefore, significant alterations of poleward heat transport by the Atlantic "conveyor" immediately imposes significant compensating adjustments on the atmospheric heat transport process and eventually on the remainder of the ocean thermohaline circulation as well. The amplitude, timing, and duration of the decreased Atlantic conveyor transport inferred to have occurred due to the GSA provides ample basis for suggesting that, in addition to hurricane variability, other regional and global climate trends occurring during the past century (e.g., "global warming") may also be largely due to the GSA and closely related effects (see Gray 1993). The nature and causes of the latter are proper issues for close study.

The following discussion reviews evidence for linkages between alterations of hurricane activity, ocean conditions in the North Atlantic and concurrent global climate events. We begin with a description of recent trends in hurricane activity and related climate trends in the tropical Atlantic. This is followed by a more extensive description of the Atlantic and global thermohaline circulation. A synthesis of these ideas is then attempted in the context of a global summary of recent concurrent climate variations. Emphasis is given to the close association between the recent 25-year decrease of intense Atlantic tropical cyclones and the concurrent long-term West African (Sahel) drought. In doing these analyses, we have focused on the need to anticipate future trends in hurricane activity. Neither new measurements nor reliable long-term indices of Atlantic conveyor transport are available for this purpose at present. Meanwhile, there is significant need to determine when the specific regional and/or global climate control features which have been inhibiting intense hurricane activity for 25 years will again trend toward conditions typical of the prior, more active mode of hurricane activity which occurred during the mid-1940s through the mid-1960s.

2.2 Decadal Climate Trends in the Tropical Atlantic

2.2.1 Decadal Variability of Atlantic Hurricane Activity

Table 2.1 provides a detailed summary of four measures of seasonal intense Atlantic hurricane activity. The hurricane categories referred to in Table 2.1 are based on the Saffir-Simpson (SS) scale of tropical cyclone intensity (see Simpson 1974; Saffir 1977). A summary of the SS storm characteristics, including central pressure, maximum sustained wind, storm surge height and estimates of the potential for coastal destruction are given in Table 2.2 for each of the five SS intensity categories. The exponential rise in estimated hurricane destruction potential for each intensity class, as indicated in the last column on the right of Table 2.2, reflects the observation that wind and storm surge destruction by hurricanes increases very sharply with increased SS intensity category (see Landsea 1993).

The total seasonal incidence of the most intense category 3, 4, and 5 hurricanes generally exhibits the greatest climatic variability, both for year-to-year and for multidecadal time frames. The decadal variability illustrated by Fig. 2.1 is also apparent in the data summary in Table 2.1 and in additional track composites shown in Fig. 2.4. The analysis of intense hurricane landfall at higher latitudes on the U.S. East Coast in Fig. 2.4 complements the analysis for all classes of hurricanes in low latitude areas of the deep tropics shown in Fig. 2.1. Time series which show the long-term trends of seasonal hurricane activity are given in Fig. 2.5. The data series in Fig. 2.5 illustrate the important differences in the seasonal incidence of total (i.e., both weak and strong tropical cyclones) activity versus the seasonal occurrence of intense (category 3-4-5) hurricanes only. Note in Table 2.1 how the ratios of earlier versus later period hurricane activity are greatest for the more intense storm activity parameters.

The category termed "named storms" in Table 2.1 and Fig. 2.5 is inclusive of all tropical storms with sustained winds greater than 34 knots. Whereas little long-term trend can be inferred for the seasonal incidence of total named storm activity in the Atlantic, shown in the left panel of Fig. 2.5, a sharp decrease of intense (category 3-4-5) hurricane activity can be inferred to have begun during the late 1960s from the right panel of Fig. 2.5. Inspection of Table 2.1 and Figs. 2.1, 2.4, and 2.5 (plus additional results given below) reveals that, in general, decadal variability is most pronounced for (1) the most intense classes of hurricanes and (2), for hurricanes originating in the deeper tropics, south of 25°N latitude. The latter point is clearly shown by the composites of hurricanes in the deep tropics in Fig. 2.1.

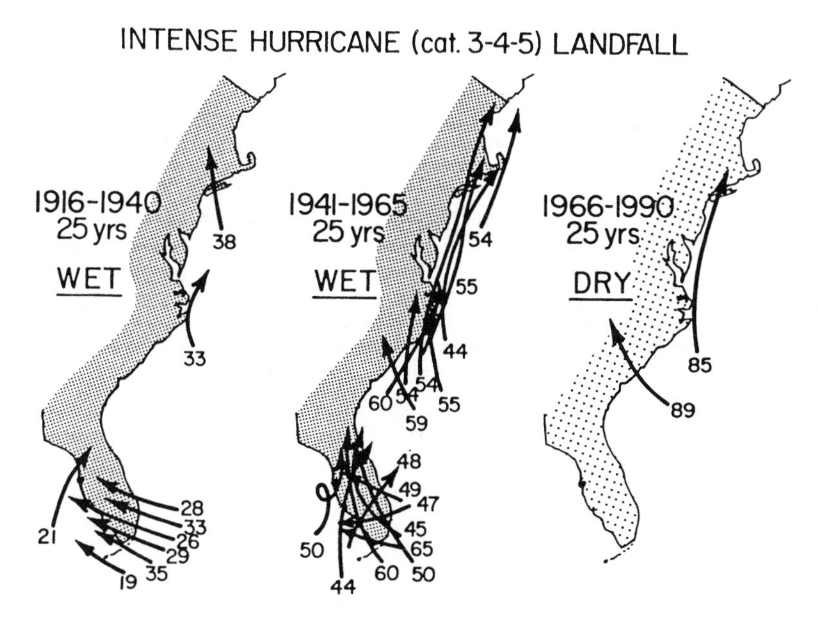

Fig. 2.4. Contrast of landfalling intense hurricanes on the U.S. east coast during two 25-year periods 1916-1940 and 1941-1965 (when the Sahel area of West Africa was relatively wet) versus intense landfalling hurricanes during the recent 25 years between 1966-1990. Numbers plotted by each track indicate the year of intense storm.

Table 2.1. Summary of Atlantic tropical cyclone statistics for 1944-1994. The numbered columns show yearly incidence of: 1) named storms (NS), 2) named storms days (NSD), 3) hurricanes (H), 4) hurricane days (HD), 5) "intense", category 3-4-5 hurricanes (IH), 6) category 3-4-5 hurricane days (IHD), 7) hurricane destruction potential (HDP), 8) net tropical cyclone activity (NTC). Mean values are shown for each column for the periods 1944-1969 and 1970-94. Ratios of the means for 1944-1969 to those for 1970-1994 are given at the bottom.

Year	NS	NSD	H	HD	IH	IHD	HDP	NTC
1944	11	53	7	27	3	4.25	73	115
1945	11	42	5	14	2	3.75	42	86
1946	6	17	3	6	1	.50	14	38
1947	9	52	5	28	2	6.50	88	106
1948	9	52	6	29	4	4.75	85	118
1949	13	62	7	22	3	3.25	64	114
1950	13	98	11	60	7	15.50	200	237
1951	10	58	8	36	2	5.00	113	119
1952	7	40	6	23	3	4.00	70	96
1953	14	64	6	18	3	5.50	59	119
1954	11	52	8	32	2	8.50	91	128
1955	12	83	9	47	5	13.75	158	195
1956	8	30	4	13	2	2.25	39	68
1957	8	38	3	21	2	5.25	67	84
1958	10	56	7	30	4	8.25	94	137
1959	11	40	7	22	2	3.75	60	97
1960	7	30	4	18	2	9.50	72	96
1961	11	71	8	48	6	20.75	170	218
1962	5	22	3	11	0	0.0	26	33
1963	9	52	7	37	2	5.50	103	115
1964	12	71	6	43	5	9.75	139	165
1965	6	40	4	27	1	6.25	73	85
1966	11	64	7	42	3	7.00	121	138
1967	8	58	6	36	1	3.25	98	96
1968	7	26	4	10	0	0.00	18	40
1969	17	83	12	40	3	2.75	110	154
Mean 1949-1969	9.85	52.08	6.27	28.46	2.69	6.13	86.42	115.27
1970	10	23	5	7	2	1.00	18	63
1971	13	63	6	29		1	1.00	65
1972	4	21	3	6	0	0.00	14	28
1973	7	32	4	10	1	0.25	21	51
1974	7	32	4	14	2	4.25	46	75
1975	8	42	6	20	3	2.25	53	91
1976	8	45	6	26	2	1.00	65	83
1977	6	14	5	7	1	1.00	19	46
1978	11	40	5	14	2	3.50	40	85
1979	8	44	5	22	2	5.75	73	94
1980	11	60	9	38	2	7.25	126	134
1981	11	60	7	22	3	3.75	63	112
1982	5	16	2	6	1	1.25	18	36
1983	4	14	3	4	1	0.25	8	31
1984	12	51	5	18	1	0.75	42	77
1985	11	51	7	21	3	4.00	61	109
1986	6	23	4	10	0	0.00	25	38
1987	7	37	3	5	1	0.50	11	47
1988	12	47	5	21	3	9.0	81	121
1989	11	66	7	32	2	9.75	108	135
1990	14	66	8	27	1	1.00	57	101
1991	8	22	4	8	2	1.25	22	59
1992	6	39	4	16	1	3.25	51	66
1993	8	30	4	10	1	0.75	23	53
1994	7	28	3	7	0	0.00	15	36
Mean 1970-1994	8.6	38.64	4.96	16	1.52	2.51	45	74.6
Ratio	1.14	1.35	1.26	1.78	1.77	2.44	1.92	1.55

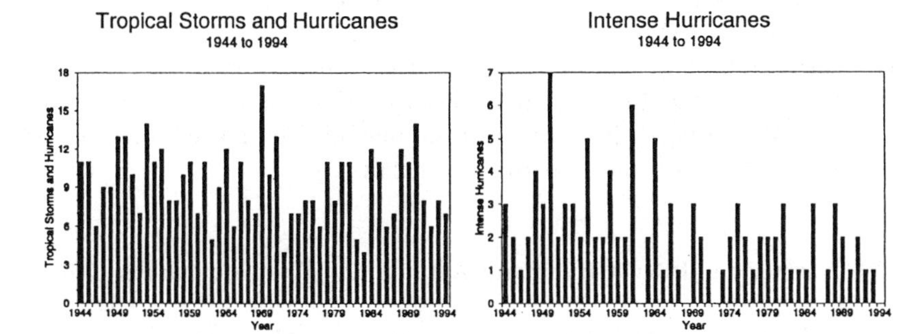

Fig. 2.5. Annual total incidence of Atlantic named storms (tropical storms plus hurricanes) between 1944 and 1994, and (right panel) Intense (category 3-4-5) hurricanes between 1944 and 1994 (left panel).

Table 2.2. Specific central pressure, wind speed, and surge criteria for the five hurricane intensity categories of the Saffir-Simpson scale. The basis for the relative potential destruction shown in the right hand column is described at length in Landsea (1991).

Saffir-Simpson Category	Range of Central Pressure (mb)	Maximum Sustained Wind Speed (m s^{-1})	Storm Surge (m)	Relative Potential Destruction
1	980	33-42	1.0-1.7	1
2	965-979	43-49	1.8-2.6	10
3	945-964	50-58	2.7-3.8	50
4	920-944	59-69	3.9-5.6	100
5	< 920	> 69	≥5.6	250

Intense hurricane activity was significantly greater during the 1950s and 1960s, in comparison with the 1970s and 1980s and the first half of the 1990s except, as discussed below, during 1988, 1989, and very recently during 1995. However, as noted in Fig. 2.5, little or no decadal scale trend was observed in the total frequency of named storms during this period (see also Gray 1990; Landsea 1991; Landsea and Gray 1992). Satellite detection of a few weaker storms in remote areas during recent years may bias these results slightly. Because the total incidence of Atlantic tropical cyclones, including tropical storms and weak and strong hurricanes, has been relatively constant during recent decades, little notice has been given to the marked decrease of the intense hurricanes and, therefore to the parallel trend in hurricane related destruction which also began to decrease during the late 1960s. The large multidecadal variations become evident only when analyzed in terms of the seasonal totals of SS category 3-4-5 hurricanes, as shown in Figs. 2.1 and 2.4. Obvious differences in Figs. 2.1 and 2.4 are noticeable for intense landfalling hurricanes striking the U.S. East Coast and Peninsular Florida. However, these differences are less apparent for landfalling intense hurricanes along the remainder of the Gulf of Mexico Coast (not shown, see Landsea et al. 1992). We believe that decadal differences are less distinct in the Gulf because a somewhat smaller percentage of Gulf hurricanes develop from easterly wave disturbances originating in West Africa. Since the advent of satellite monitoring in 1967, all intense hurricanes striking the U.S. East Coast have developed from African waves whereas some intense Gulf hurricanes have developed from local frontal boundaries and related circulations (e.g., Hurricane Alicia in 1983). Consequently, there is less of a link to decadal changes in West African rainfall and related oceanic and atmospheric conditions in the deep tropical Atlantic region

2.2.2 Covariation of Intense Hurricane Activity and West African Rainfall

The "Sahel" is an extended east-west zone in North Africa wherein generally marginal and highly variable rainfall occurs. This area lies between the Sahara Desert to the north (Fig. 2.6) and the transitional grasslands and rainforest areas further south. Rainfall variations in much of the Sahel tend to closely parallel variations of seasonal Atlantic hurricane activity. The sense of this association is that intense Atlantic hurricane activity tends to be enhanced during summers when the Western Sahel has above-average precipitation and vice versa. This association holds for both year-to-year variations and for longer-term, decadal time frames (Landsea and Gray 1992; Landsea et al. 1992). Consequently, long-term regional circulation anomalies tied to Sahel rainfall can also be related directly to variable hurricane activity; in recent decades this association has linked drought to a decrease of intense hurricanes. This covariability can also be used to make qualitative links between both of these trends and variable Atlantic thermohaline transport.

Given that the tropical Atlantic lies immediately west and downwind (in the tropical easterlies) of West Africa and that the majority of Atlantic hurricanes develop from easterly waves moving westward from West Africa, it is not surprising that variable aspects of summer monsoon rainfall in West Africa and trends in intense Atlantic hurricane activity are related. But it is surprising that they are so well related, both on seasonal and longer-term time scales. Indeed, there is also a good predictive link between early season West African rainfall (prior to August) and subsequent late summer (August, September, October) hurricane activity.

Precursor signals for seasonal hurricane activity can also be observed in West African rainfall during the late summer and fall of the prior year. This latter association presumably involves more vigorous vegetation cover and soil moisture processes early on in the following year which enhance *in situ* recycling of monsoon moisture to the atmosphere (see Gray et al. 1991). Regardless, this precipitation-hurricane association is a reflection of the strength and position of (1) the West African monsoon trough and (2) the Atlantic (Bermuda) summer center of high pressure as these two features become established during June and July. A strong summer monsoon at an anomalously high-latitude position in June-July is conducive to continued abundant West African rainfall and comparatively vigorous

Atlantic easterly wave systems during the remainder of the summer (see Gray 1990).

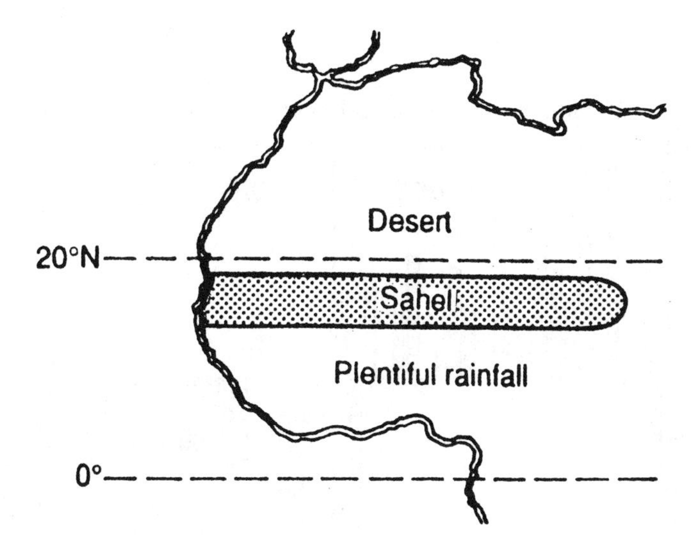

Fig. 2.6. Map of Northwest Africa showing the approximate location of the Sahel region in relation to the Sahara Desert to the north and grassland areas of more reliable rainfall farther south.

There is adequate continuous meteorological data coverage of the Western Sahel to provide reliable rainfall information back to the 1890s (see Nicholson 1989; Landsea 1991; Landsea et al. 1992). The precipitation data series in Fig. 2.7 show variations of a Western Sahel rainfall index during the last 50 years. The persistent, decades long drought which began about 1970 in West Africa and the concurrent trend to reduced intense hurricane activity evident in Fig. 2.5 appear to be closely associated manifestations of a single large-scale climate forcing factor (see Gray 1990; Gray and Landsea 1993). Presumably, as stated above, both of these trends are related to a decrease in the net northward ocean heat transport throughout much of the Atlantic basin beginning in the late 1960s. The precipitation time series in Fig. 2.7 contrasts the greater amounts of rainfall which occurred in the western Sahel throughout most of the 50-year period between roughly 1920 to 1970 versus the somewhat deficient totals especially during the recent 25-year period between 1970-1994. Shinoda and Kawamura (1994) have shown that the drought conditions in the Sahel from the late 1960s to the late 1980s were primarily due to a strong reduction in the total amount of rainfall within the West African monsoon. An equatorial shift of the rainbelt, while helping to account for the Sahel drought in the years 1968, 1973, 1985, and 1987, was in general a secondary component in comparison to the monsoon-wide rainfall reduction. It will be shown in the following section that both wet and dry rainfall variations in West Africa during the early decades of this century were also closely concurrent with complementary (increased/decreased) trends in hurricane activity and the inferred mode of the Atlantic conveyor circulation.

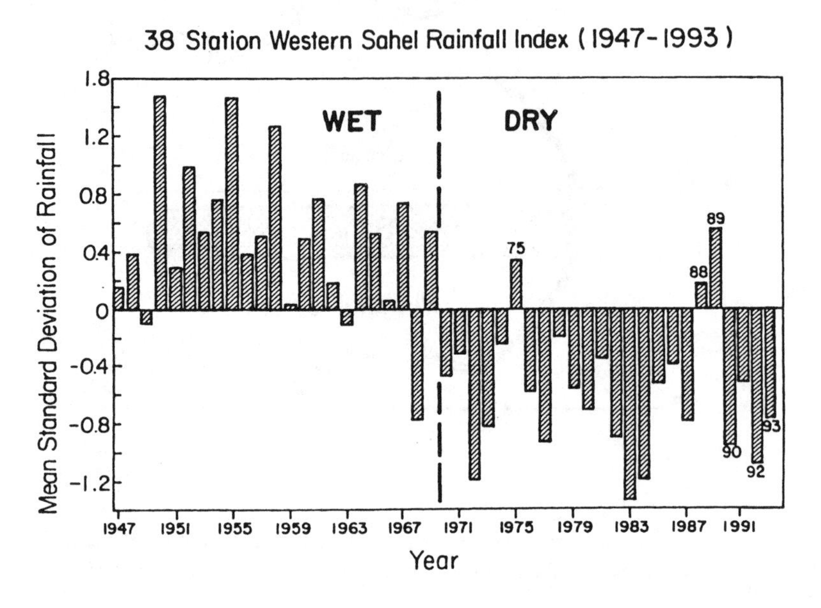

Fig. 2.7. Time series of the normalized Sahel Rainfall Index for the years 1947 to 1993 (after Landsea 1991).

Inferred weakening of the Atlantic Ocean conveyor circulation during recent decades appears to have contributed to important changes in the regional configuration of SST anomalies (both warming and cooling) which are especially evident in areas just off the northwest African Coast. The distribution of SST differences for 1950 to 1964 minus 1970 to 1984 are shown in Fig. 2.8. The areas of positive SST difference in Fig. 2.8 were warmer during 1950 to 1964 and areas of negative difference were cooler. Hence, an area of appreciable cooling occurred during 1970 to 1984 along the African coast south of about 15°N latitude and east of about 30°W longitude (positive differences) whereas coastal areas north of 15°N warmed (negative differences).

As illustrated in Fig. 2.9, the concurrent development of these surface temperature anomalies has caused collocated anomalies of surface pressure but which tend to vary in opposite directions; that is, anomalously high temperatures are associated with low surface pressures and vice versa (see Hastenrath 1990). Consequently, the long-term ocean cooling trend on the Northwest African coast has led to a gradual rise in surface pressure over parts of the subtropical Atlantic where conditions in adjacent coastal areas are directly influenced by the ocean. The resulting gradual increase of surface pressure differences due to higher pressure at Nouadhibou, Mauritania (21°N, 17°W) along the West African coast and lower pressure at the interior station at Dori, Niger (14°N, 0°W) in Fig. 2.9 illustrates this effect. The 4-mb pressure gradient change during this 35-year period is the equivalent of a geostrophic wind change of about 10 m s^{-1}. This very large change of the pressure gradient is likely an important contributor to the observed major changes in West Africa monsoon moisture advection in recent decades and hence, is likely responsible for the large decadal decrease in Sahelian summer monsoon rainfall.

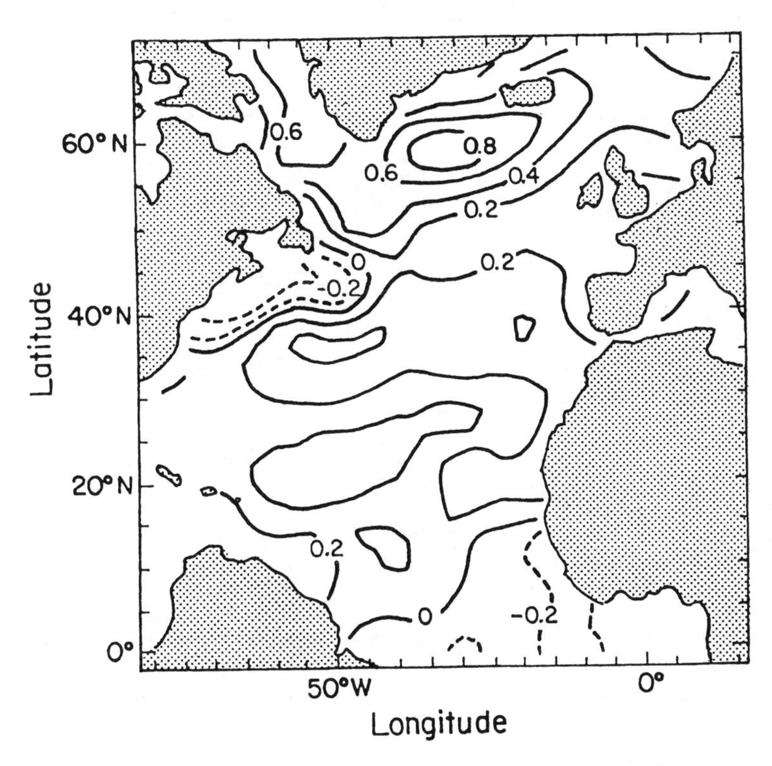

Fig. 2.8. Analysis of mean North Atlantic SST differences for warm months (June through October) for 1950 to 1964 minus 1970 to 1984. Contour interval is 0.2°C. Positive areas were warm during 1950-1964 relative to temperatures during 1970 to 1984 (after Kushnir 1994).

The protracted Western Sahel drought relented briefly during 1988 and 1989; rainfall then was more typical of seasonal values during the 1950s and early 1960s. This increased rainfall was accompanied by the development of a total of five category 4 and 5 Atlantic hurricanes during these two seasons. Similarly, the five category 4-5 hurricanes which occurred in 1988 and 1989 generated a total of 18.75 intense hurricane days (9.0 and 9.75 for 1988 and 1989, respectively) which is well above the long-term average of 2.1 intense hurricane days observed during the 25-year period of 1967-1991 (excluding 1988 and 1989). Following the 1989 hurricane season it seemed reasonable to infer that the long running multidecadal Sahel drought might be ending with far reaching implications for increased hurricane destruction during the following years. We now see that the drought did not totally end in 1988 and 1989 as drought conditions returned to West Africa during 1990 through 1993. Enhanced Sahel rainfall during 1994 and 1995 again indicates that we may now be seeing the onset of comparatively wet longer-term conditions in the Sahel and more Atlantic intense hurricane activity. The exceptionally active 1995 hurricane season (5 intense hurricanes, 11.5 intense hurricane days) is entirely consistent with this inference.

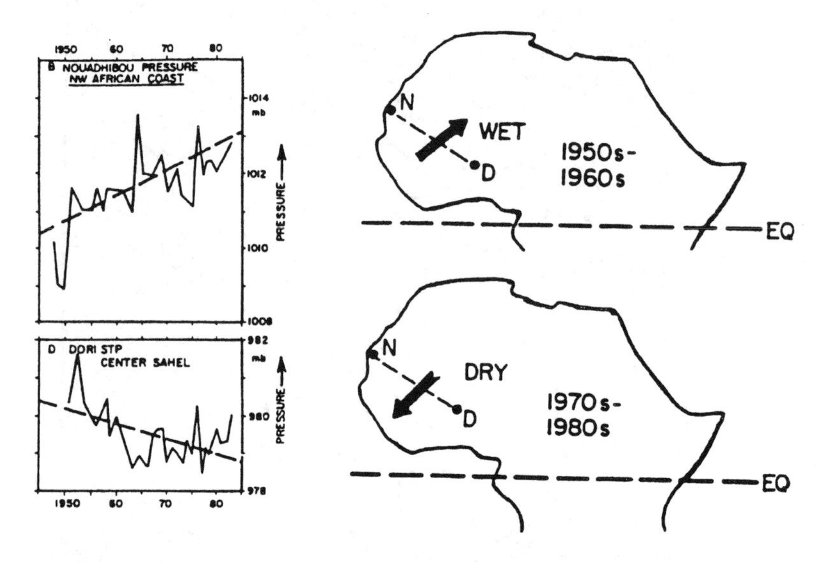

Fig. 2.9. (Left) Time series of mean annual surface (station) pressure and linear trend for 1947-1984 at Nouadhibou, Mauritania on the Atlantic coast and at Dori, Upper Volta. The net change of the geostrophic wind due to these pressure trends is illustrated in the panels on the right (after Hastenrath 1990).

The two year 1988-1989 interruption of the Sahel drought appears to have been due largely to a teleconnected response to the unusually cold Equatorial Pacific Ocean SST La Niña conditions which developed during these two years. La Niña conditions of comparable magnitude had not occurred since the middle 1950s. Teleconnected effects due to these cold SSTs contributed to weakening the upper tropospheric westerly winds within the lower Caribbean Basin and their eastward extension into the tropical Atlantic. When this upper-level circulation anomaly occurs, reduced upper tropospheric westerly winds over the tropical Atlantic become associated with the establishment of stronger upper tropospheric easterlies over West Africa and a general enhancement of the West African monsoon rainfall (Gray 1990; Landsea and Gray 1992). Decadal trends in the mode of ENSO activity also appear to be linked to Atlantic thermohaline variability. We will briefly touch on this topic again in Sect. 2.4.

2.3 The Nature and Variability of Atlantic Ocean Thermohaline Transport

The idea of an Atlantic "conveyor" provides a useful conceptual model for understanding the influence of variations in the Atlantic thermohaline circulation. This conveyor circulation is driven by the broad scale distribution of evaporation and surface layer salinity and temperature along the north-south extent of Atlantic Ocean. A simplified rendering of the Atlantic conveyor process and its likely global linkages is shown in Fig. 2.10 (cf., Fig. 2.2). Whereas the actual global thermohaline circulation is immensely more complex (see Schmitz 1995), Fig. 2.10 shows the key linkages to be considered here. Continuous north-south continental land areas flank the relatively narrow and enclosed Atlantic Ocean basin from 35°S northward to the open Arctic Ocean areas east of Greenland. This enclosed ocean configuration is unique in that it causes a large net evaporative loss of water vapor for the Atlantic Basin as a whole. Air entering the basin at high latitudes occurs mostly as westerlies which are dried during passage over the western highlands of both North and South America. Similarly, most incoming tropical air arrives as very dry easterlies out of Africa while tropical air exits the basin as moist westward moving trade winds passing over the relatively narrow and low areas of Central America. The resulting net excess of evaporation over precipitation in the Atlantic Basin creates large surface salinity values in comparison with the other large ocean areas.

A net northward transport of heat as warm upper surface layer water in the Atlantic, as noted in the chapter introduction, is engendered by the sinking of massive quantities of this comparatively high salinity water as it is chilled in the very high latitude areas of the North Atlantic. Related deep water formation processes in the Antarctic region are of lesser magnitude and are less spatially confined occurring through brine rejection during sea ice formation in several areas along the perimeter of the continent, primarily in the Ross and Weddell Seas (Toggweiler 1994; Schmitz 1995). The chilled, high salinity North Atlantic surface water descends to the deep ocean levels and moves southward, out of the basin and gradually becomes incorporated into the deep Southern Hemisphere circumpolar currents.

A more detailed view of present knowledge on the mean surface layer circulation in the North Atlantic Ocean is shown in Fig. 2.11 (from Schmitz and McCartney 1993). The numerous currents and recirculation features (gyres) shown in Fig. 2.11 represent the recent consensus regarding the best known Atlantic circulation features while smoothing over numerous lesser eddies. The circled numbers in Fig.

2.11 give estimated flow values expressed in Sverdrups (Sv; one Sv is a volume flux of 10^6 m^3 s^{-1}). The dashed lines in the northern portion of the domain indicate comparatively warm high latitude circulations. The boxes at the ends of these high latitude flow elements designate areas of subsidence and the numbers in boxes are estimated subsidence flux values in Sv. The estimates of net volume sinking in Fig. 2.11 indicate that this process is concentrated (7 Sv) in the Davis Straight east of Newfoundland and in the Norwegian Sea (3 Sv).

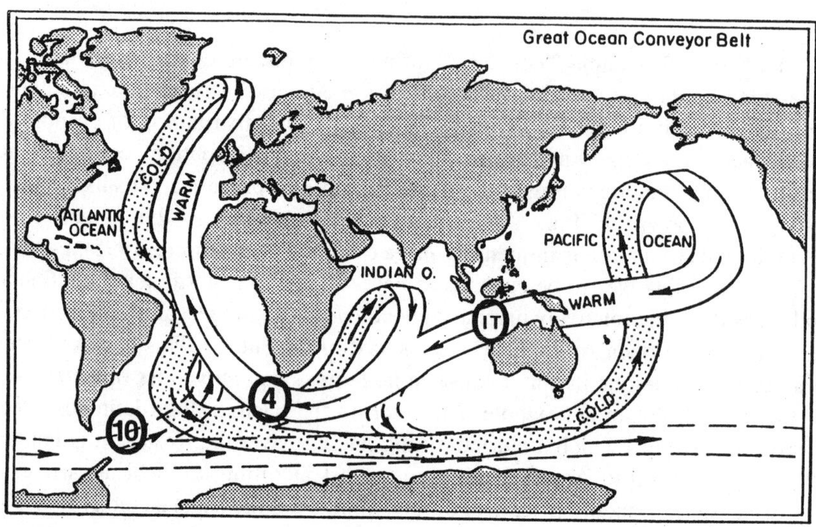

Fig. 2.10. Conceptual illustration of the Atlantic conveyor circulation. High salinity water is chilled and sinks in the far North Atlantic, promoting a compensating northward surface layer flow of warm, high salinity water (after Broecker 1991). Important components of the southern circumpolar circulations, represented by the dashed lines, have been added to Broecker's original figure (see text). The area labeled "IT" just northwest of Australia specifies the "Indonesian Throughflow". The encircled numbers represent estimated long-term mean values for upper-level current flux into the South Atlantic Ocean in Sverdrups (Sv = $10^6 m^3$ s^{-1}) from the South Pacific (10 Sv) and from the South Indian Ocean (4 Sv) which are entrained into the northward moving upper layer circulation of the South Atlantic.

Major alterations of the strength of the Atlantic thermohaline circulation in the recent 10,000 to 50,000 year time frame have been hypothesized as a key factor influencing the rates of onset and retreat of glaciation during the most recent ice-age. Abrupt discharges of fresh water from the collapse of huge pools of glacial melt over North America provide a plausible mechanism for large ocean freshenings causing extended Atlantic conveyor shut down periods observed in paleoclimate data (Broecker 1991). Although processes linked to the onset and retreat of the ice-ages represent very large variations of the thermohaline circulation on time scales of thousands of years, observations indicate that weaker, shorter term variations also occur, as shown in Fig. 2.12. The oxygen-isotope record of the Greenland ice core in Fig. 2.12 provides a proxy profile of regional North Atlantic temperature changes during the last ten thousand years. Note in Fig. 2.12 that since the final warming at the end of the last ice age (~ 10,000 B.P.), North Atlantic temperatures have remained comparatively steady but with many small multidecadal or century length temperature fluctuations of a few degrees Celsius occurring throughout the record. These smaller multidecadal and century-scale ocean temperature fluctuations are likely linked to similar but comparatively weak variations of the ocean thermohaline circulations. These smaller variations appear to be analogous to trends observed during the present century, the notable recent consequences of which have been manifest during the last 20-25 years when the North Atlantic cooled (see Kushnir 1994; Hurrell 1995).

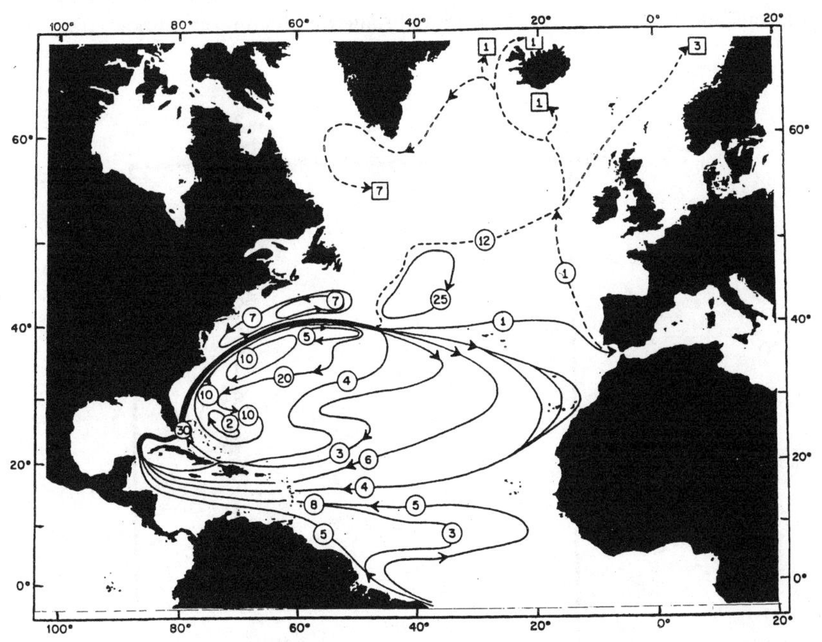

Fig. 2.11. Configuration of major ocean circulation features in the North Atlantic. Numbers enclosed in circles and boxes indicate water volume flux in Sverdrups (Sv or 10^6 m^3 s^{-1}). Dashed lines represent high latitude flows of water warmer than 7°C and boxes enclose estimates of sinking surface water flux (from Schmitz and McCartney 1993).

Direct observational evidence of persistent large-scale freshening of the ocean surface layer in the far North Atlantic during the late 1960s is shown in Fig. 2.13; this is the so called GSA event mentioned previously. Weather ship "Bravo", as noted in the caption to Fig. 2.13, was located in the area of the Labrador Sea near the "7 Sv" subsidence area shown in Fig. 2.10 where considerable deep water formation normally takes place. The effects of the GSA event in slowing the conveyor system have imposed small but nevertheless significant changes on the climate of the North Atlantic. Anomalously cold SST values due to diminished northward conveyor heat transport have altered the surface heat balance over very large areas and, thereby, have altered the large-scale atmospheric circulation. These atmospheric effects appear to include changes in (1) the regional North Atlantic-Western Europe air temperature, winds, surface pressure and precipitation distribution (see Deser and Blackmon 1993; Kushnir 1994; Hurrell 1995), (2) in the distribution of surface-500 mb thickness over much of the Northern Hemisphere (see Shabbar et al. 1990), (3) in the land-sea monsoon circulations over northwest Africa shown previously in Fig. 2.9 and (4) the vertical variability (shear) of east-west (zonal) winds throughout the deep tropics of the Atlantic and Caribbean region; these among numerous others. As discussed in greater detail below, we believe these effects are likely related to the recent quarter century Sahel drought and concurrent reduction of Atlantic Basin intense hurricane activity, as well as numerous additional closely concurrent climate trends in more remote areas.

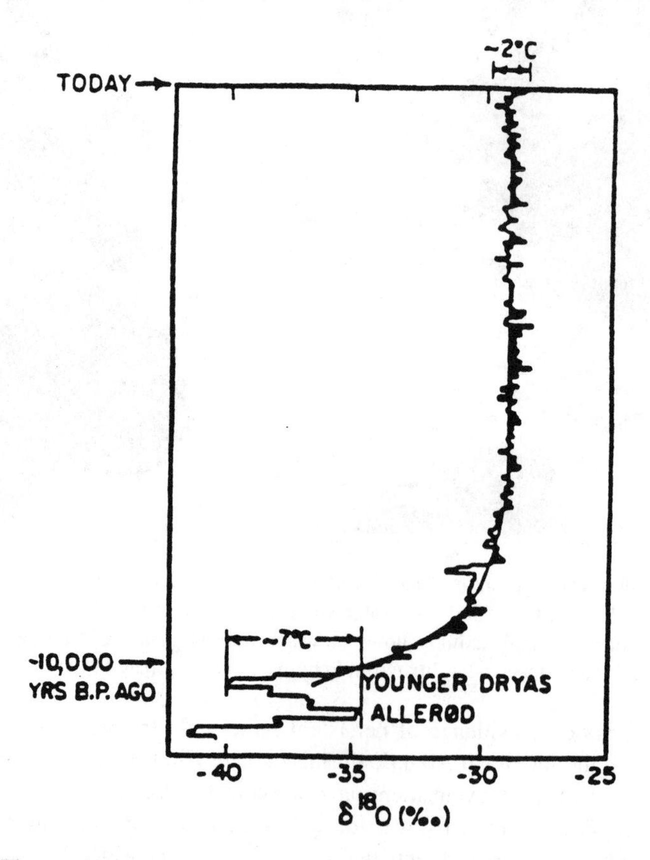

Fig. 2.12. Time variation of the fractional concentration of oxygen 18 (versus oxygen 16) in the ice core taken at Camp Century Greenland (see Dansgaard et al. 1971). Fluctuations to the left represent increased concentrations of ^{18}O in the ice, signifying the inferred variations of regional air temperature during the last 10,000 years (from Broecker 1991).

Extensive research is showing that several such variations of the Atlantic conveyor likely occurred during this century. Another weak conveyor episode appears to have occurred prior to about 1915. Reduced salinity (Dickson et al. 1990), lower SSTs and lower surface pressure (Kushnir 1994; Hurrell 1995) were observed in the far North Atlantic and accompanied reduced hurricane activity and Sahel drought during this period. An inferred stronger conveyor circulation transferred more energy into the high-latitude North Atlantic region between 1920 and 1965, followed by the present weak conveyor environment with significantly less northward ocean heat transport since the late 1960s. The upper layer freshening (salinity decrease) leading to the most recent diminished mode of the Atlantic conveyor (i.e., Fig. 2.13) was apparently caused by the anomalous advection of relatively thick Arctic pack ice (comprising a large fresh water component) through the Fram Strait (between Greenland and Spitsbergen) into the North Atlantic via the Greenland and Labrador seas during the mid-1960s (Walsh and Chapman 1990). Melting of these positive sea ice anomalies significantly lowered observed surface salinity values in the Labrador Sea area which eventually spread to much of the far North Atlantic.

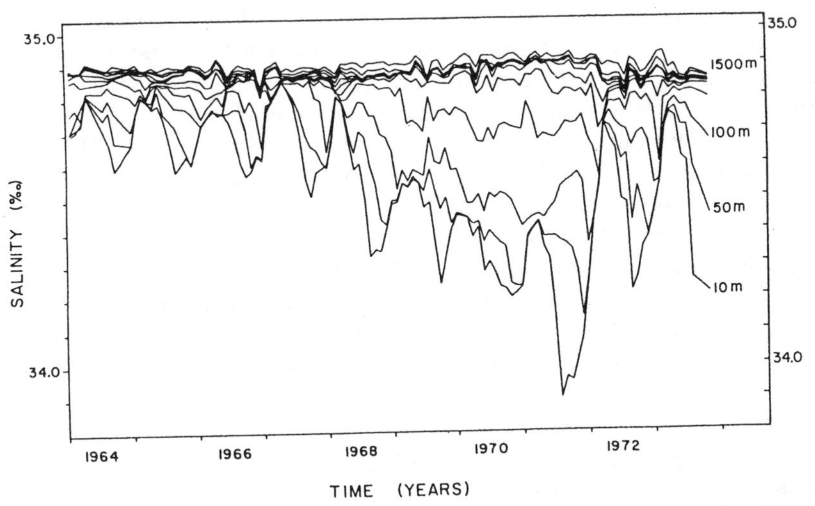

Fig. 2.13. Monthly mean salinity values taken at various depths by weather ship BRAVO showing the Great Salinity Anomaly (GSA) between 1964 and 1973 (after Lazier 1980).

Additional considerations relate to the eventual recycling of ocean deep water. As was suggested in Fig. 2.10, there are numerous potential paths whereby North Atlantic deep water (NADW) can eventually make its way back into the ocean surface layer of the South Atlantic and become reincorporated into the northward Atlantic conveyor flux. At present, the process is thought to involve various extended loops around the bottom and intermediate levels of the Indian and North Pacific ocean basins while slowly diffusing upward into the surface layer. This surface layer water eventually becomes reincorporated into the upper water of the tropical Indian Ocean and the South Circumpolar Currents (Schmitz 1995). Note in Fig. 2.10 that the estimated rate of resupply of upper layer ocean water into the South Atlantic occurs as about 10 Sv of relatively cold, low salinity water entering from the South Pacific through the Drake Passage and about 4 Sv of comparatively warm, saline water which moves westward from the Indian Ocean (Schmitz 1995).

Data on the average age of global ocean deep water suggests mean recycling times on the order of 600 years (Toggweiler 1994). However, observations indicate that vertical diffusion processes in the deep ocean are far too slow to accommodate this rate of recycling. This disparity has led to suggestions of deep upwelling of weakly stratified south circumpolar water as a major deep water removal-recycling process (Toggweiler 1994). This additional upwelling appears to be accomplished primarily by west wind driven pumping processes centered in areas of the South Atlantic just east of the Drake Passage. Hence, deep water recycling rates may also involve variable aspects of high-latitude Southern Hemisphere winds

Another important consideration is that variable amounts of warm and salty upper-level water from the South Indian Ocean enter the South Atlantic in the area west of the Cape of Good Hope. This westward transfer occurs primarily as large "retroflection rings" or eddies which break off the Agulhas Current where it turns sharply eastward to join the south circumpolar flow. Consequently, significant alterations of the upper layer salinity may occur in the South Atlantic due to variable rates of wind driven upwelling and of variable inflow of warm, saline Indian Ocean water. These trends have the potential to alter salinity of water entering the conveyor and thus, to eventually alter deep water formation rates in the North Atlantic and northward heat transfer on time scales of decades, centuries and longer.

Figure 2.14 gives a schematic summary view of four primary processes which likely influence the Atlantic portion of the global thermohaline circulation. Because of the potential associations with diverse climate trends including hurricane activity, the nature and causes of variable Arctic sea ice advection and South Atlantic upwelling and advection processes influencing North Atlantic salinity have become very important research questions. Some studies are focusing on variable precipitation and ice formation feedback processes within the arctic basin. Studies by Walsh and Chapman (1990), Mysak et al. (1990) and Serreze et al. (1992) present data and plausible hypotheses for 15 to 80 year pack-ice oscillations driven by variable interactions involving high latitude albedo and land surface hydrology processes and regional Arctic Ocean circulations. Recent data presented by Aagaard (1995) shows that average ice thickness and, thereby, net flux of fresh water passing southward through the Fram Strait and into the North Atlantic has recently been steadily decreasing (since 1991). This latter trend poses the immediate prospect of gradually increasing ocean surface layer salinity in high latitudes of the North Atlantic leading to an increased thermohaline conveyor flow. Note however, that the currently unknown effects and functional time scales of variable inputs of anomalously warm, saline water from the Indian Ocean versus relatively cold, fresh water from the South Pacific into the upper surface layers of the South Atlantic complicates the overall problem of anticipating multidecadal change.

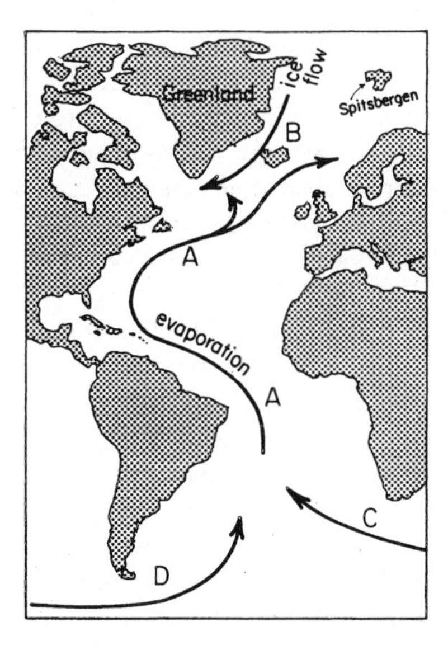

Fig. 2.14. Schematic of four factors which are believed to be primarily responsible for multidecadal variations of the strength of the Atlantic thermohaline circulation. These are (A) the rate of Atlantic Ocean evaporation, (B) the rate of ice flow (e.g., ice thickness) into the North Atlantic from the Arctic Ocean, (C) the rate of influx of warm saline water to Atlantic from the Indian Ocean, and (D) the rate of influx of relatively cold and fresh water from the South circumpolar currents. High values of processes (A) and (C) enhance the conveyor circulation, high values of processes (B) and (D) inhibit it.

2.4 Discussion and Review of Additional Related Multidecadal Climate Trends

2.4.1 Discussion

Climatic events lasting several centuries (e.g. the "climate optimum" of the late Middle Ages and the more recent "Little Ice Age") are also likely related to ocean conveyor variability, but wherein multicentury modes of interbasin ocean circulations are implicated (see Gordon et al. 1992). Regardless, the GSA-diminished conveyor transport scenario offers a plausible explanation for how the North Atlantic and much of the Northern Hemisphere became cooler in comparison with the Southern Hemisphere during the last 20 to 25 years. Significant cooling of the North Atlantic region began during the late 1960s and encompassed much of the Northern Hemisphere within a few years. The observed concurrent warming of much of the Southern Hemisphere oceans and the tropical West Pacific are illustrated in Fig. 2.15. The spatial distribution of a major mode of global SST variability during the last 100 years is shown (Fig. 2.15, upper) along with a time series (solid line in Fig. 2.15, lower) showing the time variations of the sign and amplitude of this pattern during the period (from Folland et al. 1991; also see Barnett 1984). Manabe and Stouffer (1988) obtained quite similar global SST anomaly distributions when simulating an active versus inactive Atlantic conveyor in a coupled ocean-atmosphere model. The dashed time series in Fig. 2.15, lower shows concurrent variations of Sahel rainfall. Measured rainfall variations over Northwest Africa during the last 100 years (e.g. Fig. 2.15, lower) and inferred for the last few centuries (Nicholson 1989) indicate that drought periods in the Sahel typically last 10 to 30 years, punctuated by wet periods of somewhat longer duration. The dry period prior to 1915 in Fig. 2.15, lower was similar in magnitude but not so long as the recent 1970 to present (1995) drought.

The temperature pattern in Fig. 2.15, upper represents the distribution of global SST variability extracted from an 87-year (1901-1987) data set using Empirical Orthogonal Function (EOF) analysis. The pattern is the third most powerful mode of variability (hence, EOF No. 3), following variability associated with the annual cycle and El Niño-Southern Oscillation (ENSO). Note in Fig. 2.15, upper that SST cooling associated with weak conveyor transport after 1970 also occurs over much of the North Pacific in parallel with cooling in the North Atlantic while much of the Southern Hemisphere ocean areas become warm (and vice versa when the conveyor is strong between 1920 and 1970). Very similar global difference patterns are observed as composite anomaly differences obtained by Nitta and Yamada (1989) for the 1950s versus the 1970s in a more recent (1950-1985) global SST data set.

On the basis of the time series in Fig. 2.15, lower, we may infer that the conveyor was relatively weak prior to about 1915 and after about 1965 but with a more active 1920-1964 period in between. In this context, we review other studies of regional and global data which also reflect the effects of these inferred decadal variations of the conveyor circulation. The climate record contains numerous examples of such climate trends. Notably, Kushnir (1994) used wintertime North Atlantic SST, air temperature and sea level pressure (SLP) data to specify four 15-year periods during the last 100 years which had fairly distinct climate trends and which allow two sets of composite active, versus inactive conveyor differences. Kushnir found that the periods 1900 to 1914 and 1970 to 1984 were similar in having anomalous distributions of cold SSTs and low SLP (see also Hurrell 1995) over the North Atlantic whereas 15-year periods 1925 to 1939 and 1950 to 1964 were selected as the best cases of the opposite conditions (i.e., comparatively warm SSTs and high North Atlantic SLP at high latitudes). Dickson et al. (1990) reported evidence of diminished surface layer salinity in the far North Atlantic during 1900 to 1915 which is consistent with salinity during the recent (1970-1984) GSA cold period.

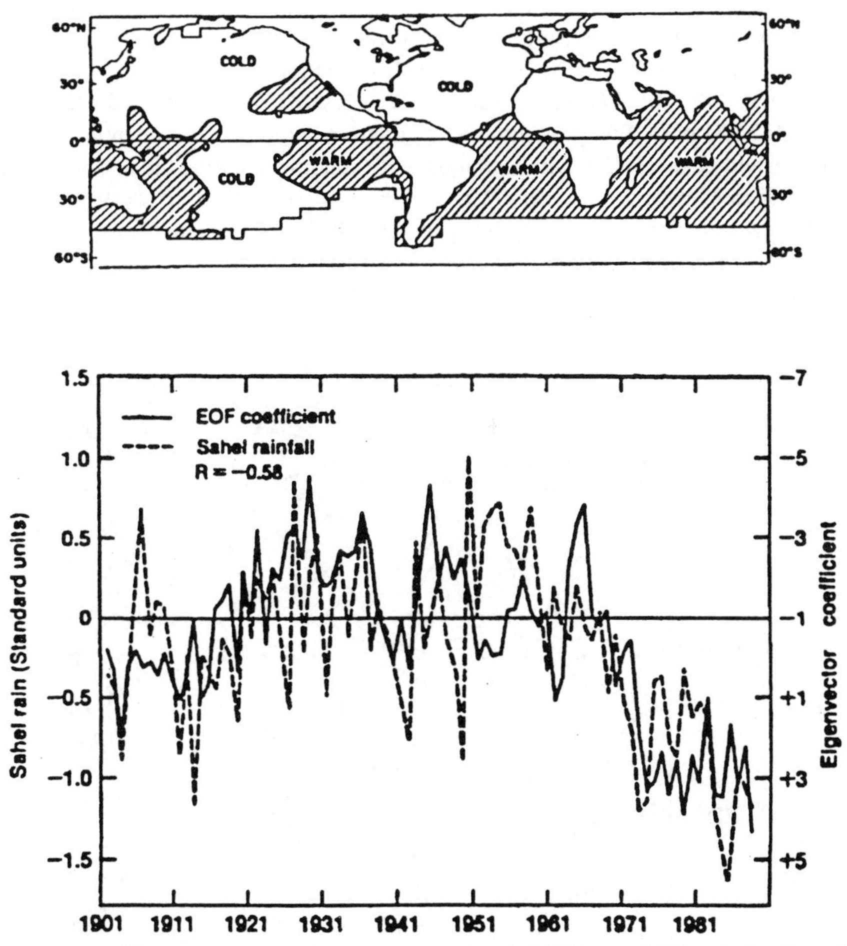

Fig. 2.15. (upper) Distribution of a major mode of global SST covariability during Boreal spring extracted from an 87 year (1901-1987) data set (data pattern extracted by EOF analysis, EOF No. 3). (lower) Time series of the variations of coefficients for SST (solid line) data pattern and for Sahel rainfall anomalies (dashed line) for the same time period; note the reversal of the sign on the right axis for the SST EOF values (from Folland et al. 1991).

Composite differences for the mean annual distribution SSTs in the North Atlantic for 1925 to 1939 (warm) minus 1900 to 1914 (cold) are shown in Fig. 2.16. A similar spatial pattern of SST differences during the summer months for 1950 to 1964 minus 1970 to 1984 SSTs was shown in Fig. 2.8 (cf. Fig. 2.15, upper, and see Kushnir 1994).

Inspection of the time series data in Fig. 2.15(lower) shows that the Western Sahel was dry between 1900 and 1914 and wet between 1924 and 1939. Similarly, trends in the Sahel during more recent periods (cf. Fig. 2.7) include wet during 1950 to 1964 versus dry during 1970 to 1984. Noting that the sign of the eigenvector coefficient on the right axis of Fig. 2.15(lower) is reversed, similar inferences can be drawn for global SSTs time series; that is, that the North Atlantic is generally warm when the Sahel becomes wet and vice versa. Hence, these results show the pre-1915 and post-1964 periods to have multiple similarities in Atlantic SST and salinity and Sahel rainfall which contrast with conditions during the intervening 50 years. Recent work by Hurrell (1995) generally verifies Kushnir's (1994) results and extends these results to long-term alterations in the Northern Hemisphere mid-latitude storm track and to resulting broadscale precipitation differences over Northern Europe.

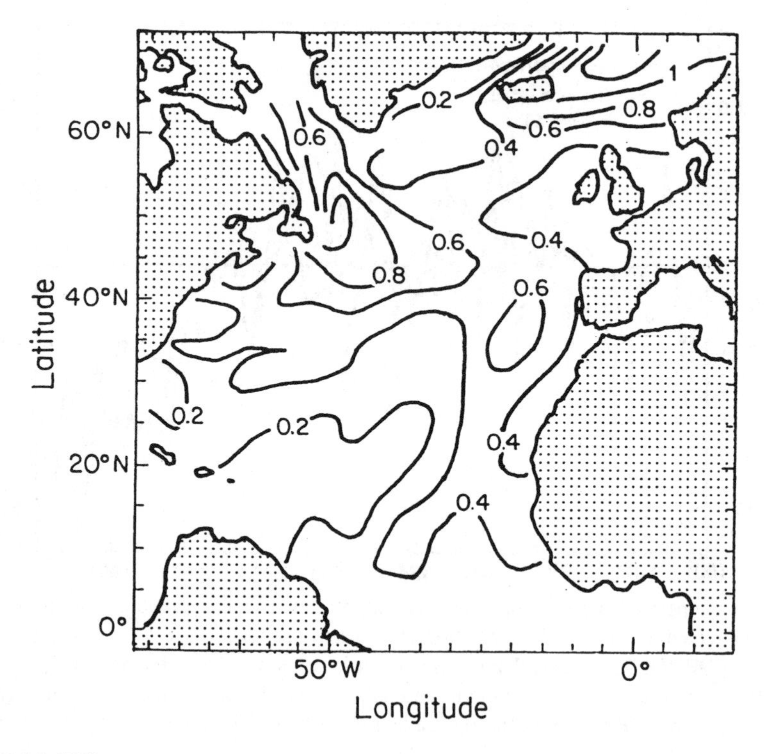

Fig. 2.16. Differences between annual mean North Atlantic SSTs for two 15-year periods (1925-1939 minus 1900-1914). Contour intervals are 0.2°C. Positive areas were warm during 1925-1939 relative to 1900-1914 (after Kushnir 1994). (Compare with Fig. 2.8.)

Details of the physical processes linking the Sahel drought with diminished hurricane activity are presented in Gray (1990), Gray and Landsea (1991), Gray et al. (1992, 1993, 1994), Landsea and Gray (1992), and Landsea et al. (1992). Basically, the reconfiguration of Atlantic SST anomalies in recent decades due to effects of the GSA weakened the summertime West African monsoon trough. As the

West African monsoon is the source of Atlantic easterly waves, it is in large part responsible for initiating nearly all intense hurricane activity in the deep tropics (Landsea, 1993). Figures 2.17 and 2.18 extend the prior analysis (i.e., Figs. 2.1 and 2.4) to comparative composites of landfalling intense hurricane tracks for the two 15-year early-20th-century periods (Fig. 2.17) and for the two more recent (Fig. 2.18) 15-year periods identified by Kushnir (1994).

A more expansive summary of the differences in hurricane activity which were illustrated in Figs. 2.17 and 2.18 is given in Tables 2.3 and 2.4. Table 2.3 shows the total number of landfalling intense hurricanes for Peninsular Florida and for the remainder of the U.S. East Coast during two 47-year periods. These two periods include 47 years of inferred strong conveyor conditions (1921-1967) versus 47 weak conveyor years comprised of two periods spanning 1900-1920 plus 1968-1992. A similar tabular summary is presented in Table 2.4 but for hurricanes in the Caribbean basin during two recent 24-year periods of comparatively weak (1970-1993) versus strong (1944-1967) conveyor conditions. The inferred strong conveyor versus weak conveyor differences for intense hurricane activity are consistently greater than two-to-one in these areas and approach a four-to-one difference for Peninsular Florida.

Table 2.3. Contrast of U.S. landfalling intense (SS 3, 4, or 5) hurricanes for two 47-year groupings of inferred weak versus strong conveyor (i.e., dry versus wet Sahel) periods. Figure 2.3 shows the locations of the two (Florida versus East Coast) regions considered.

	Peninsular Florida	East Coast	East Total
Above average Sahel rainfall and strong conveyer belt conditions; 1921-1967 (47 years)	15	12	27
Below average Sahel rainfall and weak conveyor belt conditions 1900-1920 and 1968-1993 (47 years)	4	5	9
Ratio (strong, wet/weak, dry)	3.75	2.40	3.00

Table 2.4. Contrast of the incidence of Caribbean basin hurricanes (H) and intense hurricanes (IH) during two recent 24-year periods of inferred weak versus strong Atlantic conveyor modes (cf., Figs. 2.1, 2.17, and 2.18).

	No. of Hurricanes (H)	No. of Intense Hurricanes (IH)
Above average Sahel rainfall and strong conveyer belt conditions 1944-1967 (24 years)	38	22
Below average Sahel rainfall and weak conveyor belt conditions 1970-1993 (24 years)	17	9
Ratio (strong, wet/weak, dry)	2.24	2.44

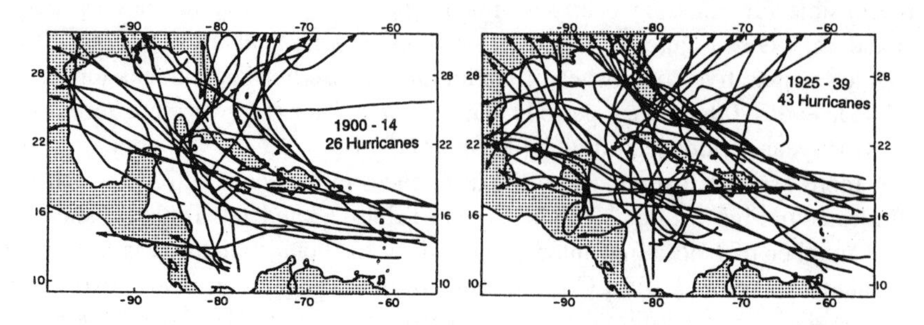

Fig. 2.17. Composite analysis of tracks for all hurricanes (category 1-5) landfalling somewhere in the Caribbean Islands, Central America, or on the U.S. East Coast during two 15-year periods: between 1900-1914 (left panel) and between 1925-1939 (right panel). The tracks represent all periods during which each storm was of tropical storm strength or greater.

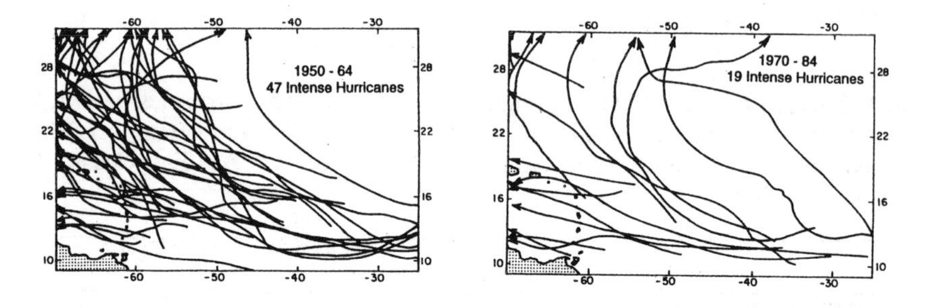

Fig. 2.18. As in Fig. 2.17 but emphasizing intense (category 3 to 5) hurricanes forming in the deep tropics (south of 20°N) east of 60°W during 1950-1964 (left panel) and during 1970-1984 (right panel).

The analyses in Fig. 2.17 were restricted to landfalling storms on the mainland or major islands so as to minimize any possible tendency for undereporting of storms in remote areas which remained at sea during either of these two early periods. As during 1950-1964, considerably greater hurricane activity occurred during the 1925-1939 inferred active conveyor (and warm North Atlantic and wet Sahel) period than during 1900-1914. The comparison of hurricane tracks shown in Fig. 2.18 is for intense (class 3-4-5) hurricanes forming in the deep tropics east of 60°W longitude during 1950-1964 versus 1970-1984. The region selected for this comparison emphasizes storms likely to have formed from African waves and thereby, helps to establish the association between hurricanes and African drought discussed previously.

As noted previously in the discussion of Figs. 2.1 and 2.4, the incidence of landfalling hurricanes on the U.S. Gulf (of Mexico) Coast shows decidedly less decadal variability (see Landsea et al. 1992). This diminished decadal variability in the Gulf likely occurs because many of these hurricanes develop from disturbances which form locally within the Gulf. This is especially true for storms developing early (June, July) and late (October) in the season. As a result, a smaller percentage of the intense hurricanes which landfall in the Gulf form from African, waves so that the incidence of intense Gulf hurricanes is less strongly influenced by variable Atlantic conveyor transport.

2.4.2 Recent Multidecadal Climate Variability in the Pacific and Indian Ocean Regions

The "warm pool" region of the equatorial West Pacific, Indonesia and the Eastern Indian Ocean is the primary ocean heat reservoir for ENSO and the associated teleconnected global climatic variability. For this reason, ENSO variability in relation to Atlantic conveyor variability, is of significant interest. As noted previously, the first two decades after 1900 appeared to have experienced climatic trends similar to those observed during the recent GSA linked periods. ENSO (and hence warm pool) linked data presented by Enfield (1989) also reflect these differences, most notably in the amount of year-to-year variability of Central and East Pacific SSTs and of the Southern Oscillation (pressure) Index (SOI; see also Trenberth 1994). During the inferred slow conveyor periods of 1900 to 1915 and 1970-present [vis-à-vis Kushnir's (1994) analysis], the amplitude of this variance in Enfield's data is consistently a factor of two greater than what occurred during the generally active conveyor period of 1920-1965. These same differences are discussed by Ramage (1983) and by Brown and Katz (1991) who noted how the significance of Sir Gilbert Walker's studies of the Southern Oscillation (e.g. Walker and Bliss 1937) was lost for 60 years, in large part because the frequency and intensity of El Niños subsided dramatically during the 1920-1965 period.

A recent trend to more and stronger El Niño events in the tropical Pacific Ocean and overall tropical Pacific and Southern Hemisphere SST warming has been observed since about 1970 (Enfield 1989; Shabbar et al. 1990). These trends may be postulated to be part of a complex sequence of global adjustments to the redistribution of ocean heat energy owing to diminished Atlantic conveyor transport and related changes in Northern Hemisphere atmospheric circulation. This hypothesis allows that some changes may occur almost immediately with the appearance of the GSA whereas other trends may not develop until several years or more later. North Atlantic variability can thus be viewed as a factor contributing to altered modes of ENSO variability in recent decades (see Trenberth 1994). Related decadal effects also appear to include the strong concurrent surface cooling trends in parts of the central and eastern North Pacific and warming in the equatorial and subtropical Northwest Pacific (e.g., Fig. 2.15, upper). A reconfiguration of the temporal and spatial variability of ENSO-linked equatorial SST anomaly patterns and changes in the well known tropical-mid-latitude teleconnections has also been observed. In the latter, we consider trend-like changes in the so called Pacific North America (PNA) pattern which links atmospheric variability in the equatorial Pacific to seasonal climate anomalies in the western North America and the Canadian Arctic (see Shabbar et al. 1990).

Michaelsen and Thompson (1992) studied various ENSO proxy indices for the last 100 years (i.e., tree-ring data, ^{18}O isotope ratios and Line Island precipitation) in comparison with standard ENSO SLP and SST indices. A listing (their Table 17.5) identifies all years since 1900 during which at least one of these indices indicated the presence of a strong or very strong El Niño. During the 30 total years identified by Kushnir (1994) as proxy "active" conveyor periods (i.e., 1925-1939 and 1950-1964), only 6 years (hence 1 year in 5) are identified by Michaelsen and Thompson (1992) as having at least one index indicating a strong El Niño; hence, evidence of some form of El Niño activity. In contrast, of the 30 inactive conveyor years identified by Kushnir (1900-1914 and 1970-1984), 15 years (or, half of all inactive conveyor years) have at least one ENSO index indicating a strong El Niño. Clearly this evidence suggests that El Niño-like conditions appear to become more pervasive during the inferred "inactive" conveyor periods. Gray (1984a) noted that teleconnected effects of El Niños are a major factor for diminished hurricane activity. Therefore, the observed trends in all the above (i.e., in ENSO activity plus North Atlantic SSTs and SLP, Sahel rainfall and global SSTs) are all broadly compatible trends and are closely coherent with observed decadal trends of Atlantic hurricane activity during the past century.

Enfield (1989) noted that strong El Niño events seem to have become more frequent in recent decades. Others have pointed out that the overall character of the development of El Niños has also changed (see Barnett 1991; Trenberth 1994). In the latter case, a distinct tendency for warm SSTs to spread eastward across the Central Pacific, somewhat in advance of the development of strong warm anomalies along the West Coast of South America, has occurred during the onset of the most recent three (1982, 1986, 1990) ENSO warming events. Gray (1984a) noted that the main feature that seemed to tie El Niños to diminished hurricane activity was enhanced vertical wind shear in the western tropical Atlantic during strong El Niños (see Fig. 2.19). That the tropical circulation has tended in recent decades to produce more strong El Niños and toward more El Niño-like conditions in general is evidenced by the trend in teleconnected Caribbean area vertical mean wind profiles shown in Fig. 2.19. The horizontal arrow in Fig. 2.19 represents the typical propagation speed of a tropical Atlantic wave disturbance moving westward into the Caribbean area. Note the much greater net vertical wind shear which occurs relative to these moving disturbances since 1970. The effect of this greater vertical wind shear is to pull apart developing tropical storms and to either prevent or seriously diminish their intensification

Recent research has also revealed a wide array of similar decadal scale climate trends occurring concurrently (or very nearly so) in the tropical West Pacific and Indian Ocean warm pool region. Gaffen et al. (1991) and Graham (1990) show upward trends for equatorial West Pacific SSTs, tropospheric temperatures, and humidity, beginning in the early 1970s. Regional tropopause heights (Reid and Gage 1993) and tropospheric thickness anomalies (Sheaffer and Gray 1995) in the warm pool area also show a distinct upward trend beginning about 1970, leveling off somewhat after 1985.

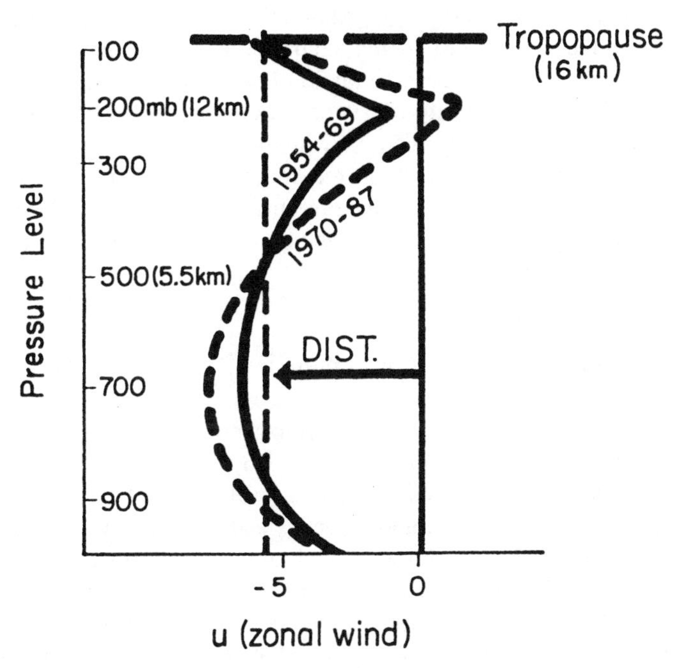

Fig. 2.19. Mean vertical wind profiles in the lower Caribbean basin for 1954-1969 (solid line) versus 1970-1987 (dashed line) adapted from Gray and Sheaffer (1991).

Some of these climate trends appear somewhat abruptly and are closely concurrent with the onset of the GSA in the mid 1960s; good examples of these being the trends in Sahel rainfall, intense hurricanes and surface pressure, winds, sea surface and air temperature, and precipitation in the North Atlantic (see Hurrell 1995). Other trends seem to have appeared more gradually with an extended transitional onset of 5 years or more. Examples of the latter delayed or more gradual trends include surface heights of major lakes, including the Great Lakes (Manchard et al. 1988) and the Great Salt Lake (McKenzie and Eberli 1987), sea level pressure-monsoon precipitation associations for the Indian monsoon (Parthasarathy et al. 1991) and conditions in the Pacific warm pool relative to ENSO variability (see Gaffen et al. 1991; Sheaffer and Gray 1995).

We note in this regard that the "global conveyor" as configured in Fig. 2.10 should not be viewed as a rigid, lock-step sort of process, periodically starting and stopping uniformly throughout its extent. A simplistic interpretation and synthesis based on such a view could be proposed wherein a disruption and subsequent slowing of the Atlantic conveyor would lead directly to a buildup of accumulated ocean surface heat energy in the warm pool due to slowing of the "Indonesian Throughflow" conveyor segment (labeled IT in Fig. 2.10) which traverses the warm pool through Indonesia west of New Guinea. Although the Indonesian Throughflow flux is known to be large (≈ 8.5 Sv) and highly variable, trends in this "throughflow" circulation appear to occur as an effect of ENSO variability (Schmitz 1995). Hence, the observed multidecadal changes of conditions in the warm pool area may more likely be due to the influence of altered atmospheric circulations (e.g., trade wind forcing) acting on the subtropical Central and East Pacific Ocean rather than as a "backing up" of impeded tropical warm water flux through the warm pool on a rigid global "conveyor belt". Consequently, slowing of the Atlantic conveyor may well be reflected in the foregoing trends in the condition of Pacific warm pool, but through associations possibly involving several anomalous large scale atmospheric circulations altering patterns of upwelling and ocean transport in remote areas of the Pacific.

2.4.3 Other Atlantic-Global Teleconnections

Ramage (1983) discusses the transient nature of various interseasonal correlation relationships which have been proposed for seasonal weather forecasting. He suggested that the multiple concurrent changes of many of these interseasonal associations (correlations) in various oceanic and atmospheric climate indices which occurred during the late 1960s should be examined to gain insight on the basic nature of global climate variability. Gray (1990) made accommodation for these effects as they relate to the association between Sahel drought and seasonal incidence of intense Atlantic hurricanes. Barnston and Livezey (1987), Chen et al. (1992), Sheaffer and Reiter (1985), and Namias et al. (1988) also made similar suggestions for accommodating concurrent long-term changes in potential seasonal forecast relationships for the tropical and North Pacific and Western North America.

The six data series shown in Fig. 2.20 provide a small but diverse sampling which is typical of numerous concurrent global climate changes that have been reported in the recent literature. As the data periods represented in Fig. 2.20 vary somewhat, a prominent arrow has been added to each time series to show the position of the year 1970 on the abscissa. For the purpose at hand, these data series and others like them can be separated into three groups as follows: (1) those trends known to be tied to hurricane variability (i.e., Fig. 2.20F); (2) trends likely to be linked with hurricane variability (Fig. 2.20A) and; (3) trends which offer interesting possibilities but require more study (Figs. 2.20B, C, D, and E).

Some specific considerations regarding the data series in Fig. 2.20 are as follows. Figure 2.20A (North Atlantic SSTs) and 2.20F (Sahel rainfall) are similar to the results already discussed in Figs. 2.8, 2.9, 2.15, and 2.16. Figure 2.20B (after Trenberth 1990) shows long-term trends in area averaged SLP for the North Pacific, the observed trend being a general lowering of high-latitude surface pressures in the North Pacific

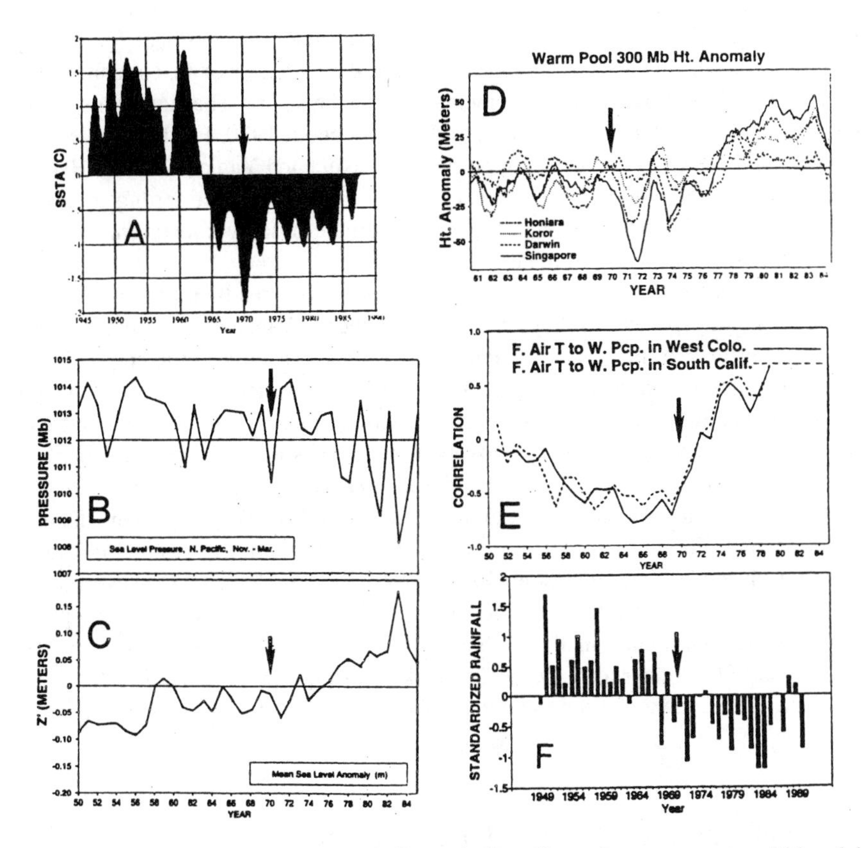

Fig. 2.20. A sampling of time series of climate indices illustrating concurrent multidecadal trends in the late 1960s and early 1970s: As the time period represented by each panel varies, a small reference arrow in each indicates 1970. (Panel A) first principal component of Atlantic zonal mean SSTs (by Kushnir; in Gordon et al. 1992; cf. Fig. 2.13b); (B) sea level pressure in the North Pacific (after Trenberth 1990); (C) sea level anomaly on the coast of central Chile; (D) pressure height anomalies in the tropical West Pacific; (E) running 10-year correlation coefficients between the primary modes of regional fall air temperature and selected winter precipitation regimes in western North America (see Sheaffer 1993); (F) time series of western Sahel rainfall anomalies (Gray et al. 1993a).

Whereas Trenberth (1990) interprets the figure as evidence of a trend in SLP beginning in 1977, these data might also be viewed as showing a trend beginning as early as 1970. Figure 2.20C shows increasing ocean surface levels in the equatorial Southeast Pacific wherein the effect is inferred to be related to decadal weakening of the Pacific trade winds and a relaxation of the residual (dynamic) westward drift of tropical surface water. Figure 2.20D shows time series of 300 mb height anomalies at five West Pacific warm pool stations including Koror, Singapore, Darwin, Majuro and Honiara. Excepting Darwin, all the data in Fig. 2.20D show a slow increase of upper-level heights (atmospheric warming) beginning about 1973. Finally, Fig. 2.20E relates to the problem of transience in empirical forecasting relationships. In this case, the input data were detrended prior to computing the running ten-year correlation coefficients shown in the figure. The sign of the fall to winter interseasonal correlations reversed sharply about 1970. Further studies and reviews of these types of trends are presented by Nitta and Yamada (1989), Hsiung and Newell (1986), and Barnett (1991), all of whom examined strong trends in Pacific SSTs and related climatic features which are broadly concurrent with the recent Atlantic (conveyor) anomaly.

The inference regarding linkages between the diverse trends in Fig. 2.20 is that the global atmospheric circulation in the Northern Hemisphere adjusted fairly rapidly to effects of the GSA on Atlantic Ocean heat transport. For example, Knox et al. (1988) noted a sharp change in Northern Hemisphere geopotential height anomalies beginning in the 1960s. The contrasting global ocean-atmosphere trend characteristics inferred for comparatively active versus inactive conveyor periods (e.g., Manabe and Stouffer 1988; Folland et al. 1991) appear to be rather systematic such that, when considered separately, both modes likely contain distinct interseasonal relationships which are useful for making simple seasonal forecasts during each mode (i.e., Gray et al. 1993, 1994; also, Fig. 2.20E). However, these prospects not withstanding, the more important opportunity may lie in adopting these and similar trends in the global climate for identifying changing decadal modes useful for longer term, decadal-scale forecasting of important climate trends including the incidence of intense hurricanes.

2.5 Summary and Thoughts for the Future

A review of the most recent 100 years of data shows that variations of hurricane activity in the tropical Atlantic appear to be linked (1) to mode-like variations of regional and global SSTs, and (2) to related trends in global air temperature, pressure anomalies and atmospheric circulations. All of these effects extend well beyond the tropical Atlantic. The preeminent effect which seems to dominate all others as a unifying process for these changes is decadal variations in the Atlantic Ocean thermohaline circulation. Figure 2.21 presents an overview of the primary regional Atlantic climatic differences associated with comparatively strong (left panel) versus weak (right panel) hurricane activity and inferred thermohaline transport conditions.

A notable decrease of upper layer salinity appeared and spread to large areas of the North Atlantic Ocean during the late 1960s and persisted for at least 10 years.

This salinity anomaly is presumed to have been caused by anomalous influx of sea ice from the Arctic. One effect of the resulting salinity anomaly was a reduction in the net rate of North Atlantic deep water formation with a collateral reduction of the compensating northward conveyor circulation of warm, salty water in the ocean surface layer (e.g., right panel in Fig. 2.21). The inferred weakening of the net northward Atlantic thermohaline transport during the last 25 years is likely the cause of the cooling of large areas of the North Atlantic Ocean and concurrent warming in much of the South Atlantic. Various observational data presented here and numerous recent numerical studies (e.g., Brewer et al. 1993; Dickson and Brown 1994; Huang 1994; Weisse et al. 1994; Weaver et al. 1994, among many others) testify to the feasibility of this concept.

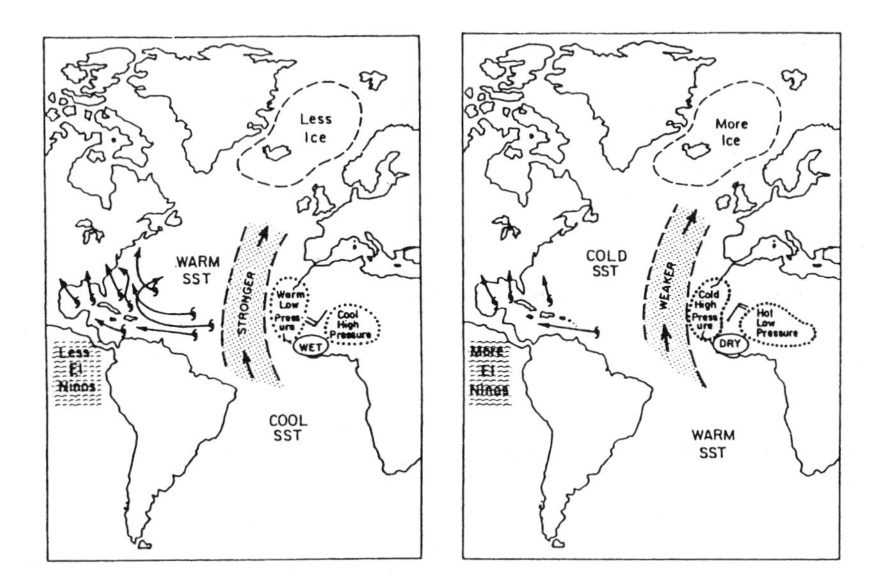

Fig. 2.21. Summary of the main regional differences between strong (left panel) versus weak (right panel) conveyor linked conditions in the Atlantic, West Africa, and East Pacific).

The effects of these changes in the regional distribution of tropical SST anomalies are particularly evident as areas of cooling along the coast of Northwest Africa. This long-term ocean cooling has caused a gradual rise in surface pressures along the subtropical Northeast Atlantic and adjacent parts of the Northwest African coast where conditions are directly affected by the ocean. These increased coastal pressure values have created an increased low level pressure gradient between the northwest African Coast and the central Sahel region. In the latter, region surface pressure has been influenced by increasingly hot, dry regional land surface conditions rather than by cooler ocean temperature changes. The combined direct effect of these two complementary changes in the regional surface pressure gradient (as shown in Figs. 2.9 and 2.20) has been a decreased southwest monsoonal surface wind circulation (note wind barbs in Fig. 2.21), a weaker monsoon and less vigorous easterly waves propagating into the tropical Atlantic. Figure 2.21 also reflects the increased tendency to El Niño conditions (and hence teleconnections) in the tropical Pacific.

It is also likely that regional feedback enhancements associated with persistent dry northeast wind anomalies (emanating out of the Sahara) have contributed to the intensity of the long-term Sahel drought. These changes have reduced moisture available for evaporation from the land surfaces, resulting in warmer summertime land surface temperatures. The hydrostatic effects of these temperatures created even lower surface pressures over the interior Sahel land areas which, when added to the positive SST induced pressure anomaly along the African coast, further alter the pressure gradient and thereby further accentuate the dry northeast wind and reduced moist southwesterly flow. Thus, once the ocean induced cooling along the West African coast initiates an increase in dry northeast winds, an interior land feedback response causes a greater enhancement of the effect.

The foregoing sequence of associations offers a plausible interpretation of how the multidecadal Sahelian drought and enhanced El Niño has been initiated by anomalously low ocean salinity thousands of miles from either effect in the far North Atlantic and why these effects have persisted so long. As the West African monsoon is the source of Atlantic easterly wave activity, these effects are at least partly responsible for the recent 25 years of greatly reduced Atlantic intense hurricane activity. Trends to more El Niño like conditions throughout the global tropics are also implicated in decreased hurricane activity. The prominent multidecadal changes in West African rainfall and intense Atlantic hurricane activity thus appear to be a natural consequence of the variable Atlantic thermohaline circulation. The nature of the teleconnected tie to increased El Niño activity needs more study, but historical and geological evidence indicate that similar trends, but of both comparable and greater amplitude and duration have occurred many times in the past. Hence, these multidecadal changes are not necessarily associated with recent anthropogenic climate alterations.

The rather loosely specified salinity effects to which we have ascribed recent inferred slowing of the Atlantic conveyor circulation are unlikely to continue indefinitely. Rather, we must anticipate that regional conditions governing the conveyor circulation will eventually assume a stronger mode. Although net Atlantic transport has recently been reduced, evaporation and related global processes have continued. Other factors left unchanged, these processes alone will gradually alter distributions of ocean salinity which, in time, will influence the intensity of the Atlantic circulation. Presumably, in this "salt oscillator" mechanism, the thermohaline circulation must increase when enhanced net rates of sinking again develop in the North Atlantic and the "conveyor" intensifies with an attendant increase in net northward heat transport by the Atlantic Ocean.

The timing of any forthcoming increases in Atlantic conveyor transport, leading to increased West African rainfall and intense Atlantic hurricane activity, remains open to question. There is some evidence from 1994-95 data that the last 25-year mode of atmospheric response may even now be changing. Paleoclimatologists are studying the changes of the conveyor belt circulation on time scales of a century to thousands of years. Shorter, decadal to century time scale conveyor belt changes are also now beginning to receive attention. It is hoped that data being collected in new research, most notably NOAA's Atlantic Climate Change Program (ACCP), will identify additional factors which reflect impending changes in the conveyor and closely linked climatic conditions. For now, more study is needed on specific atmospheric and oceanic parameters in the existing observational records which may first signal that alterations in the conveyor circulation are occurring.

Acknowledgments

We thank Richard Taft and William Thorson for their very expert assistance in the processing of massive amounts of data and Barbara Brumit and Amie Hedstrom for manuscript and data analysis assistance. We also appreciate numerous useful discussions with John Knaff. The preparation of this report and the research studies described herein were supported by the National Science Foundation.

References

Aagaard, K., 1995. The fresh water flux through the Fram Straight: A variable control on the thermohaline circulation. Preliminary extended abstracts, ACCP Principal Investigators Meeting, Miami, FL, 1995.

Angell, J.K., 1988. Variations and trends in the tropospheric and stratospheric global temperatures. *Journal of Climate*, **1**, 1296-1313.

Barnett, T.M., 1984. Long-term trends in surface temperatures over the oceans. *Monthly Weather Review*, **112**, 303-312.

Barnett, T.M., 1991. The interaction of multiple time scales in the tropical climate system. *Journal of Climate*, **4**, 269-285.

Barnston, A. and Livezey, R.E., 1987. Classification, seasonality and persistence of low frequency atmospheric circulation patterns. *Monthly Weather Review*, **115**, 1083-1126.

Bjerknes, J., 1964. Atlantic air-sea interaction. *Advances in Geophysics*, Academic Press, 1-82.

Brewer, P.G., Broecker, W.S., Jenkins, W.P., Rhines, P.B., Rooth, C.G., Swift, J.H., Takahashi and Williams, R.T., 1983. A climatic freshening of the deep Atlantic North of 50 N over the past twenty years. *Science*, **222**, 1237-1239.

Broecker, W.S., 1991. The great ocean conveyor. *Oceanography*, **4**, 79-89.

Brown, B. and Katz, R., 1991. The use of statistical methods in the search for teleconnections: past, present and future. In: Glantz, M., Katz, R., and Nicholls, N., (eds), *Teleconnections Linking World Wide Climate Anomalies*, Cambridge University Press, Cambridge, 371-400

Bryan, K. and Stouffer, R., 1991. A note on Bjerknes hypothesis for North Atlantic variability. *Journal of Marine Systems*, **1**, 229-241.

Chen, T.C., VanLoon, H. and Yen, M.C., 1992. Changes in the atmospheric circulation over North Pacific-North America since 1950. *Journal of Meteorological Society of Japan*, **70**, 1137-1146.

Dansgaard, W., Johnson, S.J., Clausen, H.B. and Langway, C.C., Jr. 1971. Climatic record revealed by Camp Century Ice Core. In: Turekian, K.K., (ed.), *The Late Cenozoic Glacial Ages*, Yale University Press, New Haven, 37-56.

Deser, C. and Blackmon, M.L., 1993. Surface climatic variations over the North Atlantic during winter: 1900-1989. *Journal of Climate*, **6**, 1743-1753.

Dickson, R. and Brown, J., 1994. The production of North Atlantic deep water. *Journal of Geophysical Research*, **99**, 12319-12341.

Dickson, R., Gmitrowicz, E.M. and Watson, A.J., 1990. Deep water renewal in the North Atlantic. *Nature*, **344**, 848-850.

Enfield, D., 1989. El Niño, past and present. *Reviews of Geophysics*, **27**, 159-187.

Folland, C., Owen, J., Ward, M. and Colman, A., 1991: Prediction of seasonal rainfall in the Sahel region using empirical and dynamical methods. *Journal of Forecasting*, **10**, 21-56

Gaffen, D.J., Barnett, T.P. and Elliott, W.P., 1991. Space and time scales of global tropospheric moisture. *Journal of Climate*, **4**, 989-1008.

Gordon, A.L., 1986. Interocean exchange of thermocline water. *Journal of Geophysical Research*, **91**, 5037-5046.

Gordon, A.L., Zebiak, S.E. and Bryan, K., 1992. Climate variability and the Atlantic Ocean. *EOS*, **73**, 161-165.

Graham, N., 1990. Decadal-scale climate variability in the 1970's and 1980's. Observations and results. Proceedings of the Fifteenth Annual Climate Diagnostics Workshop. Asheville, NE October 29 - November 2, 1990. (U.S. Dept. of Commerce PB90-198573)

Gray, W.M., 1979. Hurricanes: their formation, structure and likely role in the tropical circulation. Supplement to Meteorology Over the Tropical Oceans. Published by RMS, James Glaisher House, Grenville Place, Bracknell, Berkshire, D.B. Shaw, (ed.), 155-218.

Gray, W.M., 1984a. Atlantic seasonal hurricane frequency: Part I: El Niño and 30 mb quasi-biennial oscillation influences. *Monthly Weather Review*, **112**, 1649-1668.

Gray, W.M., 1984b. Atlantic seasonal hurricane frequency: Part II: Forecasting its variability. *Monthly Weather Review*, **112**, 1669-1683.

Gray, W.M., 1990. Strong association between West African rainfall and U.S. landfalling intense hurricanes. *Science*, **249**, 1251-1256.

Gray, W.M., 1993. Atlantic conveyor belt alterations as a possible cause of multidecadal global surface temperature change. Preprint, Fourth AMS Conference on Global Change Studies, Anaheim, CA, American Meteorological Society, Boston, 384-388

Gray, W.M. and Landsea, C.W., 1991. Predicting US hurricane spawned destruction from West African rainfall. Paper given at 13th Annual Hurricane Conference, Miami, FL, April 3-5, 40 pp. (Available from Department of Atmospheric Science, Colorado State University, Fort Collins, CO, 80523).

Gray, W.M. and Landsea, C.W., 1993. West African rainfall and Atlantic basin intense hurricane activity as proxy signals for Atlantic conveyor belt circulation strength. Preprints, Fourth AMS Conference on Global Change Studies, Anaheim, CA, January.

Gray, W.M., Landsea, C.W., Mielke, P. and Berry, K., 1993. Predicting Atlantic basin seasonal tropical cyclone activity by 1 August. *Weather and Forecasting*, **8**, 73-86.

Gray, W.M., Landsea, C.W., Mielke, P. and Berry, K., 1994. Predicting Atlantic basin seasonal tropical cyclone activity by 1 June. *Weather and Forecasting*, **9**, 103-115.

Gray, W.M. and Sheaffer, J.D., 1991. El Niño and QBO influences on tropical cyclone activity. In Glantz, M., Katz, R., and Nicholls, N., (eds.), *Teleconnections Linking World Wide Climate Anomalies*, Cambridge University Press, Cambridge, 257-284.

Hastenrath, S., 1990. Decadal-scale changes of the circulation in the tropical Atlantic sector associated with Sahel drought. *International Journal of Climatology*, **10**, 459-472.

Hsiung, J. and Newell, R.E., 1986. Factors controlling free air and ocean temperature of the last 30 years and extrapolation to the past. In: Berger, W.H. and Labeyrie, L.D., (eds.), *Abrupt Climate Change*, Reidel, Dordrecht, 67-87.

Huang R., 1994. Thermocline circulation: Energetics and variability in a single basin model. *Journal of Geophysical Research*, **99**, 12471-12485.

Hurrell, J., 1995. Decadal trends in the North Atlantic oscillation and regional temperature and precipitation. *Science*, **269**. 676-679.

Kay, P.A. and Diaz, H.F. (eds.), 1985. Problems and prospects for predicting Great Salt Lake levels. Center for Public Affairs Administration, University of Utah, Salt Lake City, UT, 67-91.

Knox, J.L., Higuchi, K., Shabbar, A. and Sargant, N., 1988: Secular variation of northern hemisphere 50 kPa geopotential height. *Journal of Climate*, **1**, 500-511.

Kushnir, Y., 1994. Interdecadal variations in North Atlantic sea surface temperature and associated atmospheric conditions. *Journal of Climate*, **7**, 141-157.

Landsea, C.W., 1991. West African monsoonal rainfall and intense hurricane associations. Dept. of Atmospheric Science Paper No. 484, Colorado State University, Fort. Collins, CO, 80523, 280 pp.

Landsea, C.W., 1993. A climatology of intense (or major) Atlantic hurricanes. *Monthly Weather Review*, **121**, 1703-1713.

Landsea, C.W. and Gray, W.M., 1992. Associations of Sahel monsoon rainfall and concurrent intense Atlantic hurricanes. *Journal of Climate*, **5**, 435-453.

Landsea, C.W., Gray, W.M., Mielke, P.W. Jr. and Berry, K.J., 1992. Multidecadal variations of Sahel monsoon rainfall and U.S. landfalling intense hurricanes. *Journal of Climate*, **5**, 1528-1534.

Landsea, C.W., Gray, W.M., Mielke, P.W. Jr. and Berry, K.J., 1994. Seasonal forecasting of Atlantic hurricane activity. *Weather*, **49**, 273-284.

Lazier, J.N.R., 1980. Oceanographic conditions of weather ship Bravo, 1964-1974. *Atmosphere-Ocean*, **18**, 227-238.

Manabe, S. and Stouffer, R., 1988. Two stable equilibria of a coupled ocean-atmosphere model. *Journal of Climate*, **1**, 841-866.

Manchard, D., Sanderson, M., Howe, D. and Alpagh, A., 1988. Climatic change and Great Lakes levels: the impact on shipping. *Climatic Change*, **12**, 107-133.

McKenzie, J.A. and Eberli, G.P., 1987. Indications for abrupt Holocene climatic change: Late Holocene oxygen isotope stratigraphy of the Great Salt Lake, In: Berger. H. and Labeyrie, D., (eds.), *Abrupt Climatic Change*, Reidel Dordrecht, 127-136

Michaelsen, J. and Thompson, L.G., 1992. A comparison of proxy records of El Niño/Southern oscillation. In: Diaz, H.F. and Markgraf, V. (eds.), *El Niño: Historical and Paleoclimatic Aspects of the Southern Oscillation*. Cambridge University Press, 323-348

Moses, T., Kiladis, G., Diaz, H.F. and Barry, R.G., 1987. Characteristics and frequency of reversals in mean sea level pressure in the North Atlantic sector and their relationship to long-term temperature trends. *Journal of Climatology*, **7**, 13-30.

Mysak, L.A., Manak, D.K. and Marsden, R.F., 1990. Sea-ice anomalies observed in the Greenland and Labrador Seas during 1901-1984 and their relation to an interdecadal Arctic climate cycle. *Climate Dynamics*, **5**, 111-133.

Namias, J., Yuan, X. and Cayan, D., 1988. Persistence of North Pacific SST and atmospheric flow anomalies. *Journal of Climate*, **1**, 682-703.

Nunez, R.H., O'Brien, J. and Shriver, J., 1992. The effects of El Niño rainfall or Chile (1964-1990). TOGA Notes, **8**, 4-7

Nicholson, S.E., 1989. African drought characteristics, causal theories and global teleconnections. *Geophysical Monographs*, **52**, 79-100.

Nitta, T. and Yamada, S. 1989. Recent warming of tropical sea surface temperature and its relationship to the Northern Hemisphere circulation. *Journal Meteorological Society of Japan*, **67**, 375-383.

OUCT (Open University Course Team), 1989. Ocean circulation. Pergamon Press, Oxford.

Parthasarathy, B., Rupakumar, A. and Munot, A.A., 1991. Evidence of secular variations in Indian monsoon rainfall-circulation relationships. *Journal of Climate*, **4**, 927-938.

Ramage, C., 1983. Teleconnections and the siege of time. *Journal of Climatology*, **3**, 223-231.

Reid, G.C. and Gage, K.S., 1993. Troposphere-stratosphere coupling in the tropics: the role of El Niño and the QBO. Proceedings of NATO advanced study institute on the role of the stratosphere in global change. Canquivannc, France, September, 1992.

Saffir, H.S., 1977. Design and construction requirements for hurricane resistant construction. American Society of Civil Engineers, New York, Preprints Number 2830, 20 pp.

Schmitz, W.J., 1995. On the interbasin-scale thermohaline circulation. *Reviews of Geophysics*, **33**, 151-173.

Schmitz, W. and McCartney, M.S., 1993. On the North Atlantic circulation. *Reviews of Geophysics,* **31**, 29-49.

Serreze, M.C., Maslanik, J.A, Barry, R.G. and DeMaria, T.D., 1992. Winter atmospheric circulation in the arctic basin and possible relationships to the great salinity anomaly in the Northern North Atlantic. *Geophysical Research Letters*, **19**, 293-296.

Shabbar, A., Higuchi, K. and Knox, J., 1990. Regional analysis of 50KPa geopotential heights from 1946 to 1985. *Journal of Climate*, **3**, 543-557.

Sheaffer, J.D. and Gray, W.M., 1996. Tendency for strong warming of the equatorial lower stratosphere prior to strong El Niños. *Journal of Geophysical Research*, (in press).

Sheaffer, J.D., 1993. Decadal scale trends in Pacific-Western North America teleconnections and implications for seasonal predictability of anomalous precipitation. Preprints, Fourth AMS Conference on Global Change Studies, Anaheim, CA, American Meteorological Society, Boston, 17-22.

Sheaffer, J.D. and Reiter, E.R., 1985. Influence of Pacific sea surface temperatures and seasonal precipitation over the western plateau of the United States. *Archive of Meteorological and Geophysical Bioclimate,* **34**, 111-130.

Shinoda, M. and Kawamura, R., 1994. Tropical rainbelt, circulation and sea surface temperatures associated with the Sahelian rainfall trend. *Journal of the Meteorological Society of Japan,* **72**, 341-357.

Simpson, R.H., 1974. The hurricane disaster potential scale. *Weatherwise*, **27**, 169-186.

Stommel, H., 1957. A survey of ocean current theory. *Deep Sea Research*, **4**, 149-184.

Street-Perrott, F.A. and Perrott, R.A., 1990. Abrupt climate fluctuations in the tropics: the influence of the Atlantic Ocean circulation. *Nature*, **343**, 607-611.

Toggweiler, 1994. The oceans overturning circulation. *Physics Today*, **47**, 45-50.

Trenberth, K.E., 1990. Recent observed interdecadal climate changes in the Northern Hemisphere. *Bulletin American Meteorological Society*, **71**, 988-993.

Trenberth, K.E., 1994. The different flavors of El Niño. Proceedings of the 18th Annual NOAA Climate Diagnostics Workshop, Nov 1-5, 1993, Boulder, CO.

Walker, G.T. and Bliss, E.W., 1937. World Weather. VI. *Memoirs Royal Meteorological Society*, **4**, 119-139.

Walsh, J.E. and Chapman, W.L., 1990. Arctic contribution to upper-ocean variability in the North Atlantic. *Journal of Climate*, **3**, 1462-1473.

Weaver, A., Aura, S. and Myers, P., 1994. Interdecadal variability in an idealized model of the North Atlantic. *Journal of Geophysical Research*, **99**, 12423-12441.

Weisse, R., Mikolajewicz, U. and Maier-Reimer, E., 1994. Decadal variability of the North Atlantic in an ocean general circulation model. *Journal of Geophysical Research*, **99**, 1241-1242.

3 Climate Variations and Hurricane Activity: Some Theoretical Issues

Kerry A. Emanuel
Center for Meteorology and Physical Oceanography
Massachusetts Institute of Technology
Cambridge, MA 02139, U.S.A.

Abstract

This chapter reviews the present state of knowledge concerning the dependence of hurricane activity on climate. This state of knowledge relies on theory, observations, and numerical models. Theory suggest that an upper bound on the intensity of hurricanes, as measured by their maximum wind speeds, would increase with global warming, with the amount of increase strongly dependent on the degree of warming of tropical oceans, of which estimates by climate models must be treated with skepticism at present. While the upper bound on intensity would increase, little is known about how the average intensity of storms might change. Storm frequency is another matter and is very sensitive to many features of the general circulation of the tropical atmosphere. Climate model simulations produce very different responses of hurricane frequency to climate change and should not be considered reliable at present, though in principle it should be possible to use such models to predict general levels of storm activity. There is no evidence whatsoever that the regions susceptible to hurricanes would undergo any net expansion or contraction.

The progress in understanding the relationship between hurricane activity and climate is impeded by the lack of resources and talent directed toward this problem.

3.1 Introduction

Large interannual and interdecadal variations in tropical cyclone activity, particularly in the North Atlantic basin, have been described in earlier chapters in this book, and have been related to other known variations in the climate system. What are the physical reasons for such relationships? Answering this question is an important step toward uncovering other relationships between hurricane activity and climate variations and toward predicting how hurricane activity might change with future changes in climate. In this chapter, I describe some of the current thinking about the physics of hurricanes and how this bears on the problem of relating hurricane activity to climate.

We are most concerned with three aspects of hurricane activity: their frequency, their intensity, and their geographical distribution. Any change in the frequency with which hurricanes strike populated land is of obvious concern. But the amount of damage increases roughly as the square of the intensity of the storms, as measured by their maximum wind speed, so in practice we are concerned more with the intense storms. If some aspect of climate variation were to lead to fewer hurricanes, but more intense ones, we might expect more losses. We would also be concerned if climate change were to cause hurricanes to be experienced in parts of the world now free from them, or to cease to be experienced in regions they now trouble. From a scientific standpoint, these issues are quite separate. The factors that control the intensity of hurricanes appear to be quite different from those that govern their frequency of occurrence, and this is reflected in the observation that some seasons produce very few but very intense storms. The 1992 season had few storms, but it produced Hurricane Andrew. The geographical distribution of hurricanes over a statistically large sample is determined by features of the large-scale circulation of the atmosphere and oceans which can, in theory, be simulated by global circulation models.

A fourth characteristic of hurricanes, their geometric size, has received less attention. The radial dimension of tropical cyclones ranges over nearly a factor of ten: the smallest observed storms can be placed entirely within the eyes of the largest. A storm whose radial dimension is twice the size of another will cause perhaps as much as four times the damage (all other things being equal) since the damage track will be twice as wide and each point within it will experience damaging winds for twice as long. (The magnitude and area covered by oceans waves and the storm surge will also be greater.) But so little is now known about the factors that determine the geometric size of individual storms that I am not able to discuss the matter here. In the following sections, I focus instead on the factors affecting the intensity, frequency, and geographical distribution of hurricanes.

3.2 Intensity

The intensity of an individual hurricane, as measured conventionally by its maximum surface wind speeds or minimum surface pressure, is affected at any given time by a large and complex array of physical processes governing the interaction of the storm with the underlying ocean and with its atmospheric environment. Few of these processes are well understood. For a given ocean temperature and atmospheric thermodynamic environment, there is an upper bound on the intensity that a storm may achieve, but very few storms achieve this bound in practice. This is demonstrated in Fig. 3.1. One approach to understanding hurricane intensity is to first try to understand what controls the upper bound on intensity, and then find out what keeps most storms from achieving this limit.

The mature hurricane may be thought of as a *heat engine*; i.e. an engine that converts heat energy into the mechanical energy of the winds. (An automobile engine does the same thing: after first converting the chemical energy of gasoline into heat when the spark plug ignites the mixture of gas and air, the heat is converted into the motion of the piston as the hot air rapidly expands.) There is one small subtlety of the atmosphere: its heat content is measured not just by its temperature, but also by the amount of water vapor in it (its humidity). In a hurricane, heat is added to the atmosphere when water evaporates from the ocean. The atmosphere's heat gain is the ocean's loss, and this is reflected by a slight drop in the ocean temperature, just exactly as one's skin temperature drops when water left over from a swim evaporates. But the tropical ocean is such a vast reservoir of heat that the drop in temperature is imperceptible.

For a heat engine to operate for any appreciable time, the heat put into the engine must ultimately be taken out again. A famous theorem of thermodynamics states that the maximum fraction of heat energy that can be converted to mechanical energy by a heat engine is given by

$$\frac{T_{in} - T_{out}}{T_{in}} \tag{3.1}$$

where T_{in} is the temperature at which heat is added to the engine, and T_{out} is the temperature at which it is removed. In a hurricane, the input temperature is the temperature of the sea surface, while the output temperature is the temperature of the air flowing out of the top of the storm. The higher the storm extends into the tropical atmosphere, the lower the outflow temperature and the greater the fraction of heat input that is converted to wind energy. In a typical hurricane, T_{in} is about 28°C (301°K) and T_{out} is around -75°C (198°K), so this fraction is roughly 1/3.

How much heat can be put into a hurricane? Because the source of heat is the evaporation of sea water, the amount of heat that can be put into a hurricane is proportional to the amount of sea water that can be evaporated. One can keep evaporating sea water into air until the air becomes saturated -- that is, until the relative humidity reaches 100%. So the amount of heat that can be put into the engine from the sea is proportional to how dry the air is just above the sea surface. An expression for this maximum heat input is

$$Lq^*(1 - H) \tag{3.2}$$

where L is a coefficient called *the latent heat of vaporization*, H is the *relative humidity* of air near the sea surface (which varies between 0 and 1), and q^* is the *saturation specific humidity*, which is a *strong* function of temperature. If we now multiply this maximum heat input by the fraction of heat convertible to mechanical energy, we get an expression for the maximum mechanical energy available to the engine:

$$E = \frac{T_{in} - T_{out}}{T_{in}} Lq^{*}(1 - H) \tag{3.3}$$

What we usually want to know is not the maximum energy generation per se, but the maximum intensity that a hurricane could achieve, as measured, say, by the maximum wind speed. To estimate the maximum wind speed, we need to understand how the energy, E given by equation (3.3), is actually used. A good assumption is that it is used up mostly to overcome turbulent dissipation in the very turbulent layer of air right next to the sea. Now the rate at which energy flows into the engine from the sea is given by

$$C_h V E \tag{3.4}$$

where C_h is called the *heat transfer coefficient*, V is a representative wind speed near the sea surface, and E is given by equation (3.3). On the other hand, the rate of dissipation of mechanical energy at the sea surface is

$$C_d V^3 \tag{3.5}$$

where C_d is called the *drag coefficient*. In equilibrium, the rate of energy input must equal the rate of energy dissipation, so equation (3.4) must equal equation (3.5), giving an expression for the maximum wind speed:

$$V_{max} = \sqrt{\frac{C_h}{C_d} E} \tag{3.6}$$

This is a useful expression, because it tells us how to calculate how strong the winds might get, if we know E and the transfer coefficients. Given the state of the atmosphere and the sea surface temperature, it is easy to calculate E. This is in fact what has been done to construct the dashed line in Fig. 3.1. The climatological state of the atmosphere and the sea surface temperature near actual hurricanes was used to estimate E, and from that, the potential intensity of the storm (expressed in this case as the minimum potential pressure) was calculated. Similarly, if we could predict how the atmosphere and oceans would change with time, we could calculate this upper bound on hurricane intensity.

This is all very well and good, but it is obvious from Fig. 3.1 that expressions like equation (3.6) are useless for predicting the intensity of individual storms, or even for predicting the average intensity of storms in a given climate; the theory simply tells us a limit beyond which, in a given state of the atmosphere and ocean, the hurricane cannot go. This is useful nevertheless, because it may tell us how intense the most intense hurricanes of a given climate are likely to be. Judging from the existing record of Atlantic hurricanes, we may expect a storm approaching its limiting intensity once every few years across the entire basin, but for an area the size of the state of Florida, this may happen only once or twice in a century. A more accurate climatology of hurricane intensity, ranked as a fraction of the limiting intensity, is needed.

There are reasons to be concerned that the anthropogenic addition of greenhouse gases to the atmosphere might lead to an increase in E and therefore to an increase in the potential intensity of hurricanes; these are discussed in a previous paper by the author (Emanuel 1987). Greenhouse gases reduce the amount of infrared radiation leaving the earth's surface, and unless there is a compensating decrease in the amount of solar radiation reaching the surface by clouds (but this is complicated because clouds also trap outgoing infrared radiation), the ocean must lose the excess heat by increased evaporation of sea water. There are only two ways to achieve this: either the evaporation potential given by equation (3.2) must increase or the average surface wind speed must increase. If the evaporation potential increases, then so does the potential intensity of hurricanes, unless the efficiency given by equation (3.1) were to decrease. In fact, the efficiency also increases when greenhouse gases are added to the atmosphere. Not only does T_{in} increase, for the reasons given above, but the temperature of the tropopause, which T_{out} approximates in the deep tropics, decreases. This is because the extra greenhouse gas at high levels leads to more efficient emission of infrared radiation and thus to more cooling.

There must be considerable doubt, however, about the magnitude of the increase in potential intensity of hurricanes accompanying increases in anthropogenic greenhouse gases. The main source of uncertainty plaguing calculations of this potential increase have to do with uncertainties in the principal feedback in the climate system: the amount of water vapor in the atmosphere. Were the water vapor content to remain fixed, doubling atmospheric CO_2 would yield a tropical sea surface temperature increase of only about 0.5°C and a barely perceptible rise in the potential intensity of tropical cyclones. Most climate model simulations give much larger increases than this, owing to a positive feedback loop involving increasing atmospheric water vapor. But the physics of the processes controlling water vapor in the atmosphere are poorly understood and even more poorly represented in climate models, and what actually happens in the atmosphere is largely unknown because of poor measurements. It is now widely recognized that improvements in understanding and predicting climate hinge largely on a better understanding of the processes controlling atmospheric water vapor.

While it is certainly useful to know the limiting intensity of tropical cyclones, Fig. 3.1 demonstrates that any attempts to relate tropical cyclone intensity distributions to the state of the climate will have to reckon with the processes that prevent most storms from reaching their potential. What are these processes? Before discussing them, it must be noted that all charts such as Fig. 3.1 have, to date, relied on *climatological estimates* of the parameters that are used to estimate the potential intensity. While it is possible, in principle, to use real-time data to estimate potential intensity, this has not been done yet in any systematic way. A particularly sensitive quantity is T_{out}, the temperature at the top of the storm. Small differences in the temperature at fixed altitudes can cause the hurricane outflow to occur at very different altitudes, and thus at different temperatures. The climatological value of T_{out} may therefore not be representative of the value that occurs in the actual storm environment.

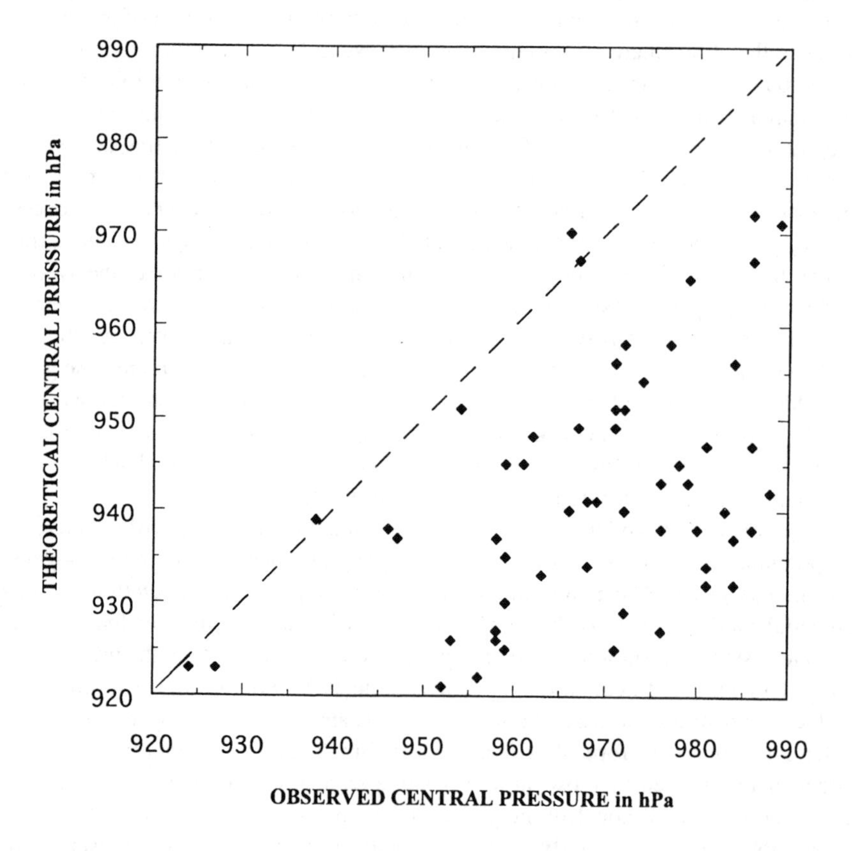

Fig. 3.1. Minimum observed central surface pressure versus calculated theoretical minimum pressure for a sample of hurricanes in the North Atlantic basin. From Schade (1994).

Among the environmental factors known to affect hurricane intensity is *vertical wind shear*. This is just the rate at which the background horizontal wind varies with altitude. Even modest magnitudes of shear prevent weak disturbances from intensifying and may also limit the intensity of mature storms. Vertical shear affects the storm dynamics in numerous ways, few of which are well understood. It acts to tilt the circulation of the storm, possibly disrupting the flow of air in, up and out through it. Also, through a complex dynamical interaction, it forces a couplet of up-down motion, straddling the storm center. This in turn causes an asymmetrical distribution of clouds and rain and may also disrupt the essential process of heat transfer from the ocean, which is particularly important under the eye wall. Finally, vertical shear causes dry air from the environment to penetrate inward closer to the core of the storm. This may lead to the formation of cold, dry downdrafts in the eyewall, partially or completely offsetting the critical moistening of the boundary later air by the ocean.

Not all dynamical interactions of tropical cyclones with their environment are detrimental to storm intensity. Tropical meteorologists in the 1940s noticed that interactions between cyclones and certain features of the flow at very high altitudes could cause intensification. Recent work by Bosart and Bartlo (1991) and by Molinari et al. (1995) have recast the analyses of such interactions in a modern dynamical framework, making the physics of the interaction somewhat more clear. It appears from this work that, under the right circumstances, the approach of a high-altitude cyclonic circulation can cause an existing tropical cyclone to intensify, or a nascent one to develop. There are also several examples of strong high-level cyclonic anomalies leading to the development of strong surface cyclones with some of the characteristics of tropical storms, even over relatively cool ocean water. One such storm occurred over the western North Atlantic in October 1992, and caused extensive coastal damage in the northeastern U.S.

Interaction of hurricanes with the underlying ocean can cause substantial reduction of the storm's intensity. This has now been well documented in modeling studies (e.g. Khain and Ginis 1991; Schade 1994). Hurricanes stir cold water up to the surface, reducing the amount of heat that flows into the storm. The magnitude of the effect depends on the thickness of the warm layer of water at the top of the ocean, on the forward speed of the storm, and on its geometric size. Typical reductions of intensity from the potential intensity are on the order of 30%.

Clearly, there are many and diverse processes that affect the intensity of individual hurricanes. We would like to know how all of these affect the statistical distributions of hurricane intensity in a given climate, but this will require a far better understanding of hurricane dynamics than we have at present. One factor limiting progress is the enormous computational demands of simulating in three dimensions the full array of scales and physics that characterize tropical cyclones. Computers are barely fast enough to simulate the interaction of a hurricane with its environment with a spatial resolution sufficient to simulate the individual cumulonimbus clouds that are the real agents of vertical heat transport in hurricanes. Another problem is social. There are a very small number of scientists working on the problem a matter, I shall take up again in the summary (Sect. 3.5).

3.3 Frequency

Very few atmospheric processes are as poorly understood as tropical cyclogenesis. In spite of years of study, it remains largely a matter of guesswork as to whether a particular tropical disturbance will become a hurricane. Thanks largely to the work of Gray (e.g. Gray 1988), we now know the atmospheric conditions that must prevail for genesis to occur, but the existence of such conditions, which are not uncommon in the tropics, is by no means a guarantee of genesis. It has been known for many years that tropical cyclones do not arise spontaneously, as do other types of storms, but must literally be triggered by disturbances of independent origin. The frequency of tropical cyclogenesis is a product of the prevalence of Gray's necessary conditions and the frequency of suitable initiating disturbances. But we do not yet know what makes one disturbance *suitable* and another *unsuitable.* Nor is there some physical constraint on the number of tropical cyclones occurring every year around the globe. As evidenced by years in which there are virtually no hurricanes in the North Atlantic, the tropical atmosphere seems to be able to live quite happily without hurricanes. Truly, these storms are accidents of nature.

Some recent developments do offer hope, however, that we may soon understand genesis well enough to be able to predict the statistical frequency of cyclone occurrence given the state of the climate. One such development is the detection by Landsea and Gray (1992) of certain strong empirical relationships between Atlantic hurricane activity and other signals in the climate system. (These are discussed fully by Gray et al., Chap. 2, this volume.) The existence of such signals clearly offers clues about the physics of genesis, but these clues have yet to be unraveled. Another development stems from some recent work with numerical hurricane simulation models and from a series of field experiments performed in the early 1990s in the western and eastern tropical Pacific. It appears that a necessary and perhaps sufficient condition for genesis is the establishment of a pillar of very humid air that extends through the entire depth of the tropical troposphere and is about 50-100 miles wide. (Normally, the tropical atmosphere is somewhat dry in middle levels.) Thunderstorms that then develop within this humid pillar do not produce the dry, cold downdrafts that characterize most such storms and which oppose the tendency of evaporation from the ocean to humidify the atmosphere. These cold, dry downdrafts are driven by the partial evaporation of falling rain, but within the humid pillar, evaporation is reduced and downdraft formation is inhibited. The formation of these humid pillars appears to be possible through a number of different mechanisms, including the lifting of the tropical boundary layer within a tropical disturbance such as an easterly wave. Humid pillars also form naturally within tropical cloud clusters, by mechanisms that have yet to be elucidated.

The problem of predicting how tropical cyclone frequency might respond to climate change can be broken into two parts: predicting how the prevalence of Gray's necessary conditions will change, and predicting how the frequency and strength of potential initiating disturbances will change. Elementary considerations suggest that anthropogenic increases in greenhouse gases will reduce the former and increase the latter. Very briefly, the strength of very large scale tropical circulations such as monsoons and the trade winds are expected to increase. (Although the pole-to-equator *surface* temperature gradient decreases, gradients at higher altitudes increase and, in the net, the strength of thermally direct circulations increases.) In general, this would be accompanied by an increase in vertical wind shear, particularly in the upper troposphere (wind shear in the lower troposphere actually decreases). This would weigh in favor of fewer cases of tropical cyclogenesis. On the other hand, the more vigorous large-scale circulation might favor more and stronger potential initiating disturbances, such as easterly waves. This would weight in favor of more tropical cyclones. Thus the problem is complex, and simple reasoning produces ambiguous results.

General circulation models (GCMs) have been used by a number of groups to explore changes in tropical cyclone activity in a double CO_2 world. To date, each of these groups has examined changes in the activity of tropical cyclones produced explicitly by the models. This approach is problematic, because neither the spatial resolution nor the physics of the models is sufficient to simulate tropical cyclones scrupulously. While the physics of mature model storms may be close to that of real hurricanes, it is very unlikely that genesis, which recent field experiments show to occur on scales as small as 30 miles, is being mimicked at all realistically by the GCMs, whose spatial resolution is more like 200 miles. For what they are worth, the GCMs produce conflicting results. The study of Haarsma et al. (1992), using the GCM run by the British Meteorological Office, shows an increase in both the intensity and frequency of tropical cyclones, but the analysis by Broccoli and Manabe (1990), using the Princeton/GFDL model, shows ambiguous results, with an increase in tropical cyclone activity if cloud-radiation feedback is not included, and a decrease in activity otherwise.

Perhaps a better strategy would be to use GCMs to assess the prevalence of Gray's necessary conditions and of potential initiating disturbances. This strategy would circumvent the need to actually simulate genesis and would be within the bounds of what the models should be capable of. At present, however, there is very little basis for taking seriously quantitative estimates of climate change produced by GCMs, if for no other reason than that there is no basis for believing that they handle water vapor correctly. But there is also good reason to be optimistic about solving the problems that plague the present generation of models, and future GCMs should prove to be valuable tools for assessing the effects of climate change on hurricane activity.

I believe that a thorough, physically based understanding of tropical cyclogenesis is a prerequisite for developing an ability to relate tropical cyclone frequency to the state of the climate. Empirical studies are enormously useful, but cannot lead to a completely general understanding of the problem. Even so, it must be admitted that such studies are far ahead of theory and modeling, which must now make an effort to catch up.

3.4 Geographical Distribution

In the current climate, hurricanes develop over tropical ocean waters where sea surface temperature (SST) exceeds about 26°C, but, once developed, they may move considerably poleward of these zones. An oft-stated misconception about tropical cyclones is that were the area enclosed by the 26°C SST isotherm to increase, so too would the area experiencing tropical cyclogenesis. Regions prone to tropical cyclogenesis are better characterized as places where the atmosphere is slowly ascending on the largest scales. Since about as much atmosphere is descending as ascending, it is hard to change the total area experiencing ascent. Thus there is little basis for believing that there would be any substantial expansion or contraction of the area of the world prone to tropical cyclogenesis. This is borne out by the GCM simulations performed by Haarsma et al. (1992), who show that while there is a substantial increase in the area enclosed by the 26°C SST isotherm is a double CO_2 environment, there is no perceptible increase in the area experiencing tropical cyclones.

It is conceivable, though, that changes in the large-scale circulation of the atmosphere would increase or decrease the rate of movement of tropical cyclones out of their genesis regions and into higher latitudes. It is also likely that changes in atmospheric circulation and SST distribution within the tropics would be associated with variations in the distribution of storms.

3.5 Summary

The theory of tropical cyclones, in its present state of development, yields some useful insights into the relationship between tropical cyclone activity and climate. There is a rigorous upper limit to the intensity that hurricanes can achieve, and this limit can be easily determined from known states of the atmosphere and ocean. Elementary considerations show that this limit increases with the amount of greenhouse gas in the atmosphere, but the magnitude of the increase that would result from the present injection of anthropogenic greenhouse gases into the atmosphere is unknown, owing to large uncertainties about feedbacks in the climate system. Moreover, very few storms approach their limiting intensity, and the processes responsible for keeping storm intensities below their limiting value are poorly understood and not likely to be well simulated by GCMs. The frequency with which tropical cyclones occur is a product of the prevalence of known necessary conditions for their formation and the frequency and strength of disturbances that have the potential of initiating tropical cyclones. Neither basic theory nor numerical climate simulation is well enough advanced to predict how tropical cyclone frequency might change with changing climate, and both give conflicting results on the change of tropical cyclone frequency on doubling atmospheric CO_2. There is no physical basis, however, for claims that the total area prone to tropical cyclogenesis would increase.

Progress in confronting the important relationship between tropical cyclone activity and climate cannot be made unless there are fundamental advances in understanding the basic physics of hurricanes. An important limitation to making such advances is social and political in nature: there are remarkably few scientists working on the problem (compared to the numbers working on, say, earthquakes, a phenomenon of comparable social significance). This is a complex matter of history and of the professional tastes that guide scientists in their choice of research problems. While it is difficult to alter such matters of taste, availability of government funding does act to encourage or discourage particular research activities. Continued funding of basic scientific research, some of which is directed toward improved prediction and warning of hurricanes, and toward evaluation of the influence of these great storms on society would be a prudent investment in the future.

References

Bosart, L.F. and Bartlo, J.A., 1991. Tropical storm formation in a baroclinic environment. *Monthly Weather Review*, **119**, 1979-2013.

Broccoli, A.J. and Manabe, S., 1990. Can existing climate models be used to study anthropogenic changes in tropical cyclone climate? *Geophysical Research Letters*, **17**, 1917-1920.

Emanuel, K.A., 1987. The dependence of hurricane intensity on climate. *Nature,* **326**, 483-485.

Gray, W.M., 1988. Environmental influences on tropical cyclones. *Australian Meteorology Magazine*, **36**, 127-139.

Haarsma, R.J., Mitchell, J.F.B. and Senior, C.A., 1992. Tropical disturbances in a GCM. *Climate Dynamics*, **8,** 247-257.

Khain, A. and Ginis, I., 1991. The mutual response of a moving tropical cyclone and the ocean. *Beitrag zur Physik der Atmosphere*, **64**, 125-141.

Landsea, C.W. and Gray, W.M., 1992. The strong association between western Sahel monsoon rainfall and intense Atlantic hurricanes. *Journal of Climate*, **5**, 435-453.

Molinari, J., Skubis, S. and Vollaro, D., 1995. External influences on hurricane intensity. Part III: Potential vorticity evolution. *Journal of Atmospheric Science*, **52**, 3593-3606.

Schade, L.R., 1994. The ocean's effect on hurricane intensity. Ph.D. thesis, Massachusetts Institute of Technology, Cambridge, 127 pp.

4 Numerical Simulation of Intense Tropical Storms

Lennart Bengtsson, Michael Botzet, and Monika Esch
Max Planck Institute for Meteorology
Bundesstrasse 55
20146 Hamburg, Germany

Abstract

A high resolution general circulation model has been used to study intense tropical storms. A five-year-long global integration with a spatial resolution of 125 km has been analysed. The geographical and seasonal distribution of tropical storms agrees remarkably well with observations. The structure of individual storms also agrees with observations, but the storms are generally more extensive in coverage and less extreme than the observed ones. A few additional calculations have also been done by a very high resolution limited-area version of the same model, where the boundary conditions successively have been interpolated from the global model.

These results are very realistic in many details of the structure of the storms including simulated rain-bands and an eye structure. The global model has also been used in another five-year integration to study the influence of greenhouse warming. The sea surface temperatures have been taken from a transient climate change experiment carried out with a low resolution coupled ocean-atmosphere model. The result is a significant reduction in the number of hurricanes, particularly in the Southern Hemisphere. Main reasons for this can be found in changes in the large-scale circulation, i.e. a weakening of the Hadley circulation, and a more intense warming of the upper tropical troposphere. A similar effect can be seen during warm ENSO events, where fewer North Atlantic hurricanes have been reported.

4.1 Introduction

Tropical cyclones are by far the most devastating of all natural disasters, both by causing loss of human life and by giving rise to large economic costs. The economic damages after Hurricane Andrew in Florida and Louisiana in August 1992 were in the order of 30 billion dollars.

Detection, prediction, and warning of tropical storms are some of the most important tasks of the meteorological services, and major efforts have been spent and are being spent to find ways of further improving this work. Of particular importance is the possibility of using a general purpose forecasting model for such forecasts. Such an opportunity is gradually becoming feasible as newer versions of general circulation models (GCM) are able to resolve many of the characteristic features of tropical cyclones. Today, forecasts from the large forecasting centers such as European Centre for Medium-Range Weather Forecasts (ECMWF) and the U.S. National Centers for Environmental Prediction (NCEP) provide crucial guidance in the operational prediction of hurricanes and typhoons (Shun 1992). It may be expected, as observational systems, data-assimilation, and modelling continue to develop, that tropical storms will be routinely predicted by large scale models in very much the same way as extratropical cyclones are predicted today.

The possibility of predicting tropical storms with a large-scale general circulation model also creates a very powerful tool with which to explore tropical storms in a systematic way, and to address questions concerning the fundamental physical processes behind the generation and manifestation of these phenomena and their role in the general circulation of the atmosphere. The specific question whether tropical storms will be more frequent or more vigorous or both as a result of future increased concentration of greenhouse gases in the atmosphere will also require a comprehensive modelling approach.

Intense tropical storms, hurricanes, and typhoons undergo large interannual variations. During the period 1958-1977, the annual number of intense tropical storms varied between 97 in 1971 and 67 in 1977 (Gray 1979). Even larger variations occur for each hemisphere and in specific ocean basins. In the Northwest Atlantic, for example, there were 14 hurricanes in 1971 but only 4 in 1972. Similarly, the early 1990s were noted for the very small number of hurricanes followed by a record large number in 1995.

Wu and Lau (1992) have studied the relation between El Niño-Southern Oscillation (ENSO) and the frequency of intense tropical storms in different ocean basins. It is suggested by Wu and Lau (1992) that El Niño acts to reduce the number of Atlantic hurricanes while the opposite is true in the cold La Niña phase. Another example of a process which may affect the number of Atlantic storms is the level of soil moisture in the Sahel region, as has been suggested by Gray (1988).

Another question that has been considered recently is whether the anticipated greenhouse warming (IPCC 1990, 1992) may influence the frequency, distribution and intensity of tropical storms. In view of the serious impact of an increase of these phenomena, which in fact may be more damaging than the overall warming itself, Broccoli and Manabe (1990), Haarsma et al. (1992), and Bengtsson et al. (1996a) have undertaken a series of scenario integrations with the present and a doubled CO_2 concentration, respectively, to investigate whether the suggestions by Emanuel (1986, 1988) hold in a more realistic context. Broccoli and Manabe's result is inconclusive, and it follows that the result is crucially dependent on the parameterization of clouds. In the case of fixed or climatological prescribed clouds, a doubling of CO_2 leads to an increase in the number of tropical storms, while in the case of predictive clouds the doubling of CO_2 leads to a decrease in the number of storms! Haarsma et al. (1992), on the other hand, found a larger number of hurricane type vortices in double CO_2 experiments.

Most recently, Bengtsson et al. (1995, 1996a) have undertaken two 5-year integrations with a very high resolution model (horizontal resolution 125 km) using calculated sea surface temperature (SST) data from a transient coupled run experiment. The result of the double CO_2 run shows clearly *a reduction in the number of storms by as much as 37%.*

The main purpose of this paper is to summarize the work of Bengtsson et al. (1995, 1996a) and to present some additional numerical experiments undertaken with a nested limited area model. Following a description of the model in section 4.2, we summarize in section 4.3 the experiments undertaken with the global model. In section 4.4 we analyse the result of the limited area model and in section 4.5 simulations of hurricanes in a double CO_2 climate.

4.2 Modelling Aspects

The present study has been undertaken with the so-called ECHAM3 model (Roeckner et al. 1992). ECHAM3 is the third-generation GCM used for global climate modelling investigations in Germany. It is a spectral transform model with triangular truncation. It has evolved from the medium-range forecasting model used at ECMWF, but significantly modified to make it suitable for climate prediction. This investigation has been undertaken with a high-resolution version of the model at wave number T106. This corresponds to an average horizontal resolution of around 125 km. The main aspects of ECHAM3 are given below.

The parameterization of sub-grid scale processes is heavily simplified, partly because of insufficient detailed knowledge (e.g. the turbulent transfer and cloud microphysics), and partly because a more accurate treatment would exceed presently available computer resources (e.g. the computation of a radiative transfer). The radiation scheme uses a formulation of the radiative transfer equations with six spectral intervals in the infrared and four intervals in the solar part of the spectrum (Hense et al. 1982; Rockel et al. 1991). Gaseous absorption due to water vapor, carbon dioxide, and ozone are taken into account as well as scattering and absorption due to aerosols and clouds. The cloud optical properties are expressed in terms of cloud water content, which is a specific variable of the model.

The vertical turbulent transfer of momentum, heat, water vapor, and cloud water is based upon similarity theory for the surface layer and eddy diffusivity above the surface layer, as in the original ECMWF model (Louis 1979). The drag and heat transfer coefficients depend on roughness length and Richardson number, and the eddy diffusion coefficients depend on wind stress, mixing length, and Richardson number, which has been reformulated in terms of cloud-conservative variables (Brinkop 1991, 1992).

The effect of orographically excited gravity waves on the momentum budget is parameterized on the basis of linear theory and dimensional considerations (Palmer et al. 1986; Miller et al. 1989). The vertical structure of the momentum flux induced by the gravity waves is calculated from a local Richardson number, which describes the onset of turbulence due to convective instability and the associated breakdown.

The parameterization of cumulus convection is based on the concept of mass flux and comprises the effect of deep, shallow, and mid-level convection on the budget of heat, water vapor, and momentum (Tiedtke 1989). Cumulus clouds are represented by a bulk model including the effect of entrainment and detrainment on the updraft and downdraft convective mass fluxes. Mixing due to shallow stratocumulus convection is considered as a vertical diffusion process with the eddy diffusion coefficients depending on the cloud water content, cloud fraction, and the gradient of relative humidity at the top of the cloud.

Stratiform clouds are predicted explicitly in accordance with a cloud water equation including sources and sinks due to condensation/evaporation and precipitation formation both by coalescence of cloud droplets and sedimentation of ice crystals (Sundqvist 1978; Roeckner et al. 1991). Sub-grid scale condensation and cloud formation is taken into account by specifying appropriate thresholds of relative humidity depending on height and static stability.

The land surface model considers the budget of heat and water in the soil, snow over land and the heat budget of permanent land and sea ice (Dümenil and Todini 1992). The heat transfer equation is solved in a five-layer model assuming vanishing heat flux at the bottom. Vegetation effects such as interception of rain and snow in the canopy and the stomatal control of evapotranspiration are grossly simplified. For a more extensive report, reference is made to Roeckner et al. (1992).

No attempt was made to modify the physical parameterization in order to handle processes of importance for the generation and development of hurricanes, specifically. Instead, the strategy has been to use a general purpose model, since we believe that the generation of hurricane-type vortices is inherently determined in models which can handle large-scale convective forcing properly, and where the numerical representation and resolution is sufficient. As has been discussed above, at an equivalent resolution around 500 km, already such developments occur. However, as will be demonstrated in this study, the realism in the simulation is only becoming evident at much higher resolutions.

4.3 The Global Storm Experiment

The development of well-organized tropical vortices is as typical a feature of a GCM as is the development of extratropical storms and occurs regularly in the

ECHAM model as well as in most other general circulation models. During the peak of the hurricane season, July to October in the Northern Hemisphere and January to March in the Southern Hemisphere, these vortices transport substantial amounts of eddy kinetic energy and moisture polewards during the hurricane season. Gray (1979) has estimated this to be of the order of 20% of the total transport.

The basic experiment described below consists of a 5-year integration with the T106 ECHAM model. The SST distribution was taken from the period 1979-1988. Only the averaged annual variations were considered. The initial atmosphere was interpolated from the ECMWF operational analyses.

A crucial part of the study is the specification of criteria for an automatic determination of hurricane-type vortices in the experiment. At our disposal was the standard archived data record consisting of all the basic quantities at all model levels, together with additional derived parameters, precipitation, fluxes, etc. For the experiment these data were stored twice daily. The criteria for identifying the storms were only based on their typical structure and intensity, and except for the fact that the search is restricted to ocean areas only, there are no conditions imposed on sea surface temperatures, time of the year or the geographical domain.

The notation "hurricane" were used for the intense tropical vortices generated by the model. This was done in spite of the fact that the model-simulated hurricanes differ from those of the real atmosphere in some respects. The major difference is one of scale in that the simulated ones extend over a larger area, another is the absence of an eye structure. However, as we will demonstrate in section 4.4, the experiments with a nested very high resolution limited area model show a distinct reduction in the size of the hurricane. The experiments also show the development of an eye structure.

A main objective was to base the identification of the vortices on dynamical and physical criteria only, thus avoiding empirical conditions on geographical distribution, sea surface temperature or specific time of the year. Since the search was limited to the generation of vortices, we restrict the search to ocean areas only, as inspection of a large number of maps did not show any land developments. Furthermore, we only considered storms with a lifetime of at least 36 hours. For a more extensive description, reference is made to Bengtsson et al. (1995).

4.3.1 Model hurricane structure

Based on the structure of typical tropical vortices taken from a study at T42 resolution (~275 km), the following criteria were found suitable:

1. Relative vorticity at 850 hPa > $3.5 \times 10^{-5} s^{-1}$
2. A maximum velocity of 15 m s^{-1} and a minimum surface pressure within a 7×7 grid point area around the point which fulfils condition 1.
3. The sum of the temperature anomalies (deviation from the mean as defined below) at 700, 500, and 300 hPa > 3°C.

4. The temperature anomaly at 300 mb > temperature anomaly at 850 hPa.
5. The mean wind speed at 850 hPa > mean wind speed at 300 hPa.
6. Minimum duration of the event, ≥ 1.5 days.

The minimum in surface pressure was determined as the center of the storm. Mean values were calculated within a 7 x 7 grid point area around the point of minimum pressure.

It was found that the criteria 3 to 5, above, specifying a vertical structure with a warm core and maximum wind at low levels, very efficiently eliminated extratropical storms which have a completely different structure. The only differences, which happen a few times a year, were very intense features at high latitudes in both hemispheres in the winter season.

We examine the vertical structure of the mature stage of a selected storm in Fig. 4.1 This storm is located west of Taiwan at location 24°N, 129°E. The following quantities have been calculated as a function of height and radius:

(i) the azimuthal mean of the tangential velocity,
(ii) the radial velocity,
(iii) the vertical velocity (in Pa s^{-1}),
(iv) the relative humidity,
(v) the vorticity (in $10^{-5}s^{-1}$)
(vi) the temperature anomaly.

The vertical cross section of the tangential wind compares well with a Pacific composite typhoon (Frank 1977). The maximum tangential wind occurs at around 850 hPa and is located at 2° latitude distance from the center. The maximum outflow takes place farther away from the center some 10-15 degrees of latitude distance at around 100 hPa. The agreement with the composite data from Frank (1977) is excellent. The same is also true for radial velocities. The maximum inflow takes place in the boundary layer around 950 hPa and the maximum outflow just above 200 hPa. The distance from the center for the inflow maximum is larger for the simulation with about the same ratio as for the tangential wind.

The vertical velocity agrees well with Frank (1977), with upwind motions within some 3° from the center and with an elongated zone of maximum winds through the whole depth of the troposphere. As in Frank's composite data there is a weak area of sinking motion about 5° from the center. Similarly the vertical cross section of relative humidity agrees nicely with Frank, and there is even in the simulation a weak indication of lower relative humidity (below 80%) at the very center of the storm.

The vorticity is clearly noted with values well above $10^{-4}s^{-1}$ within 2° of the center. Outside the core of intense vorticity there is an indication of a banded structure of weak bands of positive and negative vorticity through the troposphere. There is a maximum of negative vorticity around 100 hPa around 6° from the center stretching out through the lower stratosphere on top of the vortex. Finally, the temperature anomaly is reaching a value of over 8°C between 200 and 250 hPa. Observed values from individual intense storms are normally larger. Hawkins and

Imbembo (1976) have reported 16°C for Hurricane Inez in September 1966. We explored a few other storms as well and found similar characteristic structures.

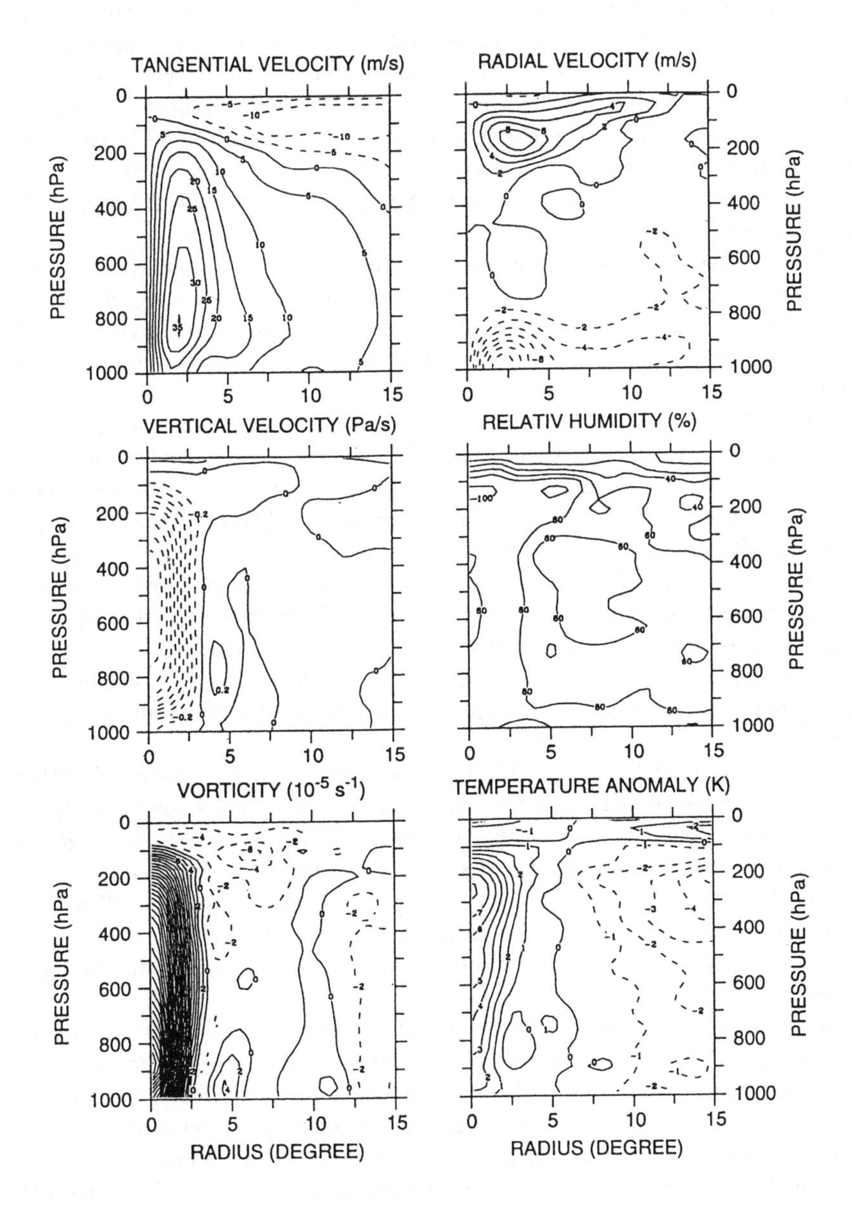

Fig. 4.1. Two-dimensional cross section (pressure height versus radial distance from storm center in degrees of latitude) of tangential wind (m s^{-1}), radial wind (m s^{-1}), vertical velocity (Pa s^{-1}), relative humidity (%), vorticity (10^{-5} s^{-1}), and temperature (K) for the maximum stage of the development at 15 August 12 UTC, year 1.

4.3.2 Geographical Distribution of Hurricanes

According to Gray (1979) there are approximately 80 tropical cyclones with a maximum sustained wind speed of 20 to 25 m s^{-1} over the globe per year. The average number of simulated tropical cyclones amounts to 83 per year, varying between 91 during the third and fifth year and 72 during the first. The observed variation over the 20-year period 1958-1978 was a maximum of 97 in 1971 and a minimum of 67 during 1977. The average observed ratio between the number of storms at the Northern Hemisphere to the Southern Hemisphere is 2.2 compared to 2.1 in the simulation.

Figure 4.2 shows the average number of simulated tropical cyclones by month all fulfilling the criteria in section 4.3.1 compared to the observed average distribution during the 20-year period in Gray's (1979) study. The agreement is good except for somewhat higher numbers of simulated storms in most months and a peculiar minimum in July.

Figure 4.3 shows the average number of storms for different ocean basins. The agreement is also here remarkably good and the number falls by and large within the range of variability as reported by Gray. There is an indication of more storms in the Northwest Pacific and Southern Indian Ocean and a slightly lower number in the Northeast Pacific and Southern Pacific.

We have examined the storms with respect to their maximum wind speed in the different ocean basins. It is interesting to note that the Northwest Pacific has by far the most powerful storm as well as the largest number of storms. The Northwest Atlantic region is particularly interesting, having a comparatively large number of very intense storms, in spite of the fact that the total number of storms is relatively small.

The highest overall wind speed is 53.1 m s^{-1} and the lowest pressure 957 hPa as selected from the archived data for every 12 h. In reality much higher wind speeds and lower minimum pressure are observed. However, these very extreme conditions can only be found over a very small area, not possible to resolve with the present model. Dell'Osso and Bengtsson (1985) and Kurihara and Tuleya (1981) have clearly demonstrated that given an adequate resolution the intensity of the storms appear to converge towards the observed values. We will discuss this further in section 4.4.

Figure 4.4 shows the cyclone tracks for all cyclones having a minimum pressure less than 1000 hPa. Figure 4.5 shows the corresponding observed tracks over a nominal 3 year period. Finally, there are a few extratropical cyclones which fulfil the criteria given in section 4.3. These are all extremely deep, highly occluded, and very slowly moving systems which occur at very high latitudes, mostly in the winter season. These events are quite rare, about 3 per year, and none of them meets the hurricane criteria more than 2 days.

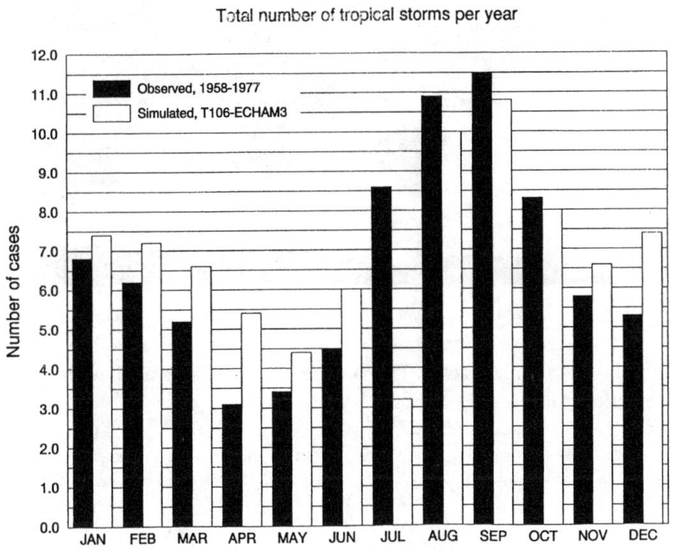

Fig. 4.2. Observed annual average number of hurricanes 1958-1977 according to Gray (1979) (solid bars) and simulated annual average number of hurricanes over five years with the T106 model (unfilled bars)..

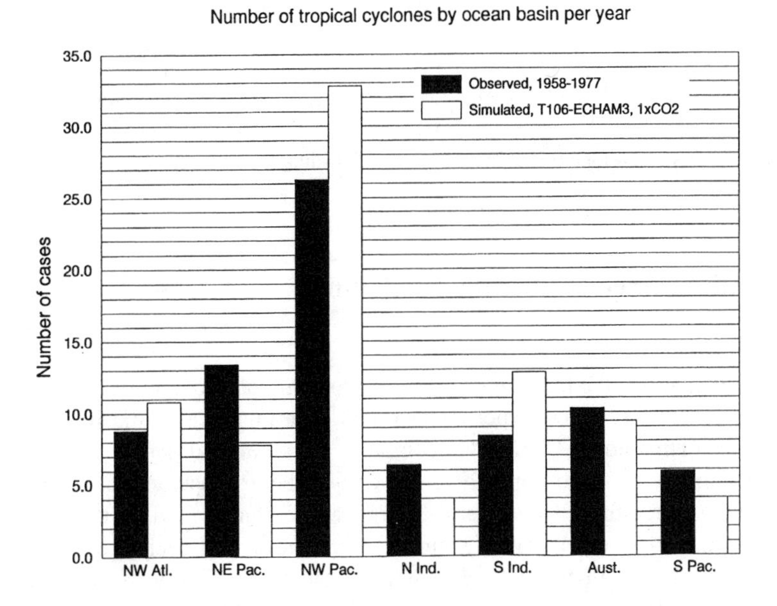

Fig. 4.3. Observed annual average number of hurricanes 1958-1977 for each ocean basin according to Gray (1979) (solid bars) and the simulated ones (unfilled bars).

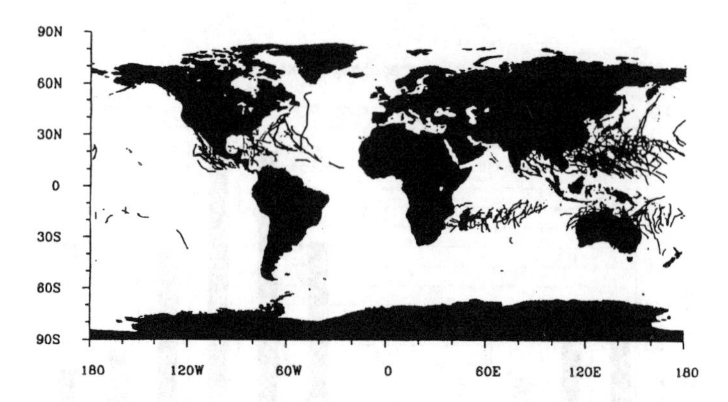

Fig. 4.4. Simulated cyclone tracks. The tracking is only indicated when the hurricanes satisfy the selection criteria. All tracks for 5 years.

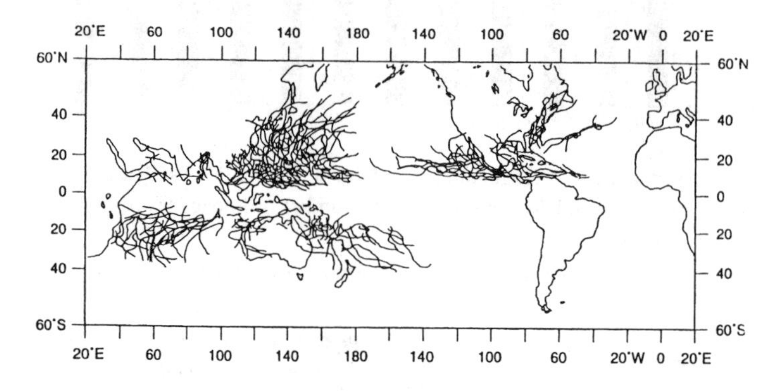

Fig. 4.5. Observed cyclone tracks over 3 years. According to Gray (1979).

4.4 Limited Area Hurricane Modelling

In order to examine the hurricanes from the global model in more details, we have also undertaken a series of experiments on an even finer scale by using a limited area model. This model HIRHAM3 combines the numerical part of the so-called HIRLAM model (Gustafsson 1993) presently used for operational short-range numerical forecasting by several European countries with the physical parameterization of the ECHAM3 model (Christensen and Meijgard 1992). The hybrid vertical coordinates and the number of vertical levels (19) is identical with the global model. In order to assure a satisfactory uniformity of the grid for arbitrary integration domains, the calculations in HIRHAM3 are carried out on a rotated latitude-longitude grid (Machenhauer 1988).

The limited area model is initialized with interpolated data from the global model. During the course of the integration the model is successively updated in a lateral boundary zone having a width of 8 grid points. The boundary data are blended with the data from the limited area model with decreasing relaxation from the boundary in order to damp possible reflection of outgoing waves. The SSTs are also interpolated onto the sea mask of the limited area model. The limited area model uses a finer and steeper orography. This may occasionally give rise to an erroneous generation of gravity waves if the boundary has steep orographic boundaries. We do not consider this to be a problem in this study since the limited areas essentially only covers ocean regions.

Here we focus on two integration areas. The first one covers a large part of the Northwest Pacific (Fig. 4.6). The integration area has 130×91 grid points and an average resolution of 0.34 lat./long. The second area has been selected for the study of hurricanes in the Atlantic sector. The area covers the Northwest Atlantic region including the Gulf of Mexico (Fig. 4.7). We have here used a somewhat finer resolution of 0.25 lat./long. but the same number of grid points. This implies a somewhat smaller area of integration. The purpose of using different resolutions for the two areas was partly to gain experimental experience but partly also to the fact that typhoons in the Northwest Pacific often have larger scales than the Atlantic hurricanes, and therefore it may be justified to apply a somewhat coarser grid in the Pacific region.

HIRHAM
North West Pacific area

Fig. 4.6. Integration area for the Northwest Pacific. A resolution of 0.34 lat./long. was used here. The coordinate system was rotated in order to give minimum distortion in the area.

Fig. 4.7. Integration area for the Northwest Atlantic. A resolution of 0.25 lat./long. was used. Otherwise, same as Fig. 4.6.

We have investigated several hurricanes and generally found that they are more spatially confined and more intense in the limited area model integrations. It is not possible to conclude anything of the general movements of the hurricanes due to the limited number of cases and the short duration of the runs (4-12 days) but there is a tendency towards stronger poleward movements. Below we describe two events, one from each region, in more detail. They have been selected from the 5-year run with the global model.

The first case is selected from August during the first year of simulation. It was one of the more intense storms of the simulation reaching a maximum wind speed of more than 52 m s^{-1}. The cyclone was first detected in the area around 15°N, 148°E, an area which is the a source region for the most powerful tropical storms on Earth. A detailed analysis of this storm has been done by Bengtsson et al. (1995). The initial pressure and low level winds for the limited area integration from August 7 of year 1 are displayed in Fig. 4.8. During the next 10 days the storm is slowly moving west-northwest under rapid intensification. There is only a minor difference in the cyclone tracks as displayed in Fig. 4.9 and a small difference in the position of landfall. The landfall in the limited area model takes place some 2° to the north. There is also a time lag of about 18 hours due to a slower movement during the first days of integration in the limited area run presumably related to a more rapid intensification.

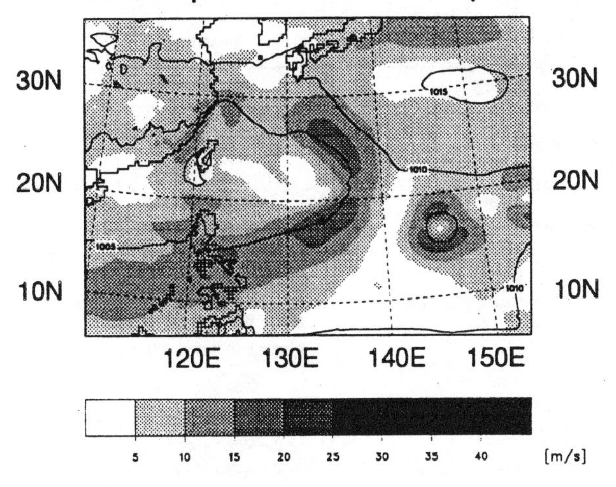

Fig. 4.8. Initial state for the experiment in the Northwest Pacific pressure (every 5 hPa) and wind speed (every 5 m s^{-1}) are indicated. The initial data are interpolated from the global T106 run. Time is 00 UTC, 7 August, year 1

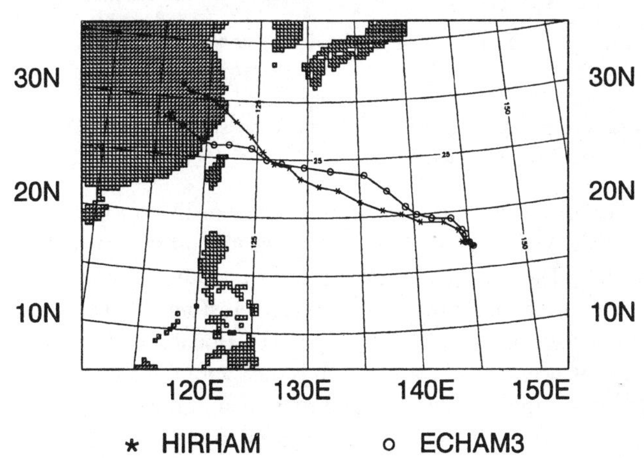

Fig. 4.9. Cyclone tracks from the global model run (open circles) and for the limited area model run, for the Northwest Pacific case. Positions are indicated for every 12 hours.

Time evolution of the North West Pacific simulation

a)

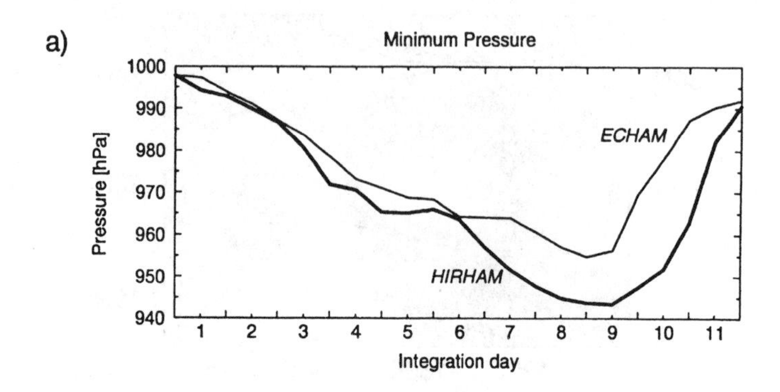

b)

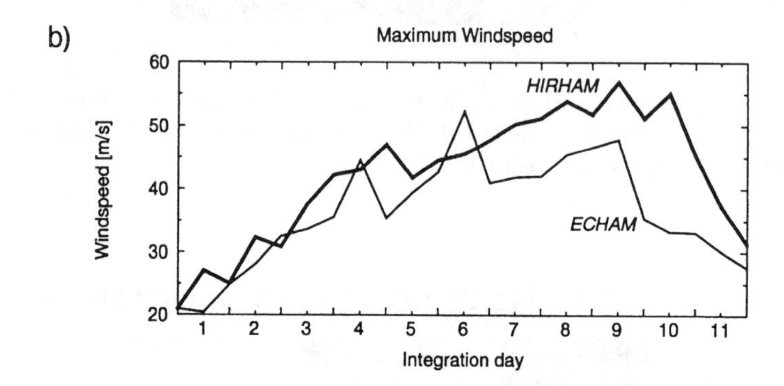

Fig. 4.10. Minimum pressure (a) and maximum wind speed (b) for the Northwest Pacific storm from 00 UTC 7 August, year 1.

The evolution of surface pressure in the center of the storm and the maximum wind speed are shown in Fig. 4.10. The intensification of the storm due to the higher resolution is clear. The minimum pressure falls by another 11 hPa and the maximum wind speed increases by a further 5 m s^{-1}. It is interesting to note the tendency of the model to generate stronger wind speed at 12 UTC (local midnight). Another typical feature of the higher resolution integration is the overall reduction in the size of the storm and the much more realistic structure with an indication of the formation of an eye (Fig. 4.11). No such feature could be seen in the integration with the global model.

The second case is selected from early November also during the first year. At the start of the integration a well organized vortex is seen north of Puerto Rico (Fig. 4.12). It was first noticed several days earlier as an easterly wave outside the African coast. During the next 6 days the hurricane moved west-northwest and then turned northeast passing over Florida (Fig. 4.13). The movement in the high resolution limited area run is slower and the hurricane is more intense. A particularly rapid intensification takes place between day 4 and 5 when the hurricane is between Cuba and Florida (Fig. 4.14). The global run does not show any such development, presumably due to the fact that the vortex is extending over too large an area, that is partly over land. A few additional case studies have been undertaken with a result very similar to the two cases reported here. All limited area experiments so far are summarized in Table 4.1.

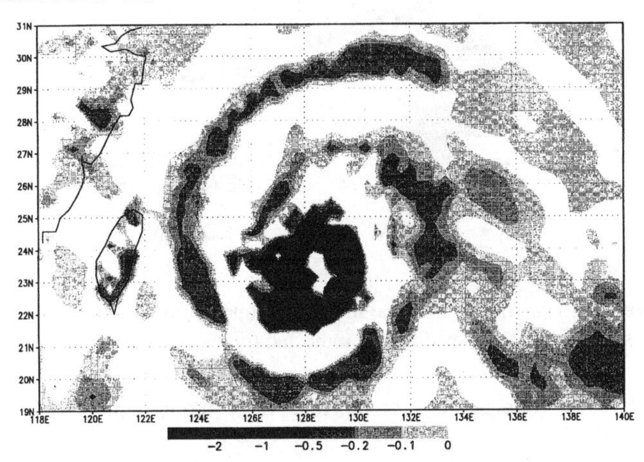

Fig. 4.11. Vertical velocity from the limited area run in the Northwest Pacific. Only the regions of rising motion are contoured, blank areas denote sinking motion.

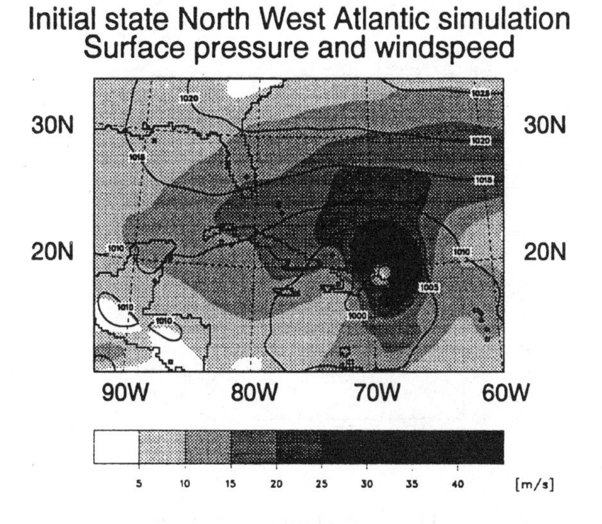

Fig. 4.12. The same as Fig. 4.8 but for the Northwest Atlantic. 00 UTC, 1 November, year 1.

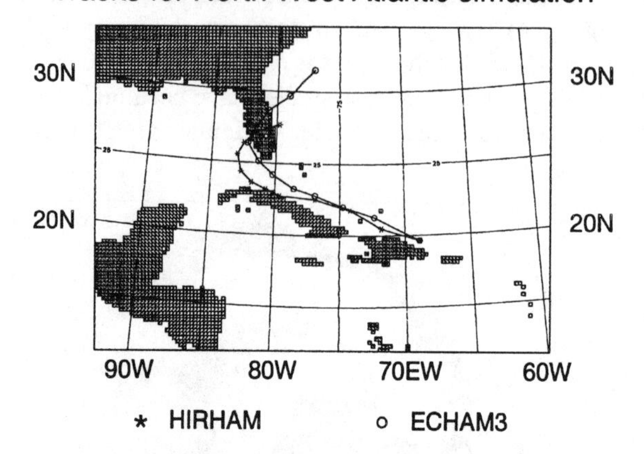

Fig. 4.13. Cyclone tracks from the global model run (open circles) and from the limited area run. Northwest Atlantic case. Positions are indicated for every 12 hours.

Time evolution of the North West Atlantic simulation

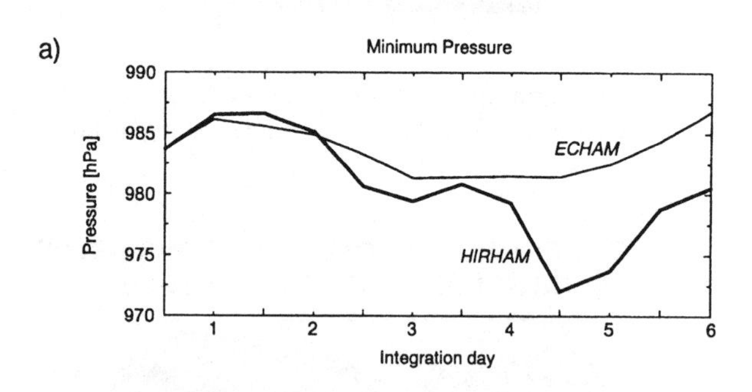

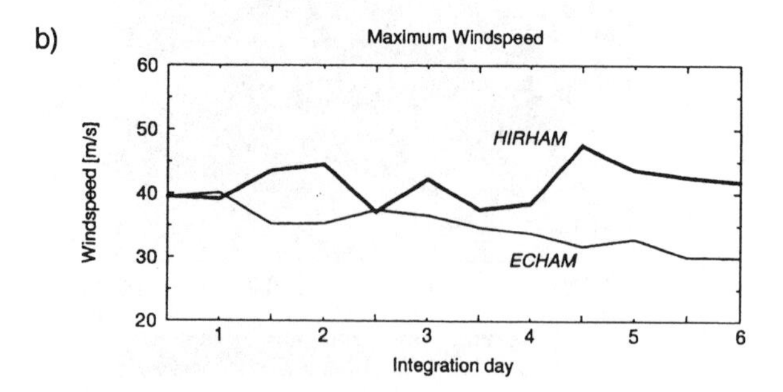

Fig. 4.14. Same as Fig. 4.10 but for the Northwest Atlantic case.

Table 4.1 Summary of limited area integrations.

	ECHAM3-T106		HIRHAM3	
NWA	resolution 125 km		resolution 28 km	
	min. pressure	max. windspeed	min. pressure	max. windspeed
1	983	37	978	44
2	981	39	972	47
NWP	resolution 125 km		resolution 44 km	
	min. pressure	max. windspeed	min. pressure	max. windspeed
3	996	36	989	42
4	981	46	948	74
NWP	resolution 125 km		resolution 38 km	
	min. pressure	max. windspeed	min. pressure	max. windspeed
5	954	52	939	57

Minimum pressure [hPa] and maximum windspeed [m/s] at 850 hPa.
NWA = North-West Atlantic
NWP = North-West Pacific

4.5 Hurricanes in a Greenhouse World

Several climate change integrations have been undertaken in recent years and considerable discussions of these experiments have taken place (e.g IPCC 1990, 1992, and 1996). However, these experiments have so far been mainly concerned with the overall global effects of climate warming and comparatively little discussion have taken place to address the more important issues of regional climate changes. A major reason has been that climate modelling and analysis have so far mostly been concerned with physically and dynamically simple models or with comprehensive models with insufficient resolution to address regional issues in earnest.

It is particularly important to explore whether extreme events like hurricanes and typhoons may be more common and more devastating or affecting larger areas in the future due to an increase in the sea surface temperature as a consequence of greenhouse warming. Empirical evidence as well as numerical simulation studies (Bengtsson et al. 1995) do indicate that the tropical storms only develop in areas where the sea surface temperature is higher than some 26°C. At first estimate therefore, an increase of such an area would have the consequences that tropical storms may affect larger areas and occur in regions where they presently do not occur.

Although several climatologists, e.g. Committee on Earth Sciences (1989) and Schneider (1990), argue that the climate warming may have started and indications exist that the intensity of tropical storms as well as extratropical storms may have increased, there are also other studies (Idso 1989; Idso et al. 1990) which claim that such signs cannot yet be detected. The long series of tropical cyclones of the North

Atlantic Ocean 1871-1992 (Neumann et al. 1993) does not show any particular trend in the number of storms over the last 60 years, but instead typical large variations from year to year. Gray (1989) has made the point that warmer sea surface temperatures do not necessarily mean increased hurricane activity, since high sea surface temperature is just one of several conditions favoring hurricane development.

There have also been a few numerical studies (Broccoli and Manabe 1990; Haarsma et al. 1992), which have been undertaken with low resolution climate models, indicating that the tropical vortices generated by the model may increase with increased greenhouse warming. But as was demonstrated in the paper by Broccoli and Manabe, their result was crucially dependent on the parameterization of clouds.

Bengtsson et al. (1996a) have undertaken a 5- year simulation of intense tropical storms with a global climate model at T106 resolution described in section 3. They used the SSTs from a previously undertaken climate change experiment (Cubasch et al. 1992). This experiment calculated the transient climate change with a coupled ocean/atmosphere model at an horizontal resolution equivalent to T21. The coupled model was integrated for 100 years from 1985 to 2085, assuming approximately 1% annual increase of CO_2 (IPCC Scenario A, business as usual according to IPCC 1990). The SST data at the time of the CO_2 doubling, including an annual cycle, were used as boundary conditions for the 5-year T106 model.

The results of the experiment by Bengtsson et al. (1996a) can be summarized as follows. The number of hurricanes in the double CO_2 case is 37% less than in the control experiment. At the Southern Hemisphere the reduction is as high as 57%. There are no changes in the distribution of storms. The statistical significance of the reduction has been evaluated by a non-parametric Mann-Whitney test (Conovar 1980), where the probability estimate is based on a ranking of the number of storms for each individual year for the present and the double CO_2 climate. Based on the Mann-Whitney test the reduction of storms is found to be significant for the Northwest Pacific, South Indian Ocean, and Australia within a 97.5-99% confidence level.

Bengtsson et al. (1996a) suggest that the enhanced vertical windshear, reduced low level vorticity as well as the reduced vertical lapse rate may contribute to the reduction. The Hadley circulation in the double CO_2 run is also weakened, particularly for the Southern Hemisphere hurricane season. This can clearly be seen in Fig. 4.15 where the upper level outgoing branch of the Hadley circulation is significantly reduced in the double CO_2 case.

It is also interesting to note the similarity in the overall temperature distribution of the upper troposphere between the double CO_2 and the control run and between different ENSO years. This is illustrated in Fig. 4.16, where we have calculated correlation between SST in the Niño region of the equatorial Pacific and the 200 hPa temperature during northern winter (December-February) (see Gray et al., Chap. 2, this volume). The figure highlights the difference in upper tropospheric temperature between warm and cold ENSO events for the period 1979-1993. The correlation pattern is strikingly similar to the temperature change pattern illustrated in Fig. 4.17. As has been pointed out by O'Brien (pers. comm.) the number of hurricanes events are significantly reduced at least in the Northwest Atlantic during warm ENSO events. (Fig. 4.18).

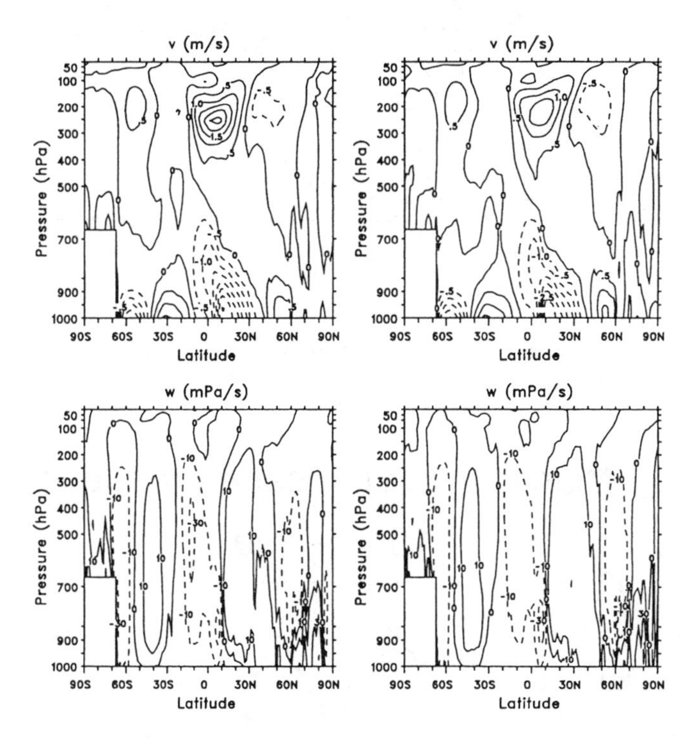

Fig. 4.15. Vertical cross section of the meridional and vertical wind during February for the control (left) and the double CO_2 case (right).

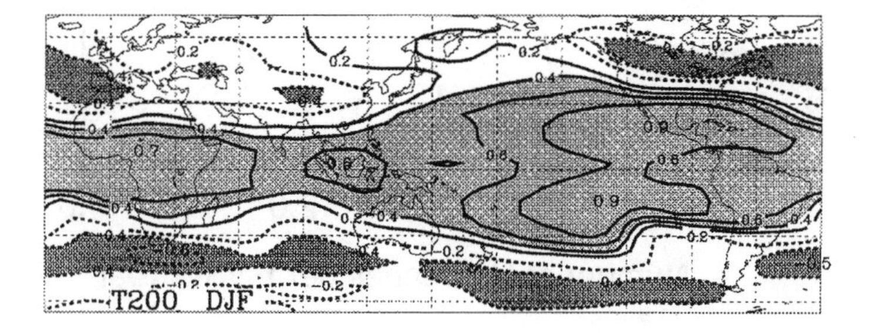

Fig. 4.16. Correlation pattern between SST from the NINO3 area and the temperature at 200 hPa. Calculations have been obtained from an ensemble of five integrations 1979-1993 using observed SST. From an experiment by Bengtsson et al. (1996b).

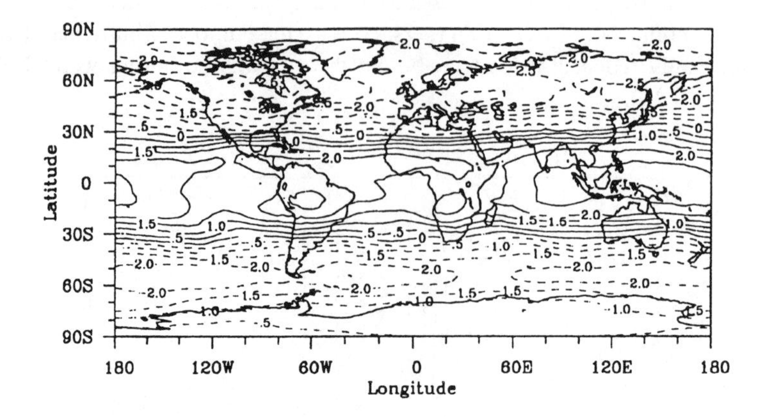

Fig. 4.17. The temperature change at 200 hPa between the control and the double CO_2 simulation (annual mean).

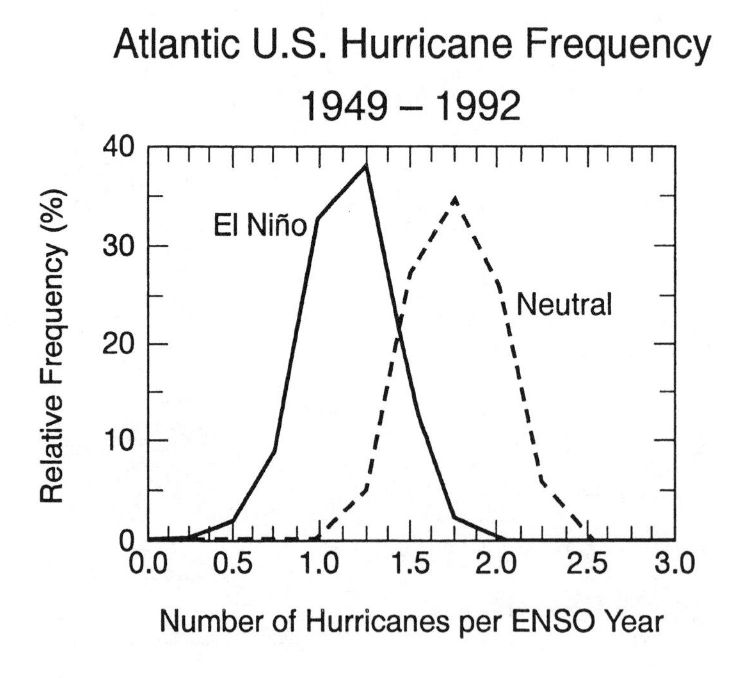

Fig. 4.18. Hurricane frequency in the Northwest Atlantic during El Niño years and normal years, respectively (courtesy of J. O'Brien).

4.6 Conclusions

This study has demonstrated the feasibility of using high-resolution climate models for simulation and investigation of hurricanes. A global model with a resolution of T106 provides a realistic structure of simulated hurricanes as well as a geographical distribution in good agreement with observations. In order to explore the effect of a further increase in horizontal resolution we have nested a limited area model into the global model using grid length of 0.34 and 0.25 lat./long. In order to minimize projection errors we have used a rotated latitude-longitude grid. The limited area experiments provide more confined developments with a smaller and more intense vortex then in the global integration. There is a clear indication of the development of an eye structure in the limited area model run.

The global experiment has been repeated using double CO_2 concentration as well as consistent SSTs obtained from a transient coupled model experiment. The overall finding is a substantial reduction in the number of storms particularly in the Southern Hemisphere. In comparison to the control experiment there are no changes in the distribution of the storms, neither in space, nor in time.The low level vorticity in the hurricane genesis regions is generally reduced compared to the present climate, while the vertical tropospheric windshear is slightly increased. Most tropical storm regions indicate reduced wind speeds and a slightly weaker hydrological cycle.

It is not possible at the present time to draw any general conclusions from the double CO_2 experiment. The result can be expected to be sensitive to the physical parameterization, which in turn is open to some criticism due to the lack of appropriately detailed empirical knowledge, for example of cloud and radiation processes. Furthermore, the SST changes were obtained from another experiment using a less advanced atmospheric model than in this study and with limited capability to reproduce coupled ocean-atmosphere modes. Nevertheless, the similarity between the warming pattern here and in warm ENSO events (having fewer hurricanes at least in some regions) adds credibility to the present result.

Acknowledgments

The authors are indebted to their colleagues at the Max Planck Institute for Meteorology. They particularly acknowledge the technical assistance of Norbert Noreiks and Kornelia Müller.

References

Bengtsson, L., Botzet, M., and Esch, M., 1995. Hurricane-type vortices in a general circulation model. *Tellus,* **47A**, 175-196.

Bengtsson, L., Botzet, M., and Esch, M., 1996a. Will greenhouse gas-induced warming over the next 50 years lead to higher frequency and greater intensity of hurricanes? *Tellus*, **48A**, 57-73.

Bengtsson, L., Arpe, K., Roeckner, E., and Schulzweida, U., 1996b. Climate predictability experiments with a general circulation model. *Climate Dynamics,* **12**, 261-278.

Brinkop, S., 1991. Inclusion of cloud processes in the ECHAM PBL parameterization. In: R. Sausen (ed.): *Studying Climate with the ECHAM Atmospheric Model. Large Sale Atmospheric Modelling*, Report No. 9, 5-14, Meteorologisches Institut, Universität Hamburg.

Brinkop, S., 1992. *Parameterisierung von Grenzschichtwolken für Zirkulationsmodelle. Berichte aus dem Zentrum für Meeres- und Klimaforschung*, Reihe A. Meteorologie, Nr. 2, Meteorologisches Institut der Universität Hamburg, 77 pp.

Broccoli, A.J. and Manabe, S., 1990. Can existing climate models be used to study anthropogenic changes in tropical cyclone climate? *Geophysical Research Letters*, **17**, 11, 1917-1920.

Christensen, J.H. and van Meijgaard, E., 1992. On the construction of a regional atmospheric climate model. DMI Technical Report, 92-14.

Committee on Earth Sciences, 1989. *Our Changing Planet: The FY1990 Research Plan.* Office of Science and Technology Policy, Washington, D.C., 183 pp.

Conover, W.J., 1980. *Practical Non-parametric Statistics.* 2nd ed. John Wiley & Sons, New York.

Cubasch, U., Hasselmann, K., Höck, H., Maier-Reimer, E., Mikolajewicz, U., Santer, B., and Sausen, R., 1992. Time-dependent greenhouse warming computations with a coupled ocean-atmosphere model. *Climate Dynamics*, **8**, 55-69.

Dell'Osso, L. and Bengtsson, L., 1985. Prediction of a typhoon using a fine-mesh NWP model. *Tellus*, **37A**, 97-105.

Dümenil, L. and Todini, E., 1992. A rainfall-runoff scheme for use in the Hamburg GCM. In: *Advances in Theoretical Hydrology: A tribute to James Dooge*, J.P. O'Kane (ed.), European Geophysical Society Series on Hydrological Sciences, 1, Amsterdam: Elsevier 129-157.

Emanuel, K.A., 1986. An air-sea interaction theory for tropical cyclones. part I: Steady state maintenance. *Journal of Atmospheric Science*, **43**, 585-604.

Emanuel, K.A., 1988. Toward a General Theory of Hurricanes. *American Scientist*, 371-379

Evans, J., 1992. Comments on: Can existing climate models be used to study anthropogenic changes in tropical cyclone climate. *Geophysical Research Letters*, **19**, 14, 1523-1524.

Frank, W.M., 1977. The structure and energetics of the tropical cyclone, Paper I: Storm structure. *Monthly Weather Review*, **105**, 1119-1135.

Gray, W.M., 1979, *Hurricanes: Their Formation, Structure and Likely Role in the Tropical Circulation. Meteorology over the Tropical Oceans.* D.B. Shaw, (ed.), Royal Meteorological Society, London, 155-218.

Gray, W.M., 1988. Environmental influences on tropical cyclones. *Australian Meteorological Magazine*, **36**, 127-139.

Gray, W.M., 1989. Background information for assessment of expected Atlantic hurricane activity for 1989. 11th Annual National Hurricane Conference, Miami, NOAA, 41 pp.

Gray, W.M., Sheaffer, J.D., and Landsea, C.W., 1996. Climate trends associated with multidecadal variability of Atlantic hurricane activity, (Chapter 2, this volume).

Gustafsson, N., 1993. HIRLAM2 Final report. HIRLAM Tech. Report No. 9. SMHI, Norrkoeping, Sweden.

Haarsma, R.J., Mitchell, J.F.B., and Senior, C.A., 1992. Tropical Disturbances in a GCM. *Climate Dynamics*, **8**, 247-257.

Hawkins, H.F. and Imbembo, S.M., 1976. The structure of a small, intense hurricane, Inez 1966. *Monthly Weather Review*, **104**, 418-442.

Hense, A., Kerschgens, M., and Raschke, E., 1982. An economical method for computing radiative transfer in circulation models. *Quarterly Journal of the Royal Meteorological Society*, **108**, 231-252.

Idso, S.B., 1989. *Carbon Dioxide and Global Change: Earth in Transition.* Tempe, AZ: IBR Press, 292 pp.

Idso, S.B., Balling, R.C., Jr. and Cerveny, R.S., 1990. Carbon dioxide and hurricanes: Implication of Northern Hemispheric warming for Atlantic/Caribbean storms. Earth in transition. *Meteorological Atmospheric Physics*, **42**, 259-263.

IPCC, 1990. *Climate Change, The IPCC Scientific Assessment.* Houghton, J., G.J. Jenkins and J.J. Ephraums (eds.), Cambridge University Press, Cambridge, 364 pp.

IPCC, 1992. *Climate Change, The Supplementary Report to the IPCC Scientific Assessment*, Houghton, J., Callandar, B.A., and Varnay, S.K., (eds.), Cambridge University Press, Cambridge, 198 pp.

IPCC, 1995. *Climate Change 1995, The Science of Climate Change,* Houghton, J.T., Meira Filho, L.G., Callander, B.A., Harris, N., Kattenberg, A., and Maskell, K. (eds.), Cambridge University Press, Cambridge, 572 pp.

JSC, 1980. *Catalogue of numerical atmospheric models for the First GARP Global Experiment, Part I*, 287-340, ICSU/WMO, Joint Scientific Committee, Geneva, April 1980.

Kurihara, Y. and Tuleya, R.E., 1981. A numerical simulation study on the genesis of a tropical storm. *Monthly Weather Review*, **109**, 1629-1653.

Louis, J.F., 1979. A parametric model of vertical eddy fluxes in the atmosphere. *Boundary Layer Meteorology*, **17**, 187-202.

Machenhauer, B., 1988. The HIRLAM Final report. HIRLAM Technical Report No. 5. DMI, Copenhagen, Denmark.

Manabe, S., Holloway, H.L., Jr. and Stone, H.M., 1970. Tropical circulation in a time-integration of a global model of the atmosphere. *Journal of Atmospheric Science,* **27**, 580-613.

Miller, M.J., Palmer, T.N., and Swinbank, R., 1989. Parameterization and influence of subgrid scale orography in general circulation and numerical weather prediction models. *Meteorological Atmospheric Physics*, **40**, 84-109.

Neumann, C.J., Jarvinen, B.R., McAdie, C.J., and Elms, J.D., 1992. Tropical cyclones of the North Atlantic Ocean, 1971-1992. Asheville, NC, November 1993 (4th revision), 193 pp.

Palmer, T.N., Shutts, G.J., and Swinbank, R., 1986. Alleviation of a systematic westerly bias in general circulation and numerical weather prediction models through an orographic gravity wave drag parameterization. *Quarterly Journal of the Royal Meteorological Society*, **112**, 1001-1031.

Ramanathan, V. and Collins, W. 1991. Thermodynamic regulation of ocean warming by cirrus clouds deduced from observations of the 1987 El Niño. *Nature*, **351**, 27-32

Rockel, B., Raschke, E., and Weynes, B., 1991. A parameterization of broad band radiative transfer properties of water, ice and mixed clouds. *Betrieb Physiches Atmosphere*, **64**, 1-12.

Roeckner, E., Arpe, K., Bengtsson, L., Brinkop, S., Dümenil, L., Esch, M., Kirk, E., Lunkeit, F., Ponater, M., Rockel, B., Sausen, R., Schlese, U., Schubert, S., and Windelband, M., 1992. *Simulation of the Present-day Climate with the ECHAM Model. Impact of Model Physics and Resolution.* Max-Planck-Institut für Meteorologie, Hamburg, Report No. 93.

Roeckner, E., Rieland, M., and Keup, E., 1991. Modelling of cloud and radiation in the ECHAM model. ECMWF/WCRP. "Workshop on clouds, radiative transfer and the hydrological cycle", 12-15 Nov. 1990, 199-222, ECMWF, Reading, U.K.

Schneider, S.H., 1990. *Global Warming. Are We Entering the Greenhouse Century?* Vintage Books, New York, 343 pp.

Shun, C. M., 1992. Performance of the ECMWF Model in Tropical Cyclone Track Forecasting over the Western North Pacific during 1990-1991, ECMWF Technical Memorandum No. 184A.

Sundqvist, H., 1978. A parameterization scheme for non-convective condensation including prediction of cloud water content. *Quarterly Journal of the Royal Meteorological Society*, **104**, 677-690.

Tiedtke, M., 1989. A comprehensive mass flux scheme for cumulus parameterization in large-scale models. *Monthly Weather Review*, **117**, 1779-1800.

Wu, G. and Lau, N.C., 1992. A GCM simulation of the relationship between tropical storm formation and ENSO. *Monthly Weather Review*, **120**, 958-977.

Section B

IMPACTS OF HURRICANE VARIABILITY

5 History of the Deadliest Atlantic Tropical Cyclones Since the Discovery of the New World

Edward N. Rappaport[1] and José J. Fernández-Partagás[2]

[1]NOAA/NWS/NHC
1169 S.W. 17th St.
Miami, FL, 33165 U.S.A.

[2]Private Consultant
825 Majorca
Coral Gables, FL, 33134 U.S.A.

Abstract

The causes, magnitude, and spatial and temporal distributions of loss of life resulting from Atlantic tropical cyclones since 1492 are documented, providing perspective on the historical threat and variability of the phenomenon. About 250 tropical cyclones were found to have taken at least 25 lives. Numerous cases of more than 1000 deaths were identified with more than 22,000 fatalities attributed to the deadliest record event. Reference data also reveal around 200 additional cases associated with loss of life that, while unquantified, could have reached 25. In total, it is estimated that one-third to one-half million people have perished in Atlantic tropical cyclones during this period.

Despite technological advances, loss of life generally increased over the past five centuries. A few exceptionally deadly storms dominate the statistics on time-scales exceeding a decade. Through the 18th century, many of the large losses occurred at sea. More recently, losses from storm surge in coastal areas and inland fresh water flooding predominate. Factors contributing to past and potential future losses include relatively poor communications systems and insufficient mitigation efforts, particularly in outlying areas of underdeveloped countries.

5.1 Introduction

The legacies of Atlantic tropical cyclones span many cultures and thousands of years. Early evidence of these storms predates extant weather records. Geologists believe that layers of sediment at the bottom of a lake in Alabama were brought there from the nearby Gulf of Mexico by storm surges associated with intense hurricanes that occurred as much as 3000 years ago (Liu and Fearn 1993). Similarly, sediment cores from the Florida west coast indicate exceptional freshwater floods during strong hurricanes more than a thousand years ago (Davis et al. 1989).

Perhaps the first human record of Atlantic tropical cyclones appears in Mayan hieroglyphics (Konrad 1985). By customarily building their major settlements away from the hurricane-prone coastline, the Mayans practiced a method of disaster mitigation (Konrad 1985) that, if rigorously applied today, would reduce the potential for devastation of economic and natural resources along coastal areas (Pilkey et al. 1984; Sheets 1990).

Many storms left important marks on regional history (Douglas 1958). In 1609, a fleet of ships carrying settlers from England to Virginia was struck by a hurricane. Part of the fleet grounded at Bermuda. The passengers became Bermuda's first inhabitants and their stories helped inspire Shakespeare's writing of *The Tempest*.

In several incidents, tropical cyclones destroyed otherwise invincible colonial armadas (Millás 1968; Hughes 1987). The French lost their bid to control the Atlantic coast of North America when a 1565 hurricane dispersed their fleet, allowing the Spanish to capture France's Fort Caroline near present-day Jacksonville, Florida. In 1640, a hurricane partially destroyed a large Dutch fleet apparently poised to attack Havana. Another naval disaster occurred in 1666 to Lord Willoughby (the British Governor of Barbados) and his fleet of seventeen ships and nearly 2000 troops. The fleet was caught in a hurricane near the Lesser Antilles. Only a few vessels were ever heard from again and the French captured some of the survivors. The 1640 and 1666 events secured, more or less, control of Cuba by the Spaniards and Guadeloupe by the French (Sugg 1968). More than two centuries later, commenting on the Spanish-American War, President McKinley declared that he feared a hurricane more than the Spanish Navy (Tannehill 1940). McKinley's concern translated to a revamped United States hurricane warning service, forerunner of today's National Hurricane Center (NHC).

Some historical events left scars. In 1495, the small town of Isabella, founded on Hispaniola by Columbus, became the first European settlement destroyed by a hurricane (Carpenter and Carpenter 1993). Other communities would suffer a similar fate. There is even conjecture that a hurricane was responsible for the mysterious disappearance of the original Roanoke Island settlement on North Carolina's Outer Banks (i.e., the "Lost Colony") in 1588 (Hunter 1982). More certainly, in 1886, the town of Indianola, Texas, was destroyed by a hurricane. It was never rebuilt. The 1900 "Galveston" hurricane severely damaged much of that city and, with it, Galveston's preeminence as the financial capital of that part of the country (Hughes 1990). Notwithstanding, subsequent extraordinary mitigation efforts, including the construction of a seawall, have allowed Galveston to become a better-protected coastal city today.

Surviving quantitative documentation about specific storms generally begins late in the 15th century during the period of New World exploration. A succession of chronologies brings the record forward to modern times (e.g., Tannehill 1940; Poey 1962; Ludlum 1963; Millás 1968).

Hebert et al. (1993) frequently update their popular statistical summary about hurricanes that affected the United States this century. Their study, which includes a tabulation of the largest United States losses of life caused by those storms, has no counterpart for earlier tropical cyclones or for casualties incurred elsewhere. In this chapter we extend their work by cataloging Atlantic tropical cyclones[1] associated with loss of life during the period 1492-1995. In addition, the temporal and spatial distributions of these storms and losses are documented, along with the recurrent causes for large disasters. The resulting data base should help place past and potential future events in historical perspective. It is also hoped that these data will make a fundamental contribution to the background information available for the study and prediction of hurricane variability on decadal time scales.

To document casualties and attendant circumstances we relied on books and articles about the weather, newspaper reports about storms, and accounts of shipwrecks. Some of these sources consulted hundreds or thousands of original documents. They provided an extensive, though admittedly not exhaustive, data base. Indeed, if current Atlantic tropical cyclone activity is representative of the past five centuries, then a staggering number of those systems (upwards of 5000!) developed during that period. Some storms were harmless. Others likely caused loss of life that was never documented, or was recorded in documents subsequently lost to deterioration with age, war, or fire (Marx 1983). It is hoped that still other cases not identified here will be uncovered in future investigations.

The catalog consists of two parts. The first, like Hebert et al. (1993), provides information about tropical cyclones responsible for at least 25 deaths. (Table 5.1 lists a subset of those, the cyclones responsible for at least 1000 deaths.) The second part identifies storms associated with loss of life that, while not quantified, could have reached at least 25, according to records of those events.[2]

5.2 Methodology

The publication *Tropical Cyclones of the North Atlantic Ocean* (Neumann et al. 1993) and the associated NHC post-analysis (rather than operational) track data set[3] served as the final authorities for Atlantic tropical cyclone histories back to 1871. Accounts of earlier storms, especially from nonmeteorological sources, were

---

[1]In this context, "Atlantic" will refer to the North Atlantic Ocean, Caribbean Sea, and the Gulf of Mexico.

[2]Both unabridged lists have been documented (Rappaport and Fernández-Partagás 1995).

[3]Available from National Climatic Data Center, Asheville, NC.

often less reliable. For example, they occasionally referred to "hurricanes" without providing elaboration. Sometimes, hurricane meant any storm of apparently exceptional ferocity (such as a powerful high-latitude storm of nontropical origin or a "severe" thunderstorm) that, perhaps, produced what we now consider *hurricane-force* winds. It is unclear in these instances whether the current requirements for a tropical cyclone were satisfied. Occasionally, however, an especially descriptive account added confidence to the interpretation, as in a summary printed in the 6 November 1761 issue of *Lloyd's List*[4]:

> Capt. Young, arriv'd at Briftol from Guadalupe, came out the 17th of Sept. in Company with a Fleet of 26 Sail, moft of them for England, under Convoy of the Griffin Man of War, who was to fee them as far as Lat. 28; but on the 27th ditto, in Lat. 22, they met with a heavy Gale of Wind, which began at the N. W. and veered all round the Compafs to the S. E. in which the Fleet were fcattered, and feveral loft their Topmafts. The next Morning he faw only nine Vessels with the Man of War; and the Captain adds, That by the Smartnefs of the Gale, and the Wind's flying about round the Compafs, he apprehends it was the Tail of an Hurricane.

Accounts that included weather observations, such as ship reports based on the Beaufort scale (Tannehill 1940), introduced in 1806, or barometric pressure measurements, also helped to clarify the nature of some rough weather events.

The present study adhered to several guidelines that minimized subjectivity and simplified the analysis. Every accepted case had a documented association with bad weather that was, or could reasonably be, related to a tropical cyclone. This requirement eliminated many cases from further consideration, even those where the remaining evidence (in the example below, the date and location of a loss of multiple ships) tempted us to attribute the disaster to a tropical cyclone:

> The Duke of Cumberland, (Captain) Ball, a Letter of Marque of Briftol, laft from the Canaries for Virginia, was loft in September laft nine Leagues to the Southward of Cape Henry; the Captain, Surgeon and twenty three Men were drowned, and 21 faved. —— about the fame time were alfo loft a Snow and a Brig, names unknown, and all the Crew of the former perifh'id. (*Lloyd's List*)

Only in obvious circumstances was a report purportedly about a tropical cyclone rejected outright (e.g., a 1791 *Lloyd's List* case of a "hurricane" over England on January 5).

Supporting data and descriptions provided by the sources were usually accepted without challenge. Unfortunately, in doing so, we likely passed along some information that either originally (or over the years) was (re)recorded incorrectly. Conflicting accounts were noted in, and by, several sources and the associated uncertainties are reflected in Table 5.1. We hope, however, that by providing all relevant reference information, the reader will gain as thorough a documentation of the event as possible.

The concept of storm track and the difference between storm motion and circulation remained obscure until Benjamin Franklin's conclusions of the mid-18th

---

[4]This account, like several that follow, is shown in an older style of English, presented by the source, where "*f*" sometimes represents "s".

century (Ludlum 1963) were extended and formalized (e.g., Reid 1841; Redfield 1936). In addition, with communications generally limited for centuries to the line of sight, storms almost always moved faster than did the information about them. This certainly contributed to the peril of people in the path of an oncoming storm. One impact on this study was to introduce uncertainty in some instances about whether contemporary storm accounts from a region referred to a single tropical cyclone or possibly to multiple systems. Cases where uncertainties persist about the number of storms involved were assigned the footnote "f" in Table 5.1.

5.3 Casualty Information

5.3.1 Losses over Open Waters

The period under study saw a large and widespread increase in Atlantic coastal population. Available records, however, suggest that the population "on" the Atlantic was the most vulnerable to storms through the 18th century. These shipborne explorers, emigrants, combatants, fishermen, traders, pirates, privateers, slaves, and tourists made up the crews and passengers on an uncounted, but enormous number of local and transatlantic sailings. Most of the ships travelled to or from the ports of Spain, France, Great Britain, and the Netherlands. They usually proved no match for the intense inner-core region of a severe tropical cyclone.

> It is doubtful if any sailing ship or any man aboard survived in this sector of a really great hurricane. (Tannehill 1995)

In fact, up to 1825,

> more than five percent of the vessels in the (West) Indes navigation were lost due to shipwrecks; the biggest part due to bad weather... (Marx 1981)

The total number of ship-related casualties associated with Atlantic tropical cyclones is unknown, but there are clues. An account of one 17th century hurricane indicates the great magnitude of some losses blamed on tropical cyclones:

> By these kind of Tempests the King of Spain hath lost at several times near to 1000 sail ship. (Ludlum 1963)

Similar disasters continued for another two centuries. Even as late as the 1830s,

> ...the annual loss of life, occasioned by the wreck or foundering of British vessels at seay, may, on the same grounds (i.e., 'the boisterous nature of the weather and the badness of the ships'), be fairly estimated at not less than One Thousand persons in each year... (Parliament Select Committee 1839)

The large number of ship losses was partially a consequence of the great number of ships that inadvertently encountered storms. An analysis of an 1845 hurricane off the U.S. mid-Atlantic coast contains, on one weather map, information from the logs of more than 50 ships within about 450 miles of the storm's center (Redfield 1846). There were likely other vessels in that area. The analyst suggested that the then-expanding electric telegraph could be used in the Atlantic ports of the United States to alert mariners of approaching bad weather. Unfortunately, occasional ship disasters related to Atlantic tropical cyclones continued into the early 1900s. Further technological advances in meteorology, communication, navigation, and the seaworthiness of ships make such losses infrequent today.

Reference materials about specific ship losses range from nonexistent to overwhelming. In some instances, where the sea claimed a lone ship or even an entire fleet, record of the cause and location of the catastrophe went down with the ship(s). Moreover, for centuries there were virtually no official records of lost ships (Cameron and Farndon 1984). On the other hand:

> if a team of one hundred researchers spent their whole lives searching through the more than 250,000 large legajos (bundles) in the Archive of the Indes (at Seville), I doubt that they could locate all the important documents concerning Spanish maritime history in the New World. (Marx 1983)

Either way, we learned little or nothing about many lost or missing crews and the circumstances behind their disappearance. For this compilation, lacking contrary evidence, the crews and passengers of ships lost over *open waters* in tropical cyclones were counted as fatalities.

5.3.2 Coastal Deaths

While losses over open waters have decreased of late, rapid growth of coastal communities over the past 500 years has meant an ever-increasing population and economic infrastructure at risk to tropical cyclones. As at sea, relatively primitive communication methods increased the possibility of disaster near the shoreline. Not until 1909 was the first *in situ* ship report of hurricane conditions (in the Yucatan Channel) received in time to assist coastal preparations (in the United States) (Garriott 1909).

There are two primary components to the danger near the shore, coastal ship losses and storm surge disasters. It is estimated that 98% of the ships lost in the Western Hemisphere before 1825 wrecked in waters no deeper than 30 feet (Marx 1983), with the unfortunate implication that many lives were lost only a short distance from the relative safety of land. Early on, many mariners lost their lives while staying too long with their vessel:

> Came to anchor in St. Thomas's harbor, and landed the mails. Here the hurricane of the 2nd (August 1837) appeared to have concentrated all its power, force, and fury, for the harbour and town were a scene that baffles all description. Thirty-six ships and vessels wrecked all around the harbour, among which about a dozen had sunk or capsized at their anchors; some rode it out by cutting away their masts, and upwards of 100 semen drowned... (Reid 1841)

In contrast, today's early warning system usually results in little or no loss of life aboard vessels that wreck on a coast or in a marina. In 1992 Hurricane Andrew, for example, only two boating-related deaths occurred in southeast Florida despite boat damage estimated at $0.5 billion (Mayfield et al. 1993). For purposes of the present study, cases with ships lost on the coast or in port were excluded from the casualty lists unless explicit documentation of sufficient loss of life was found.

Table 5.1. Atlantic Tropical cyclones causing at least 1000 deaths.

The first column indicates the areas that experienced the greatest number of deaths. For events after 1949, it also contains the name of the cyclone. The second column provides the approximate range of dates for the losses (Whenever possible, the dates were based on, or converted to, our current Gregorian calendar system which replaced the Julian calendar in the 16th century.). The third column gives the total number of deaths and the source(s) of the information, with a single source occasionally giving more than one estimate for one cyclone. (Some of the sources used the same original documents and, therefore, do not provide independent documentation.) A "+" indicates that totals from multiple sources were combined.

NAME & AREAS[a] OF LARGEST LOSS	APPROX. DATES	DEATHS AND DATA SOURCE(S)
1 MAR, STE, BAR, offshore	10-16 Oct 1780	>22000[1], 22000[2, 3,b], >20000[4]
2 Galveston (Texas)	8 Sep 1900	≥12000[5,c], >8000-12000[6,d], >8000[7]
3 FIFI: Honduras	14-19 Sep 1974	8000-10000[8], 3000-10000[9], >3000[10]
4 Dominican Republic	1-6 Sep 1930	8000[11], 4000[12,9], 2000[10,4,9,13,14,e]
5 FLORA: Haiti, Cuba	30 Sep-8 Oct 1963	8000[10], 7193[15,9], 7191[14], >7186[9]
6 Pointe-a-Pitre Bay (GUA)	6 Sep 1776	>6000[16]
7 Newfoundland Banks	9-12 Sep 1775	4000[3,f,g,h]
8 Puerto Rico, Carolinas	8-19 Aug 1899	>3433[(17, 18)+19+20,i], >3064[(13,9,14)+19+20]
9 FL, GUA, PR, TUR, MAR	12-17 Sep 1928	>3411[9+21+22+17+23], (3375-4075)[j]
10 Cuba, CI, Jamacia	4-10 Nov 1932	>3107[24+25, k], 2569[9], >2500[10], 2500[4]
11 Central Atlantic	16-17 Sep 1782	>3000[26,27,g,l,f]
12 Martinique	Aug 1813	>3000[16,f]
13 El Salvador, Honduras	4-8 Jun 1934	>3000[28], >2000[13], 506-3006[9]
14 Western Cuba	21-22 Jun 1791	3000[2,18,14,m,f], 257[29], >30[30]
15 Barbados	10-11 Aug 1831	2500[1], 1525[1], >1500[3,4,13,14,31,n]
16 Belize	6-10 Sep 1931	2500[22], 1500[9], 1500[10,4,13,14]
17 HAI, HON, offshore JAM	19-25 Oct 1935	>2150[22,o], 1168-2168[9+28], 1000-2000[4]
18 DAVID: DR, Dominica, US	29 Aug -5 Sep 1979	>2068[9], >2063[10]
19 Offshore Florida (?)	1781	>2000[32,f]
20 South Carolina, Georgia	27-28 Aug 1893	2000-2500[33], 1000-2000[34,15,9,b]
21 Eastern Gulf of Mexico	17-21 Oct 1780	2000[30]
22 Cuba	7-8 Oct 1870	2000[35], 1000[36], >800[37], ≥136[38]
23 Louisiana	1-2 Oct 1893	2000[34,13,9], 1800[15]
24 Guadeloupe, Martinique	14-15 Aug 1666	<2000[2,18,38,39,p,f]
25 Martinique	Aug 1767	1600[17]
26 Mexico	28 Aug 1909	1500[13,40], 1000-2000[36]
27 W Cuba, Straits of FL	Oct 1644	<1500[2,q]
28 Guadeloupe, Puerto Rico	26 Jul 1825	>1300[41], 500[17], 374[17,13, 42],372[14]
29 Offshore Nicaragua	1605	1300[2,r]
30 GORDON: HAI, FL, CR, DR	8-21 Nov 1994	1145[43,s]
31 Jamaica, Cuba	2-5 Oct 1780	≥1115[24+2, t], >415[2], 42[22]
32 Straits of Florida	5 Sep 1622	>1090[2, f], 590[2]
33 Gulf of Mexico	early Nov 1590	>1000[16, g]
34 Offshore Barbados	27 Sep 1694	>1000[16]
35 S Bahamas, Straits of Florida	30 Jul 1715	(>1000-<2500)[16], >1000[16], 1000[2]
36 Havana (Cuba)	15 Oct 1768	>1000[29, 38, 39, 22, 14] >100[2, u], 43[1, 44]
37 Veracruz (Mexico)	1601	1000[2, g]
38 HAZEL: HAI, US, GRE, CAN	5-13 Oct 1954	1000[10], 600-1200[11], 575-1175[9]
39 INEZ: Caribbean, Mexico	27 Sep-11 Oct 1966	1000[9]

[a] Conventional abbreviations were used for map headings (e.g., N for north or northern) and for American states and Puerto Rico. We used C for central and the following: BAR--Barbados, CAN--Canada, CI--Cayman Islands, CR--Costa Rica, DR--Dominican Republic, GRE--Grenada, GUA--Guadeloupe, HAI--Haiti, HON--Honduras, JAM--Jamaica, MAR--Martinique, STE--St. Eustatius, TUR--Turks Islands.
[b] Numerous estimates provide (sub)totals yielding a similar statistic.
[c] "The loss of life occasioned by the storm in Galveston and elsewhere on the southern coast cannot be less than 12,000 lives..." Statement of Governor Sayres on 19 Sep 1900 printed in Lester (1900).
[d] There are many estimates of the total. This one, based on the "official" summary in *Galveston in 1900* (Ousley 1900), is: 6000 in city of Galveston, 1000-1200 elsewhere on the island west of the city and more than 1000 on the mainland. Maximum estimates provided are 10000-12000. *Monthly Weather Review* indicates "Enormous loss of life...inland", as well. Most other references indicate a loss of at least 6000.
[e] At least 2000 deaths according to: Snow, E. R., 1952, *Great Gales and Dire Disasters*. Dodd, Mead and Company, New York, 263 pp.
[f] Some early storms that qualified in more than one locale may have multiple listings (Rappaport and Fernández-Partagás 1995) if the storm track is unknown.
[g] Tropical cyclone status in doubt for at least part of event.
[h] *Lloyd's List* has many accounts indicating a great many more than 500 deaths near Newfoundland. Some of the losses occurred on the northwest coast of Newfoundland and on the coast of Labrador. Hence, the total may be larger than shown by Ludlum (1963), but the storm may not have been entirely tropical, either. The dates from these sources do not match and the relationship between this entry and the other Sep 1775 storm(s) along the U.S. east coast and the storm reportedly at Hispaniola on the last days of August is not clear.
[i] Garriott (1900) and Alexander (1902) indicate thousands of additional deaths in Puerto Rico due to subsequent starvation. There were at least 50 deaths in shipwrecks along coastal Carolina according to Chapman, and to Stick, D.A., 1952, *Graveyard of the Atlantic*. The University of North Carolina Press, Chapel Hill, 276 pp. Barnes (1995) has at least 30 along the coast of North Carolina and 14 inland in that state.
[j] This range comes from references 13, 21, 22, 23, and 37, shown below in this table. For first entry Salivia (1970) has both 300 and 312 for Puerto Rico and we used the latter. Most references cite Red Cross statistics of 1836 deaths and 1870 injuries in Florida. An additional reference (4) with 1870 deaths in Florida may be in error. Flament and Martin, and Soulan (1994) have 1200 deaths for Guadeloupe. *Monthly Weather Review* adds 18 for Grand Turk and indicates others possible in Caribbean. Soulan (1994) adds 3 for Martinique. Snow (1952) has 1500-2500 deaths. Douglas (1958) has 600 in Guadeloupe, 300 in Puerto Rico, and "no one will ever know how many more than the estimated 1,800 to 2,500 died there."
[k] Cayman Islands National Archive documents indicate 101 or 102 deaths of islanders, excluding their residents lost on Cuba. Other references have smaller totals for the Cayman group.
[l] Ellms (1860) locates the disaster at 48°33'N 43°20'W, placing in doubt the tropical character of the storm. *Lloyd's List* (Oct 1782), however, has accounts of storm from the Jamaica Fleet at 43°N 48°W, and at 43°N 44°W. At the latter location, "...in a Gale of Wind from ESE...on the 16th in the Evening, when on the Morning of the 17th the Wind came out in an In*ſ*tant to N.W....the storm la*ſ*ting for two hours." A very similar account from an officer on the *Ramilies* at 42.3°N and 48.9°W is reprinted in Redfield (1836).
[m] Douglas (1958) notes a loss of 3000 people in a 1791 hurricane but gives August as the month and no location is provided.
[n] References 26 and 38 have 1477 deaths.
[o] Clark (1988) has 2150. Reference 14 has ≥2000.
[p] Alexander (1902) notes "17 sail with 2000 troops...only two were ever heard of afterwards". Other references indicate that additional ships may have survived.
[q] 13 ships carried 1500 people; 10 ships sank.

[r] Marx (1983) is probably describing the same storm when indicating no survivors of 4 wrecks resulting from "a hurricane between Serrana and Serranilla banks" in 1605.

[s] Based on 21 December 1994 Report No. 7 from the United Nations Department of Humanitarian Affairs, estimating 1122 deaths in Haiti. Earlier reports vary considerably from this figure.

[t] Seon (1986) has upwards of 1000 deaths in Jamaica, while Evans (1848) and Millás (1968) indicate 300 deaths there. Ludlum (1963) account has 200 in Savanna-La-Mar and "several white people and some hundreds of negroes killed...in the whole parish."

[u] Millás (1968) disputes accounts giving date as 25 October and deaths as more than 1000.

[1] Depradine, C.A., 1989
[2] Millás, J.C., 1968
[3] Ludlum, D.M., 1963
[4] U.S. Weather Bureau, 1940
[5] Lester, P., 1900
[6] Ousley, C., 1900
[7] Hughes P., 1987
[8] Servicio Meterológico Nacional, República de Honduras, 1975
[9] Monthly Weather Review
[10] Clark, G., 1988
[11] Final Report of the Caribbean Hurricane Seminar, 1956
[12] See references for Santo Domingo
[13] Tannehill, I.R., 1940
[14] Hurricanes in the West Indes by Country
[15] Dunn, G.E. and Miller B.I., 1964
[16] Marx, R.F., 1983
[17] Salivia L.A., 1972
[18] Alexander W.H., 1902
[19] Chapman, D.J., date unknown
[20] Barnes, J., 1975
[21] Flament, P. and Martin, R., date unknown
[22] List of disasters that occurred in the Caribbean
[23] Soulan, I., 1994
[24] Seon, K., 1986
[25] Unpublished documents of the Cayman Islands National Archive
[26] Piddington, H., 1852
[27] Ellms, C., 1860
[28] Diario de la Marina of Havana, selected issues
[29] Schomburgk, R.H., 1848
[30] Herrera D., 1847
[31] Tannehill, I.R., 1955
[32] Singer, S., 1992
[33] Ho, F.P., 1989
[34] Herbert, P.J. et al., 1993
[35] *The London Times,* selected issues
[36] *The New York Times,* selected issues
[37] Douglas, M.S., 1958
[38] Garriott, E.B., 1900
[39] Evans, J., 1848
[40] Price, W.A., 1956
[41] *Niles' Weekly Register-Chronicle*
[42] Pérez, O., date unknown
[43] Unpublished notes at the National Hurricane Center
[44] Rodríguez-Ferrer, M., 1876

Storm surge has been responsible for some of the largest losses of life associated with tropical cyclones at the coastline. Poor communication for many years left coastal communities virtually without warning of storm surge. In the United States, storm surge has been blamed for 90% of hurricane-related fatalities (American Meteorological Society 1973). Even with the many technological advances, much of the burgeoning coastal population of the Americas remains vulnerable to storm surge (Sheets 1990). There are several factors that can contribute to these potential losses: a) limited ability to forecast rapid intensification of tropical cyclones, exacerbated by b) poor planning or insufficient mitigation efforts (e.g., grid-locked roadways), c) a populace unattuned to the gravity of the threat (e.g., disbelief or apathy), and d) continuing poor electronic communications, especially in outlying areas of underdeveloped countries.

5.3.3 Inland Deaths

Inland communities are also susceptible to tropical cyclone disasters. There, fresh-water flooding from excessive rainfall, especially on elevated terrain, can lead to large numbers of deaths by drowning or from landslides and mud slides. These losses can predominate, as they did in Haiti when upwards of 1000 fatalities occurred in slides induced by thunderstorms at the fringe of Tropical Storm Gordon in 1994. Catastrophic inland loss can also occur as a result of storm surge, as it did in 1928 when Lake Okeechobee (Florida) was blown into the surrounding lowlands, killing 1836 people.

The number of inland deaths, indeed those near the coast and offshore as well, were only estimated by many of the references. Numerous entries in Table 5.1 appear rounded to the nearest ten, hundred, or even thousand. In addition, the data from many references suggest that the listed total is likely a lower threshold. Furthermore, there is evidence that casualty statistics were occasionally withheld by government officials (Pérez, date unknown).

5.4 Statistics, Analysis, and Conclusions

A total of 250 tropical cyclones were identified as causing at least 25 deaths. Table 5.1 contains information about each of the 39 most deadly cases, those reported to have caused at least 1000 deaths. Unless otherwise noted, the fatality totals discussed below refer to the first (largest) number in the third column of Table 5.1 and the remainder of the 250 cases (see Rappaport and Fernández-Partagás 1995).

The largest loss occurred in the Lesser Antilles in mid-October 1780, during "The Great Hurricane." Estimates indicate that around 22,000 deaths occurred in that storm, with a total of about 9000 lives lost in Martinique, 4000-5000 in St. Eustatius, and 4326 in Barbados. Thousands of deaths also occurred offshore. The number of fatalities during The Great Hurricane exceeds the known cumulative loss in any other year and, in fact, in any other decade (Fig. 5.1, upper panel). Interestingly, The Great Hurricane was one of three tropical cyclones to kill more than 1000 people *that month.*

The second largest loss (the largest in the United States) came as a result of the storm surge accompanying the 1900 Galveston hurricane. Although the Governor of the State of Texas estimated 12,000 fatalities just after the storm (Lester 1900), the "official" summary provides information supporting an estimate of at least 8000 lives lost (Ousley 1900).

Three other storms killed around 8000 people: 1974 Hurricane Fifi in Honduras; a 1930 hurricane in the Dominican Republic; and 1963 Hurricane Flora in Haiti and Cuba. These five deadliest tropical cyclones account for about one-third of all the deaths over the past 500 years in storms for which quantitative data on deaths has been cataloged. In fact, the 10 deadliest storms (see Table 5.1) account for almost one-half of the deaths associated with the 250 quantified cases (storms taking at least 25 lives), while representing only 4% of those cases and less than 0.2% of all tropical cyclones since 1492 (based on an estimated 5000 total storms).

Figure 5.1, lower panel, shows the number of deaths stratified by 100-year periods. The figure indicates that the number of deaths generally increased with time. The 1700s were an exception. Then, maritime losses between 1760 and 1790 dominated the relatively large total. About 70,000 deaths have occurred in the 1900s. In part, this large total is related to the increased population at risk. Beyond that, the nature of losses is evolving. In the United States, for example, the frequency with which hundreds or thousands of lives are lost has dropped significantly in the latter half of this century because storm surge losses have decreased. This occurred because of improvements in the hurricane warning *system* comprising forecasts, communications and preparedness, and mitigation practices (and to a lower frequency of strong storms making landfall). While large United States losses remain possible, that threat is chiefly limited to high population density areas where an approaching, rapidly-developing storm leaves insufficient time for thousands of people to reach safety, especially if an ineffective evacuation (e.g., an overcrowded, submerged coastal roadway) or *no* evacuation is conducted.

In some underdeveloped countries, the threat of large losses remains high because one or more of the components of the warning system is weak or nonexistent, and because of the physiography of the area that includes high terrain near the coast. Indeed, most of the largest losses during the past 50 years have occurred as a result of freshwater-induced floods, mud slides and landslides in Caribbean nations (e.g., Fifi [1974], Flora [1963], Gordon [1994]). In the case of Gordon in Haiti, for example, the land had been nearly denuded, making the treeless hillsides ripe for the 1994 catastrophic slides. Similarly, inadequate construction practices on the hills of Caracas, Venezuela contributed to a large loss of life in Tropical Storm Bret during 1993.

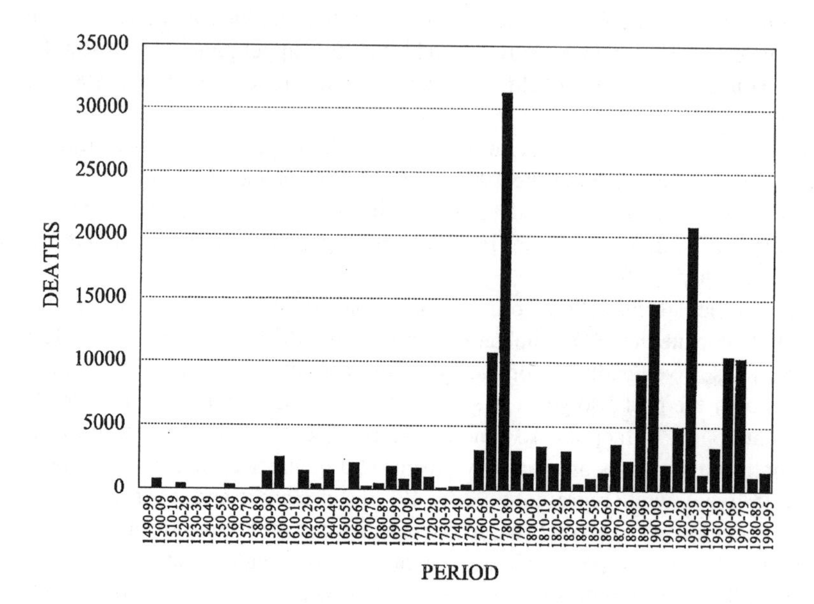

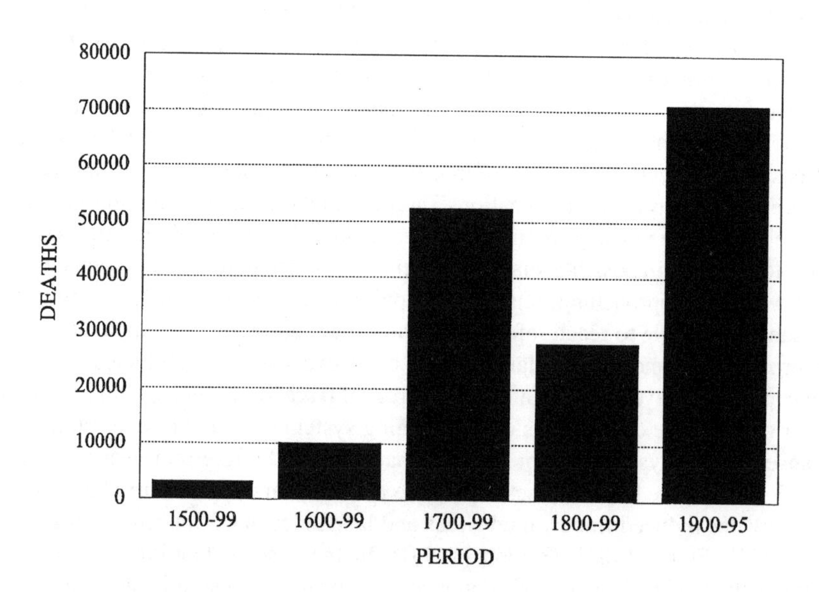

Fig. 5.1. Atlantic tropical cyclone deaths for systems documented as responsible for at least 25 fatalities, shown in (upper) 10-year periods (except for 1990-95) and (lower) 100-year periods (except for 1900-95).

The areal distribution of deaths from the 250 quantifiable cases considered here is, approximately: Greater Antilles 45,000 (29%); Offshore (entire basin) 35,000 (22%); Lesser Antilles 35,000 (21%); United States mainland 25,000 (16%); Mexico and Central America 20,000 (12%); elsewhere 1000 (<1%). For the United States mainland, the losses (using 8000 for the 1900 Galveston storm) in the 1900s are near 14,500 and in the 1800s near 8500. The Galveston storm is responsible for about one-third of the tabulated tropical cyclone-related deaths in the United States. We also found that over 90% of the offshore losses occurred more than 200 years ago (before 1790), as did all 12 offshore losses of more than 1000 people. We note that the disposition of the many casualties from shipwrecks near shore into offshore versus land losses is not certain.

Taken together, the 250 cases resulted in the loss of about 160,000 lives. The total number of deaths associated with Atlantic tropical cyclones of the past five centuries is likely much larger than 160,000. Many other lives were lost in the approximately 200 additional cases for which there could have been 25 to, perhaps, thousands of deaths in each event, but for which quantified losses could not be established. For example:

in 1553, 16 ships of the New Spain Flota were "struck by a hurricane" and not again "ever heard from." (Marx 1983)

in 1640, 36 vessels were affected, with 4 thrown on shore; "nearly all the sailors drowned, excepting 260 that were saved." (Millás 1968; italics added for emphasis)

While the total cannot be specified with confidence, we have estimated that the number of deaths in all Atlantic tropical cyclones from 1492-1995 is between one-third and one-half million people (Rappaport and Fernández-Partagás 1995).

Many factors contribute to uncertainty about the total, including relatively few references to losses in Mexico and Central America and incomplete information about losses from Spanish ships in the 1500s to 1700s and to slaves and natives of the region. There are sources that could provide more definitive information and a systematic review of that material is now underway.

Acknowledgments

Ms. Sally Haff and Mr. Robert Britter helped identify many of the reference materials used in this study.

References

Alexander, W. H., 1902. Hurricanes: Especially those of Porto Rico and St. Kitts. U.S. Department of Agriculture, Weather Bureau, Bulletin No. 32, 79 pp.

American Meteorological Society, 1973. Policy statement on hurricanes by the American Meteorological Society. *Bulletin of the American Meteorological Society.*, **54**, 46-47.

Análisis preliminar de la precipitación producida por el Huracán "Fifi" a su paso por Honduras, 1975. *Publicación No. 110, Proyecto Hon/72/006. Meteorología e Hidrología,* Sevicio Meteorológico Nacional, República de Honduras, 27 pp.

Barnes, J., 1995, *North Carolina's Hurricane History.* University of North Carolina Press, Chapel Hill, 202 pp.

Cameron, A. and Farndon, R., 1984. *Scenes from Sea and City, Lloyd's List 1734-1984. Lloyd's List,* Lloyd's of London Press Ltd, London, 288 pp.

Carpenter, S.M. and Carpenter, T.G., 1993. *The Hurricane Handbook.* Tailored Tours Publications, Inc., Lake Buena Vista, Florida, 128 pp.

Chapman, D.J. (date unknown), Our southern summer storm. Report from National Weather Service Office, Norfolk, Virginia.

Clark, G., 1988. Hurricanes of the Caribbean Sea (unpublished notes compiled by the author), National Hurricane Center, Miami.

Davis, R.A., Jr., Knowles, S.C., and Bland, M.J., 1989. Role of hurricanes in the Holocene stratigraphy of estuaries. Examples from the Gulf coast of Florida. *Journal of Sedimentary Petrology,* **59**, 1052-1061.

Depradine, C.A., 1989. Pre-1900 severe hurricanes in the Caribbean. Notes compiled for the Caribbean Meteorological Institute, St. James, Barbados.

Diario de la Marina of Havana (selected issues).

Douglas, M.S., 1958. *Hurricane.* Rinehart & Company, Inc. New York, 393 pp.

Dunn, G.E., and Miller, B.I., 1964. *Atlantic Hurricanes.* Louisiana State University Press, Baton Rouge, 377 pp.

Ellms, C., 1860. *Shipwrecks and Disasters at Sea.* I. J. Rouse, New York, pp. 428.

Evans, J. (Stormy Jack), 1848. Hurricanes 1493-1848. *Nautical Magazine*, London.

Final Report of the Caribbean hurricane seminar (1956). Published in 1958 by the Government of the Dominican Republic, Ciudad Trujillo, D. R., 395 pp.

Flament, P. and Martin, R., *Le cyclone Allen - du 29 julliet au 12 août 1980.* L'Association Météorologique de la Martinique, avec le concours, du Conseil General de la Martinique, et du Service Météorologique Antilles-Guyane, 32 pp.

Garriott, E.B., 1900. *West Indian Hurricanes.* U.S. Department of Agriculture, Weather Bureau, Bulletin H, Washington, D.C., 66 pp. (This work includes some notes of Alexander published by that author in 1902.)

Garriott, E.B., 1909. Weather, forecasts, and warnings for the month. *Monthly Weather Review,* **37**, 539.

Hebert, P.J., Jarrell, J.D., and Mayfield, M., 1993. The deadliest, costliest, and most intense United States hurricanes of this century. NOAA, *Technical Memorandum NWS NHC-31,* 41 pp.

Herrera, D., 1847. *Memoria sobre los Huracanes en la Isla de Cuba.* Imprenta de Barcina, Habana, 72 pp.

Ho, F.P., 1989. Extreme Hurricanes in the Nineteenth Century. NOAA *Technical Memorandum NWS HYDRO 43A*, Silver Spring, Maryland, 134 pp.

Hughes, P., 1990. The great Galveston hurricane. *Weatherwise*, **43**, 190-198.

Hughes, P., 1987. Hurricanes haunt our history. *Weatherwise*, **40**, 134-140.

Hunter, M.N., 1982. A watery fate for the lost colony. *The State*, 3 pp. Information supplemented by other documents of this author.

Hurricanes in the West Indies by country 1500-1979. CDPS Information Exchange, 1980, 70 pp.

Konrad, H.W., 1985. Fallout of the wars of the Chacs: The impact of hurricanes and implications for prehispanic Quintana Roo Maya processes. *Status, Structure and Stratification: Current Archaeological Reconstructions*. University of Calgary, Calgary, 321-330.

Lester, P., 1900. *The Great Galveston Disaster*. Library of Congress, 536 pp.

List of disasters that occurred in the Caribbean. Unpublished notes at the National Hurricane Center, Miami, Florida; date(s) of storms uncertain.

Liu, K. and Fearn, M.L., 1993. Lake-sediment record of late Holocene hurricane activities from coastal Alabama. *Geology*, **21**, 793-796.

Lloyd's List. Extant issues 1741-1784 and 1790-1797, Gregg International Publishers Limited (1969), Westmead, Farnborough, Hants., England.

Ludlum, D.M., 1963. *Early American Hurricanes, 1492-1870*. American Meteorological Society., Boston, 198 pp.

Marx, R.F., 1981. *Shipwrecks in Mexican Waters*. Pablo Bush Romero and Club de Exploraciones y Deportes Acuáticos de México, 76 pp.

Marx, R.F., 1983. *Shipwrecks in the Americas*. Bonanza Books, 482 pp.

Mayfield, M., Avila, L., and Rappaport, E.N., 1993. Atlantic hurricane season of 1992. *Mon. Wea. Rev.*, **122**, 517-538.

Millás, J.C., 1968. *Hurricanes of the Caribbean and Adjacent Regions, 1492-1800*. Academy of the Arts and Sciences of the Americas, Miami, Florida, 328 pp.

Monthly Weather Review, U.S. Signal Office, **1-18**, 1872-1890; U. S. Weather Bureau, **19-98**, 1891-1970; National Weather Service, **99-101**, 1971-1973; American Meteorological Society, **101-122**, 1974-1994.

Neumann, C. J., Jarvinen, B.R., McAdie, C.J., and Elms, J.D, 1993. Tropical cyclones of the North Atlantic Ocean, 1871-1992. NOAA *Historical Climatology Series 6-2*, Asheville, North Carolina, 193 pp.

Niles' Weekly Register-Chronicle (7 Oct 1815 and 27 Aug 1825), Baltimore, Maryland.

Ousley, C., 1900. *Galveston in Nineteen Hundred - The authorized and official record of the proud city of the southwest as it was before and after the hurricane of September 8, and a logical forecast of its future*. William C. Chase, Atlanta, 346 pp.

Parliament Select Committee, 1839. *Report from the Select Committee appointed to inquire into the cause of shipwrecks*, 15 August 1836, 388 pp. and *Report from Select Committee on shipwrecks of timber ships*, August 1839, 137 pp. Southampton University, Great Britain.

Pérez, O., (date unknown). Notes on tropical cyclones of Puerto Rico, 1508-1970.

Piddington, H., 1852. *The Sailor's Horn-Book for the Law of Storms*. Smith, Elder and Co., London, 360 pp.

Pilkey, O.H., Jr., Sharma, D.C., Wanless, H.R., Doyle, L.J., Pilkey, O.H., Sr., Neal, W.J., and Gruver, B.L., 1984. *Living with the East Florida Shore*. Duke University Press, Durham, North Carolina, 259 pp.

Poey, A., 1862. Table chronologique de quatre cents cyclones. Paris, 49 pp.

Price, W.A., 1956. Hurricanes affecting the coast of Texas from Galveston to Rio Grande. U. S. Department of the Army, Corps of Engineers, *Technical Memorandum No. 78*, 17 pp.

Rappaport, E.N. and Fernández-Partagás, J., 1995. The deadliest Atlantic tropical cyclones, 1492-1994. *NOAA Technical Memorandum NWS NHC-47,* Coral Gables, Florida, 41 pp.

Redfield, W.C., 1836. On the gales and hurricanes of the western Atlantic. *U.S. Naval Magazine*, 1-19.

Redfield, W.C., 1846. On three several hurricanes of the American seas and their relations to the Northers, so called, of the Gulf of Mexico and the Bay of Honduras, with charts illustrating the same. *The American Journal of Science*, **52**, 162-187, 311-334.

Reid, W., 1841. *Law of Storms.* John Weale, London, 566 pp.

Rodríguez-Ferrer, M., 1876. Naturaleza y civilización de la grandiosa isla de Cuba. Printed by J. Noguera, Madrid.

Salivia, L.A., 1972. *Historia de los Temporales de Puerto Rico y las Antillas*, Editorial Edil, Inc., San Juan, Puerto Rico, 385 pp.

Santo Domingo. Su destrucción por el huracán del 3 de Septiembre de 1930. Editado por la Empresa *Diario del Comercio*, Roques Román Hnos, 1930, 60 pp.

Schomburgk, R.H., 1848. *The History of Barbados.* Longman, Brown, Green, and Longmans. London, 1986, 695 pp.

Seon, K., 1986. Preliminary disaster catalog - Jamaica. Office of Disaster Preparedness, Government of Jamaica. Unpublished report, 31 pp.

Sheets, R.C., 1990. The National Hurricane Center—past, present, and future. *Weather and Forecasting*, **5**, 185-232.

Singer, S., 1992. *Shipwrecks of Florida.* Pineapple Press, Inc., 368 pp.

Soulan, I., 1994. Cyclones tropicaux les plus meurtriers de l'Atlantique. Personal communication, 2 pp.

Sugg, A.L., 1968. Beneficial aspects of the tropical cyclone. *Journal of Applied Meteorology*, **7**, 39-45.

Tannehill, I.R., 1940. *Hurricanes: Their Nature and History.* Princeton University Press, Princeton, New Jersey, 257 pp.

Tannehill, I.R., 1955. *Hurricane Hunters.* Cornwall Press, Inc., Cornwall, New York, 271 pp.

The London Times (selected issues)

The New York Times (selected issues)

Unpublished documents of the Cayman Islands National Archive.

Unpublished notes at the National Hurricane Center.

U.S. Weather Bureau, 1940, *Winds and Weather of the West Indian Region*, 190 pp.

6 The Frequency and Intensity of Atlantic Hurricanes and Their Influence on the Structure of South Florida Mangrove Communities

Thomas W. Doyle and Garrett F. Girod
National Biological Service
700 Cajundome Blvd.
Lafayette, LA 70506, U.S.A.

Abstract

Hurricanes are formidable forces that wreak havoc on society and nature alike. Mangrove ecosystems are especially vulnerable because they thrive in the intertidal zone of tropical regions where hurricanes originate and are most frequent. Because mangroves are found at the land-sea interface where hurricanes are often most intense, these coastal forests are subject to damage from both high wind and surge. Mangroves are also an important habitat for many colonial and migratory birds and other wildlife, along with being nursery grounds for our fisheries. The fate of mangrove habitat as influenced by hurricanes may be threatened in the future under a global warming environment that might yield more intense storms than have been previously observed. Simulation models of hurricane abiotics and mangrove community dynamics have been developed to evaluate the effects of hurricanes on mangrove habitat across the South Florida landscape. Model applications show that hurricane frequency and intensity have varied spatially across Florida's lower peninsula over the last century of record. Hindcast simulations of actual hurricane tracks and conditions seem to account for the structural composition of modern-day mangrove forests across South Florida. A recurrence interval of major storms every 30 years over the last century is the major factor controlling mangrove ecosystem dynamics in South Florida. Some climate change models predict an increase in hurricane intensity over the next century that may further alter the structure and composition of our mangrove ecosystems. Model results of climate change scenarios indicate that future mangrove forests are likely to be diminished in stature and perhaps include a higher proportion of red mangroves. This modeling approach offers the ability to assess decadal and longer time scale changes in hurricane behavior and its effects on community structure and distribution of important plant associations such as the fate of mangrove habitat.

6.1 Introduction

Mangrove forests occupy the intertidal zone of tropical coastlines worldwide. Their predominance along the land-sea interface makes these coastal forests vulnerable to sea-level rise, cyclones, and oil spill events among other disturbances, natural or anthropogenic. Because of their halophytic (salt tolerant) nature, mangroves are adapted to the added stress of waterlogging and salinity conditions that prevail in low-lying coastal environments influenced by tides. Global warming has been projected to increase sea water temperatures and lead to increases in sea level and possibly to greater hurricane intensity, which is likely to compound ecosystem stress of mangrove dominated systems.

Hurricanes are episodic climatic events of formidable force and destruction to both developed properties and natural areas. The regularity and severity of tropical storms are major determinants controlling ecosystem structure and development for coastal forests worldwide. The type, frequency, and extent of disturbance is determined not only by the dynamics of a given hurricane but also the nature and condition of the forest system and its component species. And while it is important to understand the effects and characteristics of any single storm event, it is equally important to relate ecosystem response to the cumulative effect of a history of disturbance events, both from hurricane and other sources, over many decades in order to predict long-term responses and to guide sustainable ecological resource management.

Coastal ecosystems along the Gulf of Mexico are subject to recurrent hurricanes. Tropical storm frequency along any given stretch has been estimated at one major event every 7 years. Climate change studies suggest the possibility that these storm events may become more intense in a warming global environment (Emanuel 1987). Numerous field studies have documented the susceptibility and vulnerability of neotropical mangrove species and systems to hurricane disturbance (Craighead 1964, 1971; Craighead and Gilbert 1962; Stoddart 1963; Roth 1992; Smith et al. 1994; Wunderle et al. 1992). More recent investigations by Doyle et al. (1994, 1995) of Hurricane Andrew (1992) on South Florida mangroves relate how the physical and biological elements interact to explain the varying degrees of windthrow and mortality relative to hurricane intensity, path, and direction.

Ecological models of forest growth and succession can provide a significant adjunct to empirical studies by enabling us to test hypotheses of tree and ecosystem response to long-term, multidecadal, disturbance regimes such as recurrent hurricanes. Doyle (1981) developed a forest simulation model that evaluated the role and significance of hurricane recurrence in controlling the biological diversity of Caribbean rain forests. Model results demonstrated that the prevailing hurricane frequency and damage probability were responsible for maintaining a higher species richness than would be expected in the absence of hurricanes. In this study we combine a hurricane simulation model, HURASIM, and a mangrove forest simulation model, MANGRO, to review the impact of hurricane history over the last century on forest structure of mangrove communities across South Florida. This integrated landscape modeling approach offers the ability to evaluate the temporal and spatial variability of hurricane disturbance over the last century and on the local and regional scale.

6.2 Methods

6.2.1 MANGRO Forest Model

MANGRO is a spatially explicit stand simulation model constructed for neotropical mangrove forests composed of *Avicennia germinans* (L.) Stearn (black mangrove), *Laguncularia racemosa* (L.) Gaertn.f. (white mangrove), and *Rhizophora mangle* L. (red mangrove). MANGRO is an individual-based model composed of a species-specific set of biological functions predicting the growth, establishment, and death of individual trees. MANGRO predicts the tree and gap replacement process of natural forest succession as influenced by stand structure and environmental conditions. The position of each tree is explicitly defined on a planar coordinate system with a default stand area of 1 hectare (100 m per side). Stand configuration was based on intact forest conditions with no edge effects. Canopy structure is modeled as a 3-dimensional process of crown height, width, and depth in relation to sun angle and shading by neighboring trees. Tree growth was based on growth potential for a given tree size reduced by the degree of light availability to the individual tree and species response to shade. Mortality is modeled as a self-thinning process dictated by prolonged suppression, senescence, and disturbance factors, primarily hurricanes.

6.2.2 HURASIM Hurricane Model

HURASIM is a spatial simulation model of hurricane structure and circulation for reconstructing chronologies of estimated windforce and vectors of past hurricanes. The model uses historic tracking and meteorological data of dated North Atlantic tropical storms from 1886 to 1989. The model generates a matrix of storm characteristics (i.e., quadrant, windspeed, and direction) within discrete spatial units and time intervals specified by the user for any specific storm or set of storms. HURASIM model output from Hurricane Andrew (1992) was correlated with field data to construct data tables of damage probabilities by site and species and to determine critical windspeeds and vectors of tree mortality and injury.

HURASIM recreated hurricane structure based on a tangential wind function, inflow angle offset, forward speed, and radius of maximum winds. Data input for the model includes tracking information of storm position (latitude and longitude) every 6 hours or less and maximum sustained wind speed. The model offers a suite of mathematical functions and parameter sets for the tangential wind profile taken from other hurricane studies (Boose et al. 1994; Bretschneiger and Tamaye 1976; Harris 1963; Kjerfve et al. 1986; Neumann 1987). The radius of maximum winds is determined from the reported maximum sustained wind input and a set of empirical equations. The inflow angle offset is user specified between 20 and 70 degrees, the default is 45 degrees for overland applications.

Model output is user-specified for point or area applications. Point applications involve a predetermined site location given in latitude and longitude. Area applications constitute a landscape matrix of defined geographic boundaries and resolution. Storm path and trajectory was interpolated for each location either on an hourly basis or fraction thereof. Time intervals of storm reposition and speed for this study were generated every 10 minutes. Profiles of estimated wind conditions for each site location are stored by year and storm. Minimum conditions of windspeed or distance can be set to parse the data output if warranted. Maximum wind estimates were retained from each storm greater than hurricane strength (i.e., 30 m s^{-1}) to be used in the MANGRO for projecting degree of forest impact and mortality.

6.2.3 South Florida Landscape Application

Both the MANGRO and HURASIM model simulations were projected onto a compartmentalized landscape of South Florida at a scale equal to a 7.5-minute quadrangle. A total of 41 cells of an uneven matrix design were identified across the lower peninsula of Florida that contained distributions of mangrove habitat (Fig. 6.1). Each cell represented an intact forest condition approximated by an independent simulation of the MANGRO model. Initial forest conditions were set with a normally distributed population of mature mangrove trees approximating a mean stand diameter of 50 cm and a mean stand age of 125 years. Seedling ingrowth was maintained stochastically at 1 plant per square meter every year to maintain a fully stocked stand for continuous recruitment. Equal probability was given to seedling recruitment by any species for any land area. Tree growth was constrained by size and light availability. A logistic growth function was constructed with observed data and generalized for all species taken from Doyle et al. (1995) and Craighead (1971) (Fig. 6.2). Craighead (1971) surmised that 1-m-diameter mangrove trees in the Lower Everglades were 250 years old. Stand data from 10, 30, and 60-year-old stands were used to approximate an empirical growth curve (see Fig. 6.2). Light availability was calculated for the crown zone and height of each tree as a function of light penetration through the canopies of taller neighbors. Light attenuation was calculated with Beer's Law based on a maximum leaf area index of 5.5. Growth was limited by shade tolerance (red > black > white mangrove) based on a set of light response curves (Fig. 6.3) constructed from a concurrent experimental study. While salinity is a major factor in controlling mangrove growth and succession, it was not included as an environmental parameter or growth effect in this modeling exercise. Sea-level and salinity conditions were assumed to be optimum for this simulation for all quadrangles whether seaward or not. Mortality was modeled as a stochastic process of hurricane occurrence and windspeed derived from damage probability curves developed from observed data of Hurricane Andrew (1992) effects (Doyle et al. 1994, 1995) (Fig. 6.4). If in any given year of the simulation a predicted hurricane windforce exceeded 30 m s^{-1}, a probability was derived based on windspeed from which a percentage of the standing crop or tress were stochastically removed from the forest simulation. Trees that failed to maintain 30% of potential growth accumulated an increasing probability of death by suppression such that less than 1% were likely to survive 10 years. Tree death was assumed to be complete with no retention of stump function and size for resprouting. While the white and black mangrove are known to sprout, there were no available data on the rate of regrowth from sprouting.

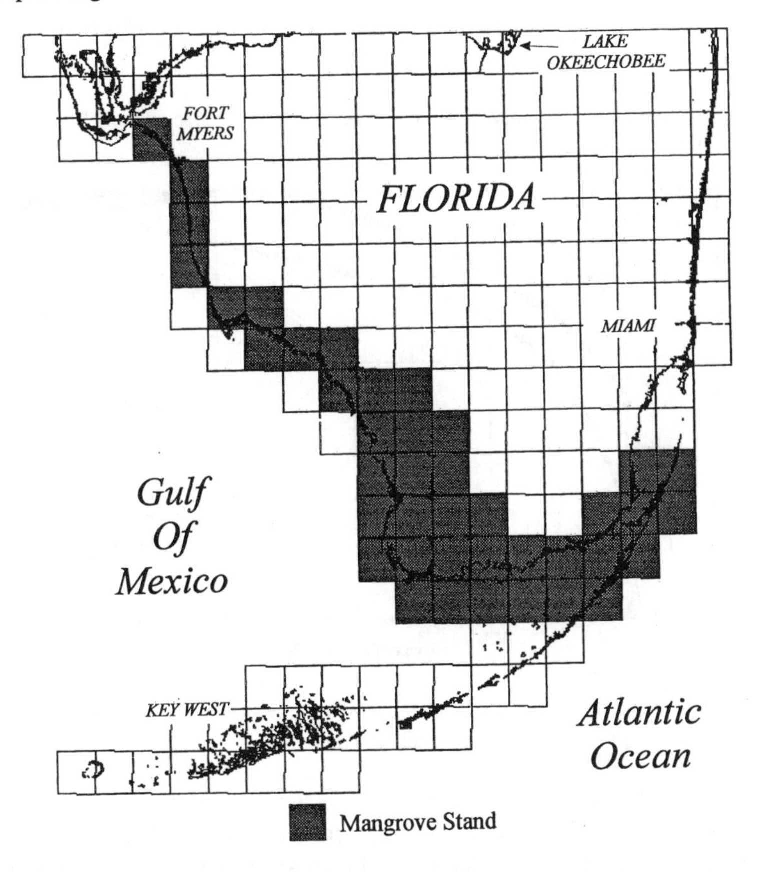

Fig. 6.1. Landscape matrix of South Florida, USA, illustrating distribution of mangrove habitat for 7.5-minute quadrangles (10-13 km on a side) included in the spatial simulation.

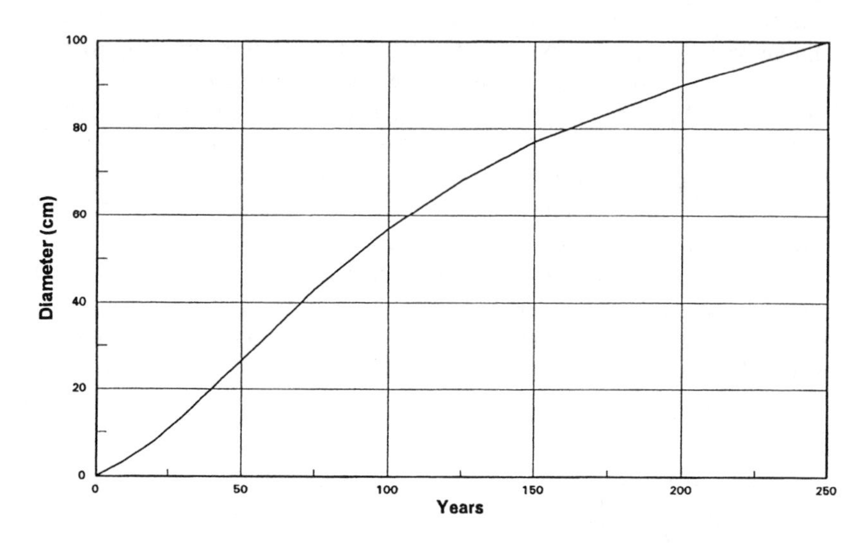

Fig. 6.2. Maximum diameter growth curve as a function of tree size and age for neotropical mangrove species.

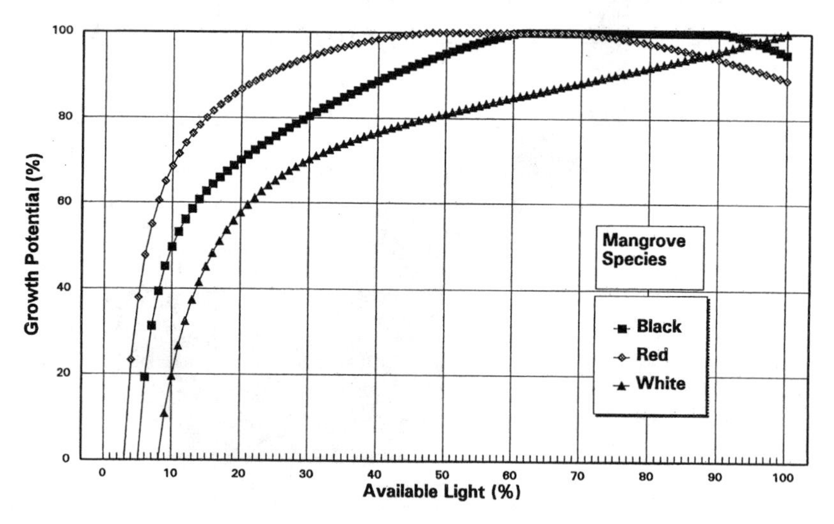

Fig. 6.3. Light response curves of red, black, and white mangrove species as a function of maximum potential growth and percent available light.

Four treatment effects were implemented including a no-hurricane simulation contrasted with a low, moderate, and high mortality effect that increases with corresponding increases in windspeed (see Fig. 6.4). A hindcast simulation for the period 1886 through 1989 was achieved by passing hurricane and site specific information from the HURASIM model to the associated MANGRO simulation for common cells. A cumulative assessment of hurricane impact was achieved by averaging stand attributes and size for the entire simulated landscape and time interval from 1890 to 1989.

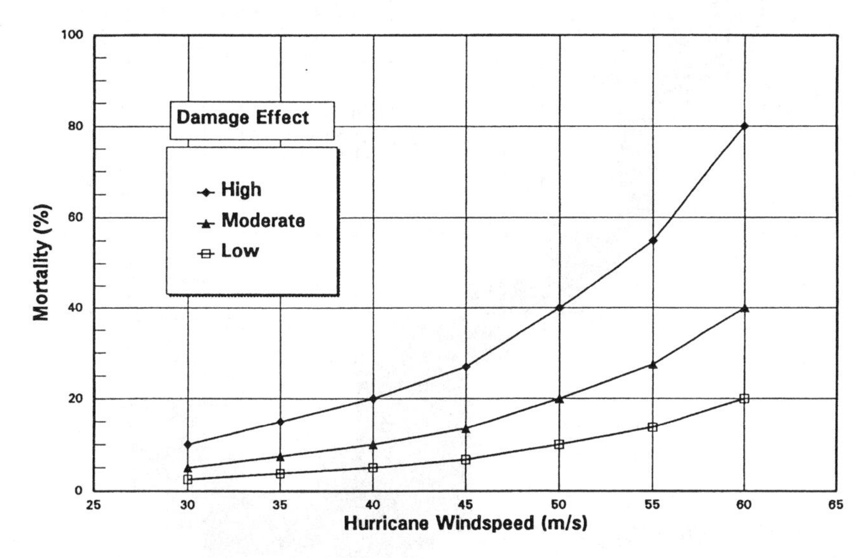

Fig. 6.4. Damage probability curves showing the percentage of mangrove trees killed at given predicted windspeeds under expected hurricane impact (moderate) and for extreme (high) and nominal (low) impact scenarios.

6.3 Results and Discussion

Model applications show that hurricane frequency and intensity have varied spatially across Florida's lower peninsula over the last century of record. For example, the distribution in the return frequencies of storms (i.e., number of storms per century) exceeding hurricane strength (30 m s^{-1}) across the simulated landscape varied geographically (Fig. 6.5). Storm frequency ranged from a minimum of 12 along the southwest coast and a maximum of 21 on the eastern shore. As a rule, hurricane frequency appears to be greatest on the east coast and least for west coast sites. Predicted storm strength likewise varied geographically over the period of record. Maximum predicted storm intensities for any one storm by quadrangle across South Florida ranged from 46-59 m s^{-1}, exhibiting a greater impact on the lower southwest coast than sites north and east (Fig. 6.6). These results suggest that the hurricane disturbance regime across South Florida includes a complement of sites with relatively low and high frequency and low- and high-intensity storm activity.

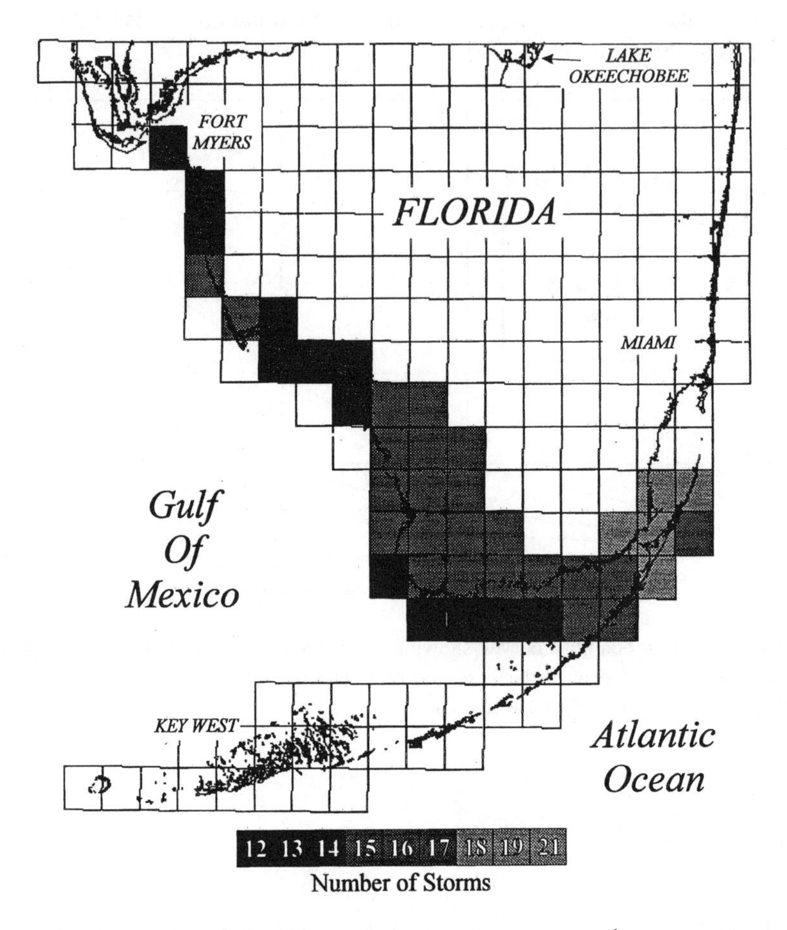

Fig. 6.5. Frequency of strikes by hurricane strength storms (> 30 m s^{-1}) by quadrangle across South Florida predicted by HURASIM model for mangrove habitat for the period 1886-1989.

Differences in stand structure were readily apparent between all hurricane impact trials and the no-hurricane control simulation over the century of hurricane history across South Florida. The change in mean stand diameter by decade predicted with each damage scenario is illustrated in Fig. 6.7. The no-hurricane trial showed an overall increase in stand stature for lack of catastrophic disturbance phenomena. All other damage scenarios contributed to declines in stand size though varying in the rate and magnitude of change from decade to decade. As expected, high damage impacts yielded a greater turnover of forest resources cumulatively and with each event. Change in forest structure was appreciably reduced by a few, large storms than by all remaining storms collectively. Over 70% of the forest reduction was attributed to 4 storms or less that occurred on regular intervals in the 1910s, 1935, and 1960, respectively. The intensity and trajectory of these storms impacted a greater proportion of the landscape and with sufficient windspeed to exceed average damage probability. Forest regrowth is appreciably slow compared to the regularity and frequency of hurricane impact. In some cases, hurricanes were as frequent as several per decade such that the loss of a single dominant tree may have easily exceeded the growth gains of new recruits.

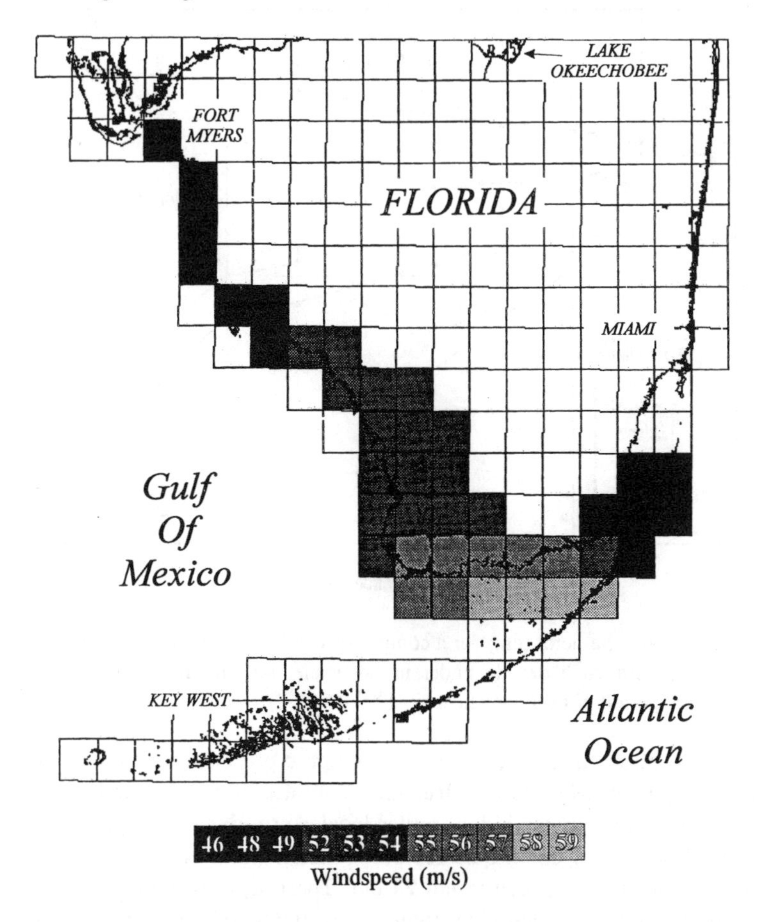

Fig. 6.6. Maximum sustained windspeed predicted for all North Atlantic tropical storms by quadrangle across South Florida by HURASIM model for mangrove habitat for the period 1886-1989.

The periodicity of hurricane impact, particularly of the strongest storms, appreciably contributed to the chronic downsizing of stand condition. Recurrent hurricanes that strike in subsequent years are less likely to reduce the overall forest stature than events that are more evenly distributed in time and space. The presence of large stump diameters of mangrove species in the Lower Everglades apparently killed by the 1935 and 1960 storms that are still evident along the southwest coast

(Doyle et al. 1995) and the relative absence of surviving trees of similar size suggests that forest disturbance from hurricanes may have been greater over this century than in the past. While the actual frequency of storms may or may not be different between centuries, the trajectory and periodicity of combined storms of high intensity may offer a plausible explanation for predicted stand conditions.

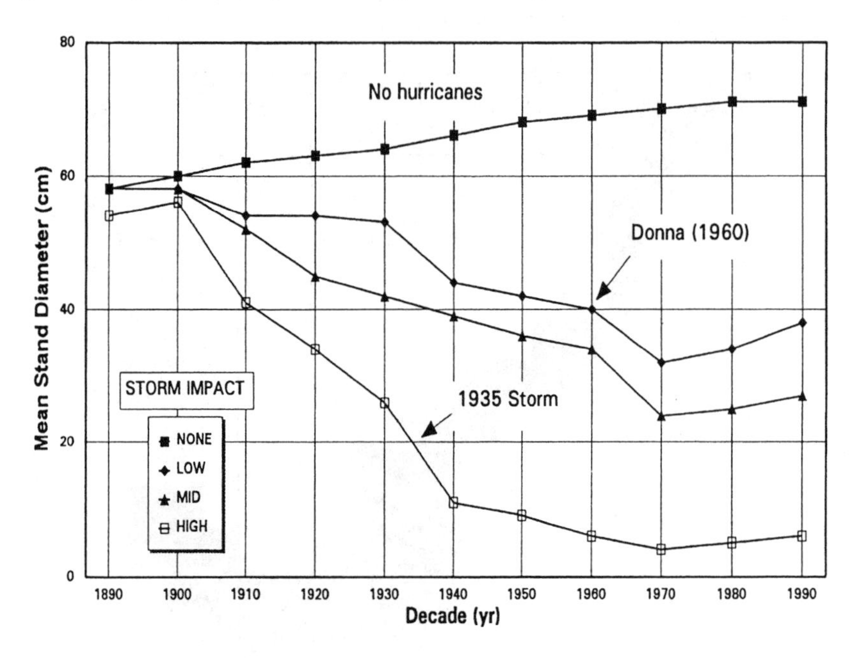

Fig. 6.7. Mean stand diameter (cm) for a composite sampling of all 41 stand simulations of mangrove habitat for each decade under no-hurricane, low, moderate, and high impact scenarios for the period of hurricane history 1886-1989.

Available stand level data were used to compare present forest conditions with simulated output of 1989 forest structure. Contrast of mean stand diameter of observed and simulated stands for similar locales are given in Table 6.1. Results show that present field conditions represent rather diminutive stands ranging in mean stand diameters between 9 and 24 cm. These figures were not statistically different from predicted stand measurements for the moderate damage probability scenario ($p < .05$). The high damage scenario resulted in a forest condition of extremely small tree sizes, akin to clearcutting with each hurricane event. The no-hurricane and low damage scenarios yielded stand measurements three or more times greater than the equivalent field condition which in part demonstrates the role hurricanes play in controlling mangrove stand structure across South Florida.

The agreement in actual and predicted forest size and structure for the moderate or expected damage scenario calibrated from Hurricane Andrew (1992) data provides some verification of model function and utility. The high damage scenario offers a measure of what impact increased hurricane intensity under a warming global environment might impose on future forests. Model results suggest that not only will forest stature be reduced significantly but that a shift in species composition might be expected with persistent and shade tolerant red mangroves notwithstanding other site and environmental conditions. Because light is the primary factor controlling tree growth and stand succession in this application, red mangroves by virtue of their greater tolerance to shade persisted in greater numbers in the subcanopy classes. Other factors such as propagate transport and salinity might actually limit red mangrove distribution and abundance relative to the other species in nature than what is considered in this version of the MANGRO model.

Table 6.1. Comparison of mean stand diameter (cm) for actual (field) and simulated stands circa 1990 from complementary grid cell locations for each hurricane impact scenario. Simulated data are displayed in columns 2 through 5; actual data are in column 6.

Site Number	Simulated None	Simulated Low	Simulated Mid	Simulated High	Actual Data
1	71.6	41.4	23.4	5.9	15.9
4	70.2	40.5	11.7	4.3	15.4
7	71.3	32.3	14.4	1.35	13.8
9	71.3	26.5	14.8	3.29	14.5
13	73.3	38.5	17.9	3.05	16.7
14	73.8	37.1	20.1	1.35	18.1
16	70.3	31.3	18.1	3.12	24.3
19	67.9	24.4	10.4	3.18	9.3
21	76	35.4	14	3.06	18.9
22	67.9	33.5	9.9	1.96	22.2
25	70.5	30.6	32.2	2.41	13.7
27	74.1	34.8	10.5	2.34	11.5
33	74.4	28	15.7	2.34	12.1
40	73.5	32.2	12	1.96	11.9

6.4 Conclusion

Hindcast simulations of actual hurricane tracks and conditions seem to account for the structural composition of modern-day mangrove forests across South Florida. The periodicity of major storms every 30 years in this century may be the most important factor controlling mangrove ecosystem dynamics in South Florida. Global climate change models predict an increase in hurricane intensity over the next century that may further alter the structure and composition of this mangrove landscape. Model results of climate change scenarios (high damage probability) indicate that future mangrove forests are likely to be diminished in stature and perhaps include a higher proportion of red mangroves. The integrative modeling approach of combining physical models like HURASIM with biological models like MANGRO offers the ability to assess large-scale and long-term processes of climate-related phenomena on our natural ecosystems. Decadal and longer time scale changes in hurricane behavior and regularity may be much more significant in shaping mangrove community structure and distribution on the landscape than can be evaluated by field studies alone.

References

Boose, E.R., 1994. Hurricane impacts to tropical and temperate forest landscapes. *Ecological Monographs,* **64**, 369-400.

Bretscheingder, C.L. and Tamaye, E.E., 1976. Hurricane wind and wave forecasting techniques. In: *Proceedings of the 15th Coastal Engineering Conference*, held Honolulu, Hawaii, July 11-17, (ASCE), Chapter 13, 202-237.

Craighead, F.C., 1964. Land, mangroves, and hurricanes. *Fairchild Tropical Garden Bulletin*, **19**, 1-28.

Craighead, F.C., 1971. *The Trees of South Florida*, vol. 1, University of Miami Press, Coral Gables, Florida 212 pp.

Craighead, F.C. and Gilbert, V.C., 1962. The effects of Hurricane Donna on the vegetation of southern Florida. *Quarterly Journal of the Florida Academy of Sciences*, **25**, 1-28.

Doyle, T.W., Wells, C.J., Krauss, K.W., and Roberts, M., 1994. The Use of Videography to Analyze the Spatial Impact of Hurricane Andrew on South Florida mangroves, 222-227. In: Proceedings of GIS/LIS 94 Annual Conference held October 27-30, 1994, Phoenix, AZ. Published by American Congress on Surveying and Mapping, Bethesda, MD.

Doyle, T.W., Smith, T.J. III, and Robblee, M.B., 1995. Wind damage effects of Hurricane Andrew on mangrove communities along the southwest coast of Florida, USA. *Journal of Coastal Research,* **18**, 144-159.

Emanuel, K.A., 1987. The dependence of hurricane intensity on climate, *Nature*, **326**, 483-485.

Harris, D.L., 1963. Characteristics of the Hurricane Storm Surge, U. S. Dept. of Commerce, Technical Paper No. 48. Washington, D.C., 138 pp.

Kjerfve, B., Magill, K.E., Porter, J.W., and Woodley, J.D., 1986. Hindcasting of hurricane characteristics and observed storm damage on a fringing reef, Jamaica, West Indies. *Journal of Marine Research,* **44**, 119-148.

Neumann, C.J., 1987. The National Hurricane Center risk analysis program (HURISK). NOAA Technical Memorandum NWS NHC 38, Washington D.C., 56 pp.

Rorth, L.C., 1992. Hurricanes and mangrove regeneration: Effects of Hurricane Joan, October 1988, on the vegetation of Isla del Venado, Bluefields, Nicaragua. *Biotropica*, **24**, 375-384.

Smith, T.J.,III, Robblee, M.B., Wanless, H.R., and Doyle, T.W., 1994. Mangroves, hurricanes, and lightning strikes, *Bioscience*, **44**, 256-262.

Stoddart, D.R., 1963. Effects of Hurricane Hattie on the British Honduras Reefs and Cays, October 30-31, 1961. *Atoll Research Bulletin*, **95,** 1-142.

Wunderle, J.M., Lodge, D.J., and Waide, R.B., 1992. Short-term effects of Hurricane Gilbert on terrestrial bird populations on Jamaica., *The Auk*, **109,** 148-166.

7 A Socioeconomic Analysis of Hurricanes in Puerto Rico: An Overview of Disaster Mitigation and Preparedness

Havidán Rodríguez
Department of Social Sciences
University of Puerto Rico-Mayagüez
Mayagüez, Puerto Rico 00680

Abstract

The island of Puerto Rico has been severely affected by a significant number of hurricanes within the past 100 years. Hurricanes San Ciriaco (1899), San Felipe (1928), San Nicolás (1931), San Ciprián (1932), Santa Clara (1956), Federico (1979), and Hugo (1989) have had devastating social and economic effects on the island. The problems of hurricanes are further exacerbated by the topography of Puerto Rico, which includes areas that are highly susceptible to flooding and landslides.

As a result of the changing demographic patterns in Puerto Rico, there has been a significant increase in population density, in the proportion of the elderly and physically disabled population, and an increasing concentration of residents in areas that are at high risk due to floods and/or landslides. These factors, combined with the dramatic increase in modern, high-rise infrastructure (i.e., hotels and condominiums) in the coasts of the island, contribute to our increasing vulnerability to disasters.

In order to prepare better for disasters and to coordinate the responsibilities and efforts of governmental organizations during periods of crisis, Law Number 22 (1976) and the Executive Order of the Governor of Puerto Rico (1993) were established. However, past experiences with disasters, such as Hurricane Hugo, show that the results of the planning, response, and recovery efforts of these organizations have been mixed. The lack of human and economic resources has presented serious problems in the management of disasters among organizations on the island.

The research literature, and data collected from disaster management officials in Puerto Rico, show that there is an urgent need to educate organizational leaders as well as the general population if we are to obtain an "appropriate" response during disaster situations. Moreover, existing disaster mitigation, preparedness, and response strategies must be revised and evaluated before the next hurricane strikes.

7.1 Introduction

Given their topographical and climatological conditions, the Caribbean islands are highly susceptible to natural disasters (e.g., hurricanes, floods, landslides, and earthquakes). Over the past century, these islands have experienced a variety of natural disasters, resulting in dramatic loss of life and property. The island of Puerto Rico has not been an exception.

Much of the land in Puerto Rico is characterized by coastal or riverline flood areas and steep mountains. Therefore, it is no surprise that a significant proportion of the population resides in areas which are extremely hazardous due to the high probability of flooding and/or landslides. Indeed, the Federal Emergency Management Agency (FEMA 1985) estimates that 47% of the Puerto Rican population resides in flood hazard areas. This represents a potentially catastrophic hazard threat to life and property.

7.2 An Overview of the Socioeconomic Impact of Hurricanes in Puerto Rico

Throughout its history, the island of Puerto Rico has been severely affected by a large number of hurricanes. Table 7.1 shows the most important hurricanes that have approached or have impacted the island, from 1508 to 1989. During the past 100 years, the island has been affected by 13 hurricanes. Further, 43 tropical storms and hurricanes have passed within 75 miles of Puerto Rico (Palm and Hodgson 1993).

On 8 August 1899 Puerto Rico experienced one of the most devastating hurricanes in recent memory. Hurricane San Ciriaco, with an estimated wind speed of 100 mph, has been called the "last great hurricane of the 19th Century" in Puerto Rico. Certainly, it was the hurricane which claimed the largest number of victims on the island; 3369 persons perished, primarily as a result of floods. The economic costs to Puerto Rico as a result of hurricane San Ciriaco are estimated at twenty million dollars.

San Felipe (13 September 1928) is known as the "second most important hurricane" to strike Puerto Rico. At an estimated 160 mph it was more powerful than Hurricane San Ciriaco. San Felipe claimed the lives of 312 persons and left 83,000 homeless. The economic losses associated with San Ciriaco were estimated at $50 million. Another important hurricane to hit Puerto Rico was San Ciprián on 26 September 1932. It is estimated, that at 120 mph, this hurricane resulted in 225 deaths. Hurricane San Ciprián destroyed the islands infrastructure leaving 25,000 families stranded, without a home, and a large segment of the population was left without basic services (González and Chaparro 1990). This hurricane resulted in approximately $30 million of economic loss to the island.

Table 7.1. Hurricanes that have approached or impacted Puerto Rico, 1508-1989.

HURRICANE	DATE
San Roque	16 August 1508
San Laureano	4 July 1515
San Francisco	4 October 1526
San Francisco	4 October 1527
Santa Ana	26 July 1530
San Hipólito	22 August 1530
San Ramón	31 August 1530
San Leoncio	September 12, 1535
San Pio	11 July 1537
San Bartolome	24 August 1568
San Mateo	21 September 1575
San Leoncio	12 September 1615
San Nicomedes	15 September 1626
------------	September 1642
------------	August 1657
San Zacarias	6 September 1713
San Candido	3 October 1713
Santa Regina	September 7, 1718
San Agustín	28 August 1722
Santa Rosa	30 August 1730
Santa Rosa	30 August 1738
San Leoncio	12 September 1738
San Esteban	3 August 1740
San Vicente	11 September 1740
San Judas Tadeo	28 October 1742
San Agapito	18 August 1751
San Jenaro	19 September 1766
San Marcos	7 October 1766
San Adrian	8 October 1766
San Cayetano	7 August 1767
Del Carmen	16 July 1772
San Agustín	28 August 1772
San Ramón	31 August 1772
San Pedro	1 August 1775
Santa Regina	7 September 1776
San Antonio	13 June 1780
San Paulino	10 October 1780
San Calixto	14 October 1780
San Lupo	25 September 1785
San Roque	16 August 1788
Santa Rosalia	4 September 1804
San Mateo	21 September 1804
San Vicente	11 September 1806
San Jacinto	17 August 1807
San Esteban	2 September 1809
San Liborio	23 July 1813
Santa Juana	21 August 1813
San Jose De Cupertino	18 September 1816
San Mauricio	22 September 1818
San Mateo	21 September 1819
San Pedro	9 September 1824
La Monserrate	8 September 1824
Santa Ana	26 July 1825
San Jacinto	17 August 1827
San Hipolito	13 August 1835
Los Angeles	2 August 1837
San Cipriano	16 September 1840

Table 7.1. (Cont.)

HURRICANE	DATE
San Vicente	11 September 1846
San Agapito	18 August 1851
San Lorenzo	5 September 1852
San Mauricio	22 September 1852
San Evaristo	26 October 1853
San Narciso	29 October 1867
Santa Juana	21 August 1871
San Felipe	13 September 1876
San Serafin	12 October 1876
San Erasmo	25 November 1878
San Rufo	28 November 1878
San Gil	1 September 1888
San Martin	3 September 1889
San Magin	19 August 1891
San Remigio	1 October 1891
San Serafin	12 October 1891
Del Carmen	16 July 1893
San Roque	16 August 1893
San Gil	1 September 1893
San Ramón	31 August 1896
San Eustaquio	20 September 1898
San Ciriaco	8 August 1899
San Juan Bautista	29 August 1899
San Ignacio	23 October 1900
San Cirilo	7 July 1901
San Vicente	11 September 1901
San Zacarias	6 September 1910
San Emigdio	5 August 1915
San Tiburcio	11 August 1915
San Hipolito	22 August 1916
San Pedro	9 September 1921
San Liborio	23 July 1926
San Felipe	13 September 1928
San Nicolas	10 September 1931
San Ciprián	26 September 1932
San Calixto	14 October 1943
San Mateo	21 September 1949
Santa Clara	12 August 1956
Donna	5 September 1960
Frances	2 October 1961
Edith	26 September 1963
Cleo	23 August 1964
Faith	26 August 1966
Inez	28 September 1966
Beulah	9 September 1967
Eloisa	15 September 1975
David	30 August 1979
Federico	2 September 1979
Allen	4 August 1980
Gert	8 September 1981
Hugo	17 September 1989

Source: Lastra-Aracil, Jorge J. (1983). El ABC Sobre Desastres Naturales (Huracanes). Carolina, Puerto Rico: Imprenta El Impresor.

Santa Clara ("Betsy"[1]) is remembered by many residents on the island because it occurred quite recently (12 August 1956) and due to its negative impact on the island economy. Santa Clara claimed the lives of 16 residents, destroyed close to 15,000 homes, and resulted in damages estimated at $40 million.

On September 1975 the island of Puerto Rico was visited by tropical storm Eloise. Eloise resulted in severe floods in many parts of the island. Primarily as a result of flooding, 44 deaths were recorded. Property damage was estimated at over $125 million.

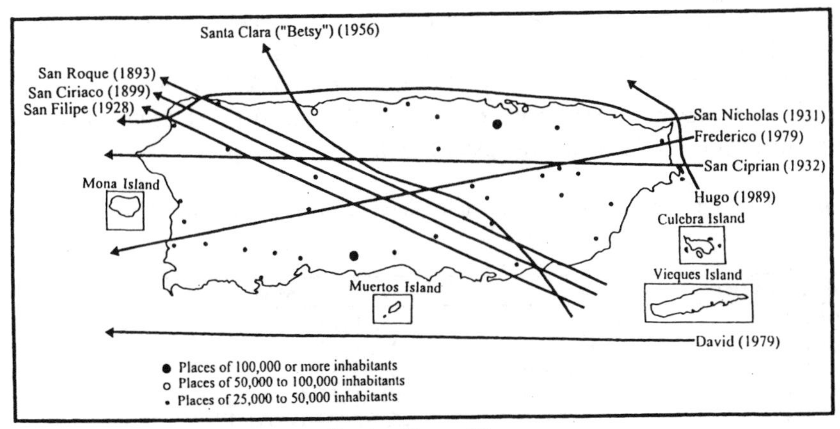

Fig. 7.1. Nine "Great" hurricanes affecting Puerto Rico.

In 1979, two hurricanes passed through Puerto Rico, hurricanes David (29 August) and Federico (4 September). David and Federico caused close to $25 million worth of damage and resulted in seven deaths.

The most recent hurricane, Hugo (17-18 September 1989), had devastating social and economic effects on the island (Aguirre and Bush 1992). Hurricane Hugo, with estimated wind speeds of up to 120 mph when it reached Puerto Rico, devastated the municipalities of Vieques and Culebras. Hugo then proceeded to enter the island through the northeast coast (see Fig. 7.1).

As a direct result of Hugo, thousands of people lost their homes, two persons drowned, seven persons died of other causes after the storm had passed, 35 municipalities were left without electricity, and the San Juan area was left without water for nine days. Hurricane Hugo was described as the costliest hurricane in the history of Puerto Rico, resulting in approximately $1 billion in economic losses.

[1]Prior to 1953 the hurricanes that occurred in Puerto Rico received the name of the saint that corresponded to the day that this physical event occurred. After 1953, the hurricanes were given the names which were established by the National Oceanic and Atmospheric Administration (NOAA).

Hurricanes San Ciriaco (1899), San Felipe (1928), San Ciprián (1932), Santa Clara (1956), Eloise (1975), David and Federico (1979), and Hugo (1989), among others, remind us of the vulnerability of Puerto Rico to this type of disaster. Moreover, as Gray and others have suggested, we might enter a period of more intense hurricanes, similar to Hugo, in the coming decades (see Gray et al. 1996, Chap. 2, this volume).

7.2.1 Floods in Puerto Rico

Flood disasters, primarily as a result of hurricanes and tropical storms, have also been very common in Puerto Rico. Since 1970, the president of the United States has declared ten major disasters in Puerto Rico (See Table 7.2). In October 1970, flooding in Puerto Rico resulted in 18 deaths and over $65 million in economic losses. In September 1975 floods on the island resulted in 34 deaths and $125 million in damages (Palm and Hodgson 1993).

Table 7.2. Presidential disaster declaration for flooding in the Commonwealth of Puerto Rico

DECLARATION	DATE	COST TO FEMA[a]
296-DR	October 1970	$14,573,834[a]
455-DR	November 1974	12,056,237[a]
483-DR	September 1975	44,749,088[a]
579-DR	September 1979	102,047,551[a]
736-DR	May 1985	25,221,000 (Projected)[a]
746-DR	October 1985	65,000,000 (Projected)[a]
768-DR	May 1986	-
805-DR	December 1989	-
842-DR	September 1989	-
-	January 1992	-
Total cost to FEMA		$269,647,710

[a] Total costs to FEMA only include these amounts

Source: Federal Emergency Management Agency (1985); Estado Libre Asociado de Puerto Rico (1991).

One of the most devastating floods in Puerto Rico occurred during 4-7 October 1985 when a tropical wave swept through the island resulting in 22 inches of rain within a 24-hour period. As a result, the "most lethal landslide in US history" (FEMA 1985) occurred in the squatter settlement of Mameyes in the municipality of Ponce leaving a death toll of 127. The intense rains resulted in the overflow of the Coamo River Dam sweeping away a span of the Las Americas Expressway and another 29 people died. Furthermore, in the squatter settlement of Las Batatas in El Tuque-Ponce, another 17 lives were lost due to flash floods and the lack of evacuation warnings (FEMA, 1985). It was estimated that over 46,000 island residents, in 30 municipalities, were affected by this storm (Rodríguez et. al. 1993).

Most recently, the "Three Kings Day Flood" (5-6 January 1992) took disaster mitigation authorities and the Puerto Rican population by surprise. As a result of heavy rainfalls, some areas in Puerto Rico received up to 20 inches of rain within a 24-hour period resulting in one of the island's worst floods. The death toll reached 20 persons. Approximately 6000 families totally or partially lost their homes and/or their personal belongings. Close to 600 persons were evacuated to 17 shelters throughout the island. On 15 January 1992, the governor of Puerto Rico petitioned the president of the United States to declare Puerto Rico a major disaster area.

The information presented in the preceding section reflects the vulnerability of Puerto Rico to disasters such as hurricanes, floods, and landslides. It is noteworthy that further natural disasters in Puerto Rico, as in other parts of the world, will be more damaging than in the past (Quarantelli 1985; Curson 1989; Aptekar 1994). Population growth, high population density, high population concentration in hazard-prone areas (i.e., coastal or riverline flood zones and steep mountainous areas), and the dramatic increase in modern high-rise infrastructure all contribute to the higher vulnerability of the population when a meteorological event occurs.

7.3 Changing Demographic Composition of Puerto Rico and Disaster Vulnerability

Puerto Rico is the smallest island of the Greater Antilles, measuring approximately 100 miles long and 34 miles wide (3435 square miles). However, Puerto Rico has the fourth largest population in the Caribbean after Cuba, Haiti, and the Dominican Republic. It is important to note that, from 1950 to 1970, Puerto Rico experienced significant demographic and socioeconomic changes. During this period, Puerto Rico was transformed from a rural/agrarian to an urban/industrial society. Furthermore, the island experienced a dramatic decline in its fertility and mortality rates and an increase in life expectancy.

During the 1940's the total fertility rate (TFR), or the average number of children for women in their reproductive ages, in Puerto Rico was 5.9 but currently it is near replacement level (2.2). The mortality rate in Puerto Rico declined from 14.5 in 1940-50 to 7 in 1993. Consequently, life expectancy on the island increased from 61 in 1949 to 75 in 1993. As a consequence of declining fertility and mortality rates, Puerto Rico experienced a significant increase in the proportion of its elderly population. In 1990, approximately 10% of the island population was 65 years of age and over (see Fig. 7.2). The increase in the elderly population on the island, among other factors, resulted in an increase in the proportion of the physically disabled population.

Figure 7.3 shows that in 1990 12.3% of the population on the island aged 16-64 reported a physical or mental handicap. Moreover, 45%, of those aged 65 and over, reported a physical and/or mental handicap.[2] Further, Fig. 7.4 shows that approximately 6% of the population aged 16-64 had a mobility limitation[3] compared to 29% for those 65 years of age and over.

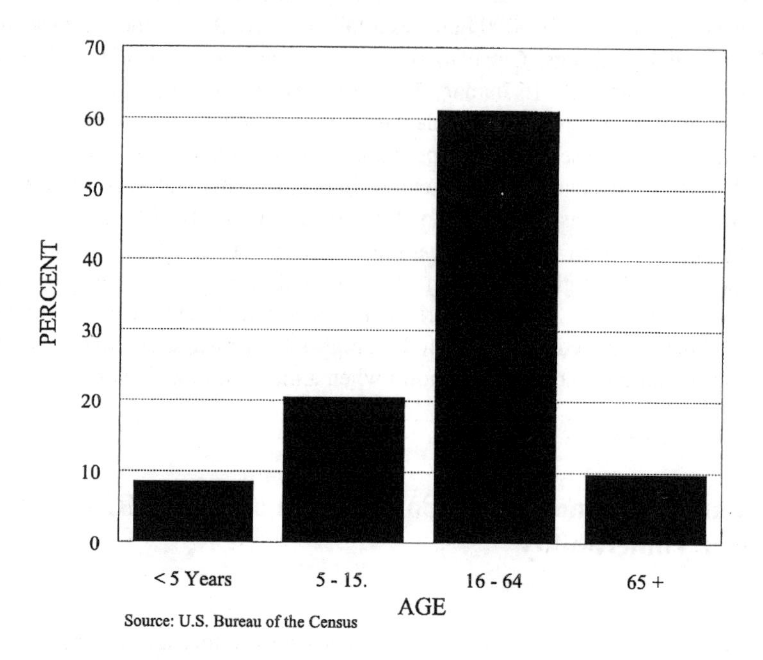

Fig. 7.2. Population of Puerto Rico, 1990: Age distribution.

The above information shows that there is a significant proportion of the Puerto Rican population that is elderly and/or has a physical or mental handicap, and/or some type of mobility limitation. This is a group that is vulnerable to disaster situations and may require special attention, in terms of evacuation and sheltering, during a period of crisis such as a hurricane.

Figure 7.5 shows that the total population of Puerto Rico increased from 2.3 million in 1960 to 3.5 million in 1990. Consequently, the population density in Puerto Rico has soared. Puerto Rico's population density of 1060 persons per square mile makes it one of the most densely populated countries in the world.

---

[2]The U.S. Bureau of the Census defines a physical or mental handicap/disability as a "health condition lasting six or more months and which limited the kind or amount of work they [persons] could do at a job or business".

[3]The U.S. Bureau of the Census defines a mobility limitation as a health condition, with a duration of six or more months, which makes it difficult for a person to leave his/her home on his/her own.

The island of Puerto Rico has also experienced a rapid process of urbanization since the 1950s. Figure 7.6 shows that the urban population on the island increased from 44% in 1960, to 56% in 1970, to 67% in 1980, and to 71% in 1990. Unfortunately, population settlement has occurred in areas of high risk such as those subject to floods and/or landslides. Ayala-Padró (1991) indicates that over 160,000 families reside in 90 communities which are located in flood-prone areas in Puerto Rico. Almost every municipality on the island has a significant segment of its population residing in areas which are highly vulnerable to flooding (FEMA 1985). The problems of these residents are further exacerbated by the fact that many of these areas are also susceptible to landslides. FEMA (1985) indicates that the "mountainous terrain and tropical climate combine to make Puerto Rico one of the most landslide-prone areas in the United States." This problem has been exacerbated by the land-use patterns of the Puerto Rican population and the construction industry.

Larsen et. al (1991) shows that, since the 1970s, there has been a significant growth in the islands construction industry. Indeed, in 1960 the number of residences in Puerto Rico approached 522,000, but this figure dramatically increased to 994,000 by 1980 (Estado Libre Asociado de Puerto Rico 1988). The continued construction of residences, roads, and other types of structures in mountainous, flood-prone areas and the coasts of the island increases the populations vulnerability to disasters. Jarvinen (1991) argues that increases in coastal population density have resulted in the potential for flood scenarios far worse than those occurring during Hurricane Hugo.

Other problems that the island currently confronts are its high poverty and unemployment rates. Data from the U.S. Bureau of the Census shows that in 1989 only 47.1% of the population, 16 years and over, was in the labor force. The unemployment rate for Puerto Rico during this period was 20.4%. Puerto Rico is characterized as having one of the highest, if not the highest, unemployment rate of all industrialized nations. Further, census figures show that close to 60% of the population was living below the established poverty levels and 31% of all households received income from public assistance (U.S. Bureau of the Census 1993). Poverty increases our vulnerability to disasters. Disasters occur more frequently in poor countries causing more suffering among the poverty stricken population. This segment of the population, due to lack of resources, resides in high risk areas, and has less protection against and experiences greater difficulties in recovering from disasters (Anderson 1991; Aptekar 1994).

In summary, as a result of the aforementioned demographic and socioeconomic changes, Puerto Rico can be characterized as an urban/industrial society with a high population density and a disproportionate population concentration in hazard-prone areas. Further, the increase in life-expectancy has resulted in a large segment of the population that will require special attention during a disaster situation (e.g., the elderly and those with physical and mental disabilities). The problems of the island are further exacerbated by its high unemployment and poverty rates. Finally, the construction of modern, high-rise infrastructure in hazard-prone areas has increased the vulnerability of the island to disasters.

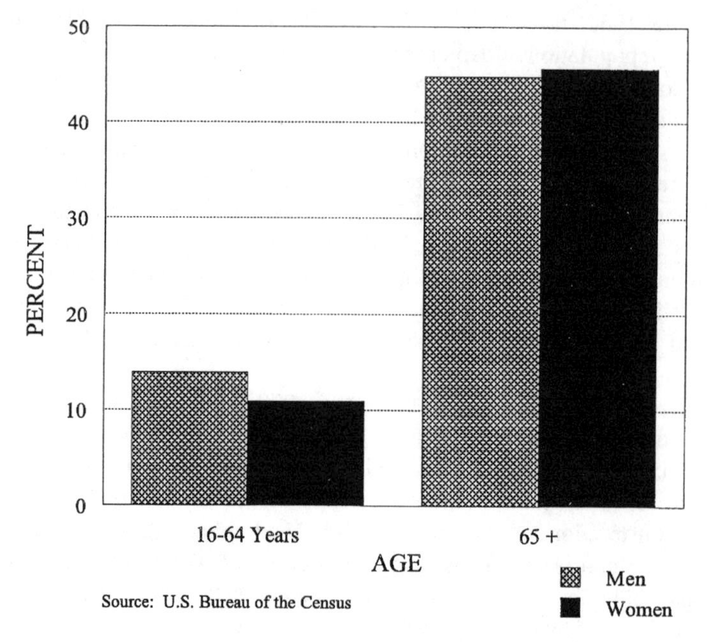

Fig. 7.3. Population with a physical/mental disability, Puerto Rico, 1990.

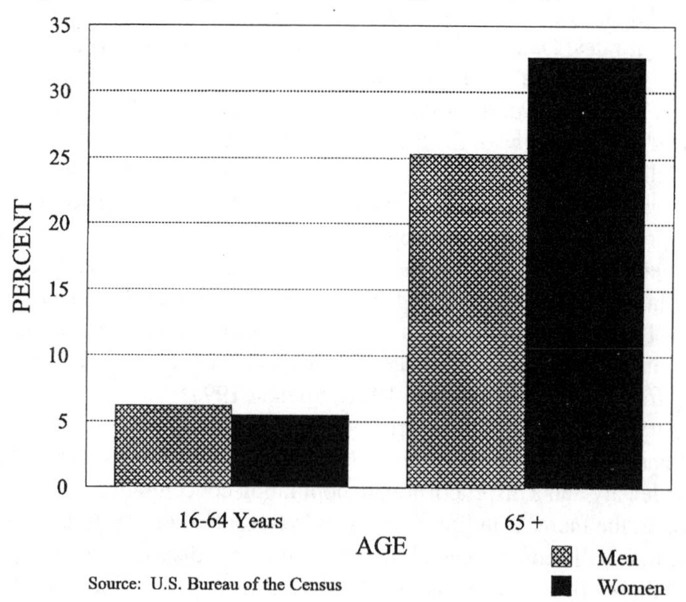

Fig. 7.4. Persons with mobility limitations, Puerto Rico, 1990.

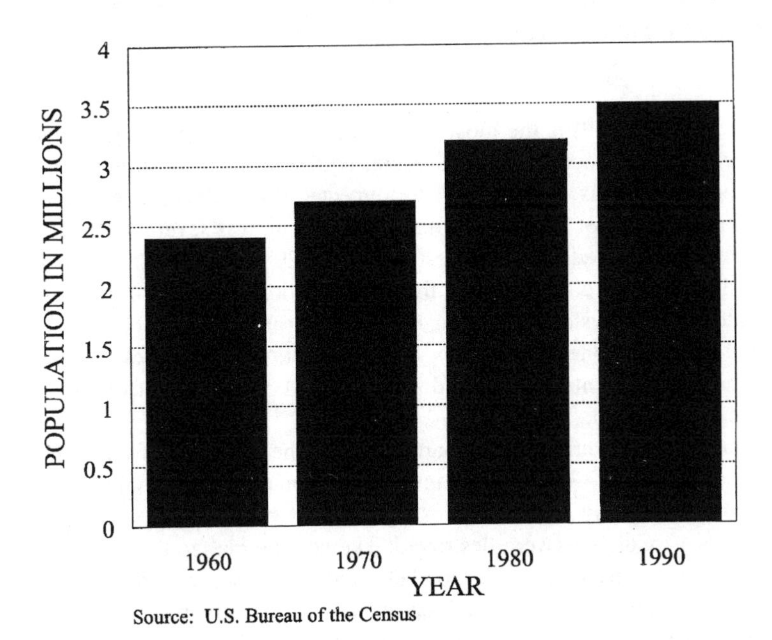

Fig. 7.5. Population of Puerto Rico, 1960-1990.

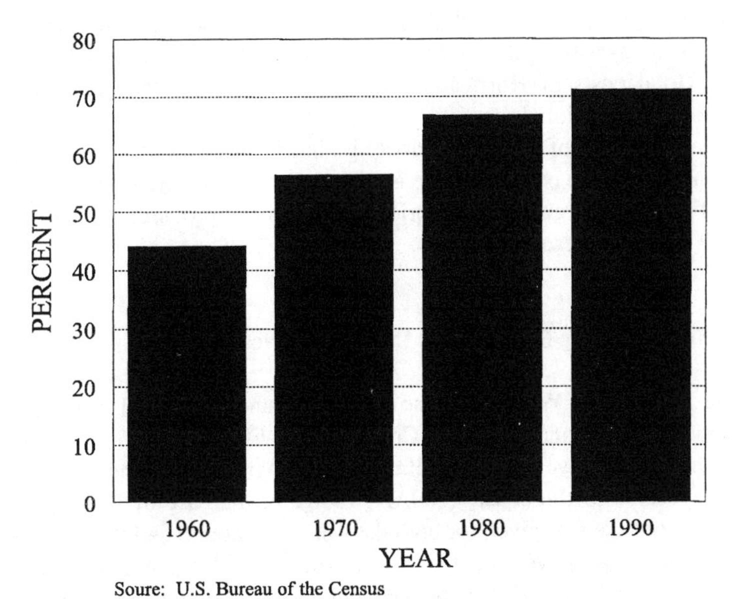

Fig. 7.6. Urban population, Puerto Rico, 1960-1990.

7.4 Disaster Mitigation and Preparedness in Puerto Rico: The Case of Hurricane Hugo

Hurricane Hugo (1989) is the most recent and the most economically devastating hurricane to strike Puerto Rico. As a consequence of the destructive nature of Hugo, the island was temporarily disconnected from the outside world. The International Airport of San Juan, Luis Muñoz Marín, was severely affected by the tremendous wind speeds of Hugo, estimated at 120 mph. The San Juan seaports were also closed, therefore, limiting the transportation into and out of these areas. The United Postal System in Puerto Rico was paralyzed for approximately four days and the telephone system was left nonfunctional. The lack of electricity, water, and other essential services disrupted the daily lives of over one million residents on the island.

The major transportation routes, particularly in the northeast of the island, were cluttered with debris. Therefore, movement from one municipality or area to another was extremely difficult. Close to 80% of the housing structures in the municipality of Culebras were destroyed. The commercial zones in Santurce and Hato Rey, the tourist area of El Condado, and the Historic Old San Juan were described as "battle zones" by the local media; the islands tropical rainforest (El Yunque) was all but destroyed; segments of the islands agriculture (e.g., coffee plantations and other minor crops) were severely affected (González and Chaparro 1990). The Department of Natural Resources of Puerto Rico (Departmento de Recursos Naturales, DRN) estimated that the agricultural sector would need about three years to recover from the severe damages caused by Hurricane Hugo (DRN 1991). The total costs of Hurricane Hugo to the Puerto Rican economy were estimated to be one billion dollars.

As with all analyses of natural events (e.g., hurricanes, floods, landslides, etc.) that affect a society, we can divide the experiences of Puerto Rico with Hurricane Hugo into two broad or general categories: successes and problems. The following section will focus on these two aspects.

7.4.1 The Success Story Prior to and During Hurricane Hugo

Hurricane Watch and Warnings. The ideal hurricane watch and warning times for the National Hurricane Center in Coral Gables, Florida is "36 hours (watch) before the hurricane's eye makes landfall and approximately 24 hours (warning) before the eye crosses the coast" (NOAA, 1990). In the case of Puerto Rico an ample amount of time was given for both the watch (63 hours) and the warning (42 hours). Therefore, the government, other disaster-related organizations, and the population of Puerto Rico had more than enough lead time to prepare for this event.

Communication. Aguirre and Bush (1992) indicate that the local Weather Service Forecasting Office (WSFO) operations and their communication with other agencies and the population at large were successful. Throughout the hurricane watch and warning there was constant communication between the weather service, emergency managers, and government officials.

The Governor of the island also played a key role during Hugo. In conjunction with Civil Defense officials and a meteorologist from WSFO, the Governor of Puerto Rico made a special hurricane broadcast through the Emergency Broadcast System on Sunday, 17 September. The Governor briefed the population on the dangers that the island would confront as a consequence of Hugo. He also emphasized the need to take immediate preventive measures to reduce the loss of life and property. If disaster warnings are to be effective and are to be taken seriously by the population they must be credible and must convince the population that there is imminent danger. The presence of the Governor of Puerto Rico in this emergency broadcast proved to be critical in mobilizing the population to take immediate action.

Previous Experiences with Disasters. Previous experience with disasters can serve as a mechanism to show the population that their lives can be seriously disrupted by such an event. Past disaster experiences can contribute to greater disaster awareness and preparedness among the general public, government officials, and disaster management organizations.

Disaster managers in Puerto Rico had confronted at least one major event in the recent past which resulted in one the worst landslides in U.S. history. As mentioned previously, 180 persons died in Puerto Rico as a result of a tropical storm (4-7 October 1985) that passed through the island. Government officials did not want to see this tragedy repeated.

In order to achieve "adequate" public response, Hugo was compared to hurricanes San Felipe (1928) and Santa Clara (Betsy, 1956). As discussed previously, these hurricanes were socially and economically devastating to the island and most Puerto Ricans had knowledge about these hurricanes either through personal experience or by folk-stories that had been passed on. The comparison of Hugo with these two hurricanes served to reinforce the need to take appropriate protective action (NOAA 1990).

One month prior to Hugo, Puerto Rico was threatened by Hurricane Dean which approached the island but then turned north and did not impact the same. However, this experience served disaster managers and the population at large as an exercise in disaster preparedness.

Evacuation. As a result of effective and efficient warning systems and good organizational coordination and communication strategies, the evacuation of over 30,000 persons in Puerto Rico proved to be very successful. The evacuation of individuals throughout the island, particularly in the northeast, started in the early hours of Sunday, 17 September and was completed approximately eight hours later

(Hazards Management Group 1993). Again, the population was convinced of the imminent danger of the hurricane and the pending need to evacuate to safer grounds and to government established shelters. The constant flow of information, provided by the Civil Defense, on the need to evacuate and the location of shelters proved to be crucial for one of the largest and most successful disaster evacuations in Puerto Rican history.

Other Preparedness Measures Prior to Hurricane Hugo. Disaster preparedness involves communication within and between organizations and the available information must be communicated to the population at large. Furthermore, there must be disaster drills, rehearsals, meetings, training, and regular updating of obsolete information (Quarantelli 1985, 1987; Britton 1987; Gillipsie and Streeter 1987). The available information shows that, prior to Hurricane Hugo, disaster managers in Puerto Rico were very active in terms of preparedness activities. NOAA (1990) provides an extensive overview of the activities undertaken by disaster-related agencies prior to Hugo that serve as good indicators of their level of disaster preparedness.

The months preceding Hugo were characterized by frequent preparedness meetings by WSFO and other government agencies; hurricane drills and workshops were conducted; the Department of Natural Resources (DRN) held its hurricane conference during the summer of 1989; the San Juan evacuation plan was reviewed; and hurricane awareness programs were regularly presented in one of the most important television networks, among other activities (NOAA 1990; Aguirre and Bush 1992).

In summary, effective hurricane warnings, constant communication and coordination between disaster-related agencies and the general population, previous experiences with disasters, and frequent disaster-related activities held by disaster managers in the months prior to Hugo had a significant impact on disaster preparedness and management in Puerto Rico.

7.5 The Problems during and after Hurricane Hugo

Although the successful strategies prior to and during Hurricane Hugo in Puerto Rico are quite evident, there were also significant problems during this event that merit our attention.

7.5.1 The Carraizo Pumping Plant

One of the greatest problems confronted by the population of San Juan was the lack of water for nine days. The lack of electricity did not allow the Carraizo Dam floodgates to open during Hurricane Hugo. Further, the emergency generator did not work (DRN 1991). Consequently, the Carraizo Dam, which provides drinking water to San Juan, overflowed and flooded all the machinery (Aguirre and Bush 1992). It is noteworthy that some of the Carraizo floodgates were inoperable and maintenance of the dam and of its equipment was poor (FEMA 1985).

What is significant about the aforementioned event is that it could have been avoided. In 1985, after the Mameyes landslide, a FEMA team found several disturbing problems with the Carraizo Dam which include "an earthen left bank which may be subject to erosion, an inadequate spillway, rust deterioration of auxiliary floodgates, and a number of other serious maintenance operation problems" (FEMA 1985). However, nothing was done to correct this situation and then the disaster happened.

7.5.2 Shelters

It has already been mentioned that over 30,000 people were evacuated as a result of the impending danger from Hurricane Hugo. Although the evacuation of the population was a success, the utilization of and the services provided in the shelters were not. The lack of sanitary facilities, beds, food, drinking water, electricity, and health services resulted in so problematic a situation that the Governor of Puerto Rico had to intervene in order to correct these deficiencies. (Aguirre and Bush 1992).

After a short visit to different shelters, the Governor threatened to fire government employees who were not fulfilling their duties and were not providing adequate services to the victims of this disaster. He also criticized the lack of cooperation of other agencies involved in disaster relief. Aguirre and Bush (1992) argue that the problems confronted with the shelters were the result of a lack of communication and coordination between the agencies responsible for the management of shelters.

Another significant problem with the structures used as shelters was that the majority of them were school buildings. Given that many shelters were used for extended periods of time, a significant part of the educational system on the island was paralyzed and the semester of an estimated 150,000 students was disrupted. The DRN (1991) indicates that, by the first week of October, 7361 persons were still living in 158 shelters in 48 municipalities; by December, 65 persons were still residing in 9 shelters in 6 municipalities. The agencies responsible for the management of shelters and the Department of Social Services, among others, need to establish adequate strategies to relocate the sheltered population to more permanent settlements once the period of imminent danger is over.

7.5.3 Lifeline Agencies

Lifeline agencies, such as those responsible for the distribution and maintenance of water and electrical services throughout the island, were severely criticized as a result of power and water shortages that affected at least 35 municipalities. Maintenance problems and lack of adequate resources and personnel were quite evident during the recovery period following Hurricane Hugo.

Essential health services in the most important Medical Center in Puerto Rico and the Veterans Administration Hospital, among others, were severely affected by the lack of water and electricity. Again, these problems could have been avoided if adequate mitigation and preparedness strategies had been undertaken by these agencies.

7.5.4 Problems for Disaster Managers

Disaster managers confronted serious problems that affected their ability to adequately respond to hurricane Hugo. Lack of trained personnel and resources, such as obtaining adequate equipment, had a significant impact on preparedness and response. Furthermore, other emergency personnel confronted difficult conditions while dealing with disaster situations. For example, NOAA (1990) indicates that although WSFO personnel were provided rooms in the International Airport Hotel they were confronted with lack of water, rest rooms, and air conditioning while having to work extremely long periods of time.

In summary, during the aftermath of Hurricane Hugo, both emergency management personnel and the population at large experienced serious problems such as lack of adequate personnel, equipment, electricity, and water, among others. However, if necessary precautions such as the maintenance and replacement of obsolete and non-functional equipment had taken place, some of the problems arising during the aftermath of Hurricane Hugo could have been avoided. In some aspects, the island was fortunate that Hugo did not follow the route of San Felipe (1928), San Ciprián (1932), or Santa Clara (1956), which practically crossed through the center of the island. The results would have been devastating.

7.6 The Role of the Puerto Rican Government in Disaster Preparedness and Mitigation

As a consequence of past disaster experiences, and in order to provide the people of Puerto Rico with the necessary mechanisms to confront future emergency and disaster situations, the Government of Puerto Rico established Law Number 22 (1976) as its public policy concerning disaster management. This law designates the Civil Defense as the key State agency responsible for disaster mitigation, preparedness, and response. Furthermore, the Executive Order of the Governor of Puerto Rico (1993) focuses on the coordination of executive functions during emergency or disaster situations. The Executive Order designates 16 organizations that must coordinate the primary functions of the Civil Defense during periods of crisis.

Past disaster events, such as the Mameyes Landslide (1985), Hurricane Hugo (1989), and the "Three Kings Day Floods" (1992) show mixed results regarding the planning, response, and recovery efforts of disaster related organizations. As mentioned previously, the "Three Kings Day Floods" took weather forecasters, disaster managers, and the Puerto Rican population by surprise; the results were devastating for the victims. Further, referring to the 1985 floods in Puerto Rico, FEMA (1985) indicates that "it is evident that there has been a lack of enforcement of the Commonwealth's regulations concerning development in the floodplain."

Rodríguez and Troche (1994), in a study of disaster management among organizational officials in the municipality of Mayagüez, suggest that disaster management agencies are confronting serious problems that can have adverse effects on disaster mitigation, preparedness, and response. For example, they indicate that the organizational officials interviewed reported the following difficulties:

1. Lack of communication equipment, transportation, appropriate shelters, and trained and qualified personnel to respond to disaster situations.

2. The positions of key personnel in government agencies are "appointed by the governor" ("posiciones de confianza"). Therefore, when the island experiences changes in the top governmental official (i.e., the governor), as a result of local elections, the majority of these administrators are replaced by those of the political party in control. For example, when one disaster manager was asked to indicate how frequently would his employees participate in disaster drills, workshops and conferences he replied that it depends on 3 November 1993 (election day in Puerto Rico). Aguirre and Bush (1992) indicate that government agencies in Puerto Rico "are vulnerable to dramatic changes that take place in executive and managerial positions as part of government employment distribution after election day." This situation does not allow for the continuity and development of disaster preparedness plans. Another concern was that the personnel who replace them are not necessarily the most experienced and trained in disaster mitigation and preparedness.

3. Personnel responsible for accomplishing specific tasks in a disaster situation, according to the organizations disaster plan, may not be aware of their specific duties and responsibilities. One organization official expressed his frustration in the following manner: "there are very good and elaborated plans, but when an emergency situation occurs the plan is locked up and the personnel does not know how it functions." Furthermore, one respondent indicated that "although there is an Executive Order, the government does not force the agencies to comply with the law."

4. A final concern among these organization officials was that there is lack of disaster awareness among the general population. This is a problem that warrants the immediate attention of government officials and disaster managers in Puerto Rico.

7.7 Conclusions and Recommendations

7.7.1 Communication

In order to better prepare for disasters, there must be an improvement in inter- and intra-organizational communication. The disaster literature suggests that inter-organizational communication and coordination in non-disaster situations is an important predictor of successful communication during times of crisis (Nigg 1987; Aguirre and Bush 1992). That is, if organizations have good communication in their day-to-day functions, the communication will most probably be good during a disaster situation (Nigg 1987). Further, Gillipsie and Streeter (1987) indicate that inter-organizational relations are positively associated with disaster preparedness.

In an attempt to improve inter-organizational communication and relationships, municipal and regional Civil Defense agencies can schedule frequent meetings with their emergency committees, which includes at least 16 organizations identified by the Executive Order of the Governor (1993). One of the objectives of the meetings could be to re-evaluate and develop new and more effective disaster strategies on the basis of past experiences. The knowledge and information developed in these meetings can then be passed on to other organizations and the general public. Coordination and communication between organizations responsible for the management of shelters in Puerto Rico could have resulted in adequate services to the victims of Hurricane Hugo.

7.7.2 Structural Mitigation

The island of Puerto Rico continues to confront serious problems that can result in a catastrophe if another hurricane of the magnitude of Hugo or Andrew strikes the island. For example, the Carraizo Dam continues to experience significant difficulties that warrant our immediate attention. The DRN (1991) indicates that as a result of increased sedimentation the dam's holding capacity continues to decline. Given that Carraizo supplies water to the San Juan area, the AAA (water utility company) personnel "will continue to be extremely reluctant to reduce the water levels of the dam" (DRN 1991), as is the usual procedure prior to hurricane or flood warnings. Further, increased sedimentation contributes to a higher probability of flooding in adjacent areas. This factor alone increases the probability of another emergency situation with the Carraizo Dam.

During 1994, the island of Puerto Rico was severely affected by a "drought." As a consequence, a water rationing plan was put into effect throughout the island affecting 29 municipalities (Puerto Rico Planning Board, 1994) and over one million island residents.[4] The AAA indicates that the primary problem was that the island received only two-thirds of its yearly precipitation. While other experts agree that there has been a shortage of water, they indicate that this has also been the case in previous years. Further, it has been suggested that the increased sedimentation in the dams of Puerto Rico and the lack of adequate maintenance have been the primary causes of the so-called "drought." Moreover, it is estimated that close to 243 million gallons of water, or 40% of the water "produced" by the AAA, is either lost and/or stolen on a daily basis in Puerto Rico. The severe problems that the AAA is currently confronting will be exacerbated if another hurricane strikes Puerto Rico. It is quite clear that the AAA needs to consider the development and implementation of a rigorous maintenance plan for its equipment and facilities.

7.7.3 Political Appointments

The appointment of directors of disaster management agencies directly involved in disaster situations by officials of the elected political party presents serious problems for the development and implementation of mitigation and preparedness plans. If the political party in power changes every four years this could lead to the restructuring of government agencies, at least in the top hierarchy, which would result in interruptions, lack of follow-up and/or total abandonment of previous mitigation and preparedness strategies. Disaster management positions need to be filled on the basis of merit, qualifications, and experiences and not based on political affiliation. During Hurricane Hugo "Civil Defense officials mentioned that commonwealth officials discriminated against their municipalities because the current governor's political party did not carry their municipalities during the last general elections" (Aguirre and Bush 1992). Indeed, these types of political changes and policies put in jeopardy the lives of millions of Puerto Ricans by creating the appropriate conditions for the development of a disaster.

7.7.4 The Demographic Component

Existing disaster mitigation, preparedness, and response strategies must be re-

---

[4]The Puerto Rico Planning Board (1994) indicates that the drought started affecting the island on August 1993 but it was in May 1994 that the water rationing program began. Although the water rationing was stopped prior to the Christmas holidays, the AAA started another rationing program in January 1995.

evaluated before the next disaster strikes. Disaster plans must take into consideration the changing demographic characteristics of the Puerto Rican population. The increasing proportion of the elderly and physically disabled population needs to be carefully considered in the existing disaster plans and the utilization of shelters. This population has specific health needs that can jeopardize their lives if not considered in the evaluation of shelters and the services provided. It is recommended that the Civil Defense and other disaster management agencies conduct local censuses to identify the disabled and chronically ill population, their mobility status, health conditions, and medical needs, among others. This information would be of great value for the development of adequate shelters and for the evacuation procedures during a disaster situation.[5]

7.5 Education and Public Awareness

Finally, there is an urgent need to educate organizational leaders, as well as the general population, if we are to obtain "appropriate" responses during disaster situations. There needs to be a massive and widespread public campaign on disasters, their structural, socioeconomic, and psychological impact. The need to develop adequate preparedness and response strategies in order to reduce the loss of life and property must be emphasized. The population concentrated in high risk areas must be informed about their vulnerability and the strategies to be followed in the event of a hurricane, flooding, and/or landslide, among others. Palm and Hodgson (1993) indicate that "the current knowledge of home locations with respect to flood hazard zones in Puerto Rico was very low. Apparently, little effort has been made to educate homeowners in flood zones about such risks, and those undertaken have been largely unsuccessful." Moreover, they indicate that the general disaster preparedness levels of Puerto Rican homeowners is very low (1992).

The available information shows that there is an urgent need to inform and educate the Puerto Rican population regarding disaster vulnerability, mitigation, and preparedness. Regularly scheduled radio and television broadcasts by the Civil Defense, the Department of Natural Resources, and FEMA, among others, can serve as valuable resources to prepare the population for any type of disaster. These organizations need to disseminate the information to the population in a simple yet efficient format. The population needs to be informed about the necessary strategies and plans that need to be developed prior to the occurrence of a physical event so that when the hurricane strikes the population can be prepared.

---

[5]The Director of the Civil Defense in the municipality of Carolina has informed the researchers at the Disaster Research Laboratory at the University of Puerto Rico-Mayagüez, that this agency has conducted this type of census.

Town meetings, workshops, seminars, and disaster-related conferences directed at the general population can be another effective measure for disaster preparedness. Public awareness is essential to achieve public safety during disaster situations. Further, community members need to become active agents in the identification of risk areas and in the development and implementation of disaster mitigation and preparedness plans. In essence, we have to make the population aware that they have a responsibility in designing and implementing disaster preparedness strategies. However, in order to mobilize the population, organizational and government "leaders must become trusted, credible information sources, and they must convey a message to the audience that will move them to action" (Palm and Hodgson 1992).

Aguirre (1993) indicates that "effective community disaster managers have a regular, comprehensive preparedness program, updated programs with defined duties and responsibilities, have good communication equipment at their disposal, and spend a great deal of their time educating the public through the mass media and other means." Clearly, all these activities represent additional economic costs and demand strong commitment by the government and other disaster management agencies in Puerto Rico. However, these can be effective mechanisms to reduce the social and economic costs associated with events such as hurricanes. Moreover, disaster related costs associated with lack of or inadequate preparedness can and will be much greater and will result in excessive human suffering. Although we may not be able to prevent natural events (e.g., earthquakes, tornadoes, hurricanes, or landslides) we can prevent or reduce the impact or consequences of such events through adequate mitigation and preparedness.

References

Aguirre, B., 1993. Planning, warning, evacuation, and search and rescue: a review of the social science research literature. Unpublished Manuscript.

Aguirre, B. and Bush, D., 1992. Disaster programs as technology transfers: the case of Puerto Rico in the aftermath of Hurricane Hugo. *International Journal of Mass Emergencies and Disasters,* **10**, 1.

Anderson, M.B., 1991. Which costs more: prevention or recovery? In: Kreimer, A. and Mchan, M. (eds.), *Managing Natural Disasters and the Environment.* World Bank, 17-27.

Aptekar, L., 1994. *Environmental Disasters in Global Perspective.* G.K. Hall and Co., New York, 198 pp.

Ayala-Padró, G., 1991. Sistema de Aviso sobre Inundaciones Repentinas. In: Bernier, L.S. and Franco-Villafañe, O. (eds.), Proceedings from the Conference on Hurricanes, June 5, 1991. Commonwealth of Puerto Rico: Department of Natural Resources, 25-44.

Britton, N.R., 1987. Towards a reconceptualization of disaster for the enhancement of social preparedness. In: Dynes, Russell, Bruna de Marchi and Pelanda Carlo (eds.), *Sociology of Disasters: Contributions of Sociology to Disaster Research.* Franco Angeli, Milan, Italy, 31-35.

Curson, P., 1989. Introduction. In: Clarke, J.I., Curson, P., Kayastha, S.L., and Nag, P. (eds.), *Population and Disaster.* Basil Blackwell Ltd., Oxford, 1-23.

Department of Natural Resources (DRN), 1991. Plan de Mitigación de Riesgos Naturales del Estado Libre Asociado de Puerto Rico. Commonwealth of Puerto Rico.

Estado Libre Asociado de Puerto Rico, 1976. Ley Número 22. Established on June 23, 1976, Commonwealth of Puerto Rico.

Estado Libre Asociado de Puerto Rico, 1988. Comision Especial de Seguridad y Protección Civil: Informe Final. Senate of Puerto Rico, 10th. Legislative Assembly, 4th. Ordinary Session, Commonwealth of Puerto Rico.

Estado Libre Asociado de Puerto Rico, 1990. Informe sobre la Resolución del Senado N'mero 280. Comisión Especial del Senado de Puerto Rico: Comisión Especial del Huracan Hugo. 11th. Legislative Assembly, 3rd. Ordinary Session, Commonwealth of Puerto Rico.

Estado Libre Asociado de Puerto Rico, 1991. Plan de Mitigación de Riesgos Naturales del Estado Libre Asociado de Puerto Rico. Commonwealth of Puerto Rico.

Estado Libre Asociado de Puerto Rico, 1993. Orden Ejecutiva del Gobernador del Estado Libre Asociado de Puerto Rico Para Establecer la Coordinación de Funciones Ejecutivas para Casos de Desastres o Emergencias. Boletín Administrativo Número OE-1993-23, Commonwealth of Puerto Rico.

Federal Emergency Management Agency, 1985. Federal Interagency Flood Hazard Mitigation Team Report for Puerto Rico. In: Response to the Oct. 10, 1985 Disaster Declaration (FEMA-746-DR-PR).

Gillipsie, D.F. and Streeter, C.L., 1987. Conceptualizing and measuring disaster preparedness. *International Journal of Mass Emergencies and Disasters*, **5**, 155-176.

Hazards Management Group, Inc., 1993. Behavioral Assumptions for Hurricane Planning for Puerto Rico's South Coast. Prepared for the U.S. Army Corps of Engineers, Jacksonville District.

Jarvinen, B.R., 1991. An Evacuation Study in Puerto Rico with the New SLOSH Model. In: Bernier, L.S. and Franco-Villafañe, O. (eds.), Proceedings from the Conference on Hurricanes, June 5, 1991. Commonwealth of Puerto Rico, Department of Natural Resources, 87-96.

Larsen, M.C., Torres-Sánchez, A.J., and Simon, A., 1991. Relación entre Lluvia y Deslizamientos de Tierra en la Cordillera Central de Puerto Rico. In: Bernier, L.S. and Franco-Villafañe, O. (eds.), Proceedings from the Conference on Hurricanes, June 5, 1991. Department of Natural Resources, Commonwealth of Puerto Rico, 97-116.

Lastra-Aracil, J.J., 1983. *El ABC Sobre Desastres Naturales (Huracanes)*. Imprenta El Impresor, Carolina, Puerto Rico.

National Oceanic and Atmospheric Administration, 1990. Natural Disaster Survey Report: Hurricane Hugo, September 10-22, 1989. United States Department of Commerce.

Nigg, J.M., 1987. Communication and Behavior: organizational and individual response to warnings. In: Dynes, Russell, Bruna de Marchi and Pelanda Carlo (eds.), *Sociology of Disasters: Contributions of Sociology to Disaster Research*. Franco Angeli, Milan, Italy, 103-118.

Palm, R.I. and Hodgson, M.E., 1993. *Natural Hazards in Puerto Rico: Attitudes, Experiences, and Behavior of Homeowners*. University of Colorado, Boulder, Colorado. Institute of Behavioral Science, Program on Environment and Behavior, Monograph 55.

Puerto Rico Planning Board, 1994. Análisis del Impacto de la Sequía en la Economía de Puerto Rico Durante los Años Fiscales 1994 y 1995. Commonwealth of Puerto Rico.

Quarantelli, E.L., 1985. The Need for Planning, Training, and Policy on Emergency Preparedness. Disaster Research Center, University of Delaware, Preliminary Paper #101.

Quarantelli, E.L., 1987. What should we study?: questions and suggestions for researchers about the concept of disasters. *International Journal of Mass Emergencies and Disasters*, **5**, 7-32.

Quarantelli, E.L., 1987. Criteria Which Could be Used in Assessing Planning Preparedness Planning and Managing. Disaster Research Center, University of Delaware, Preliminary Paper 122 pp.

Rodríguez, H. and Troche, M., 1994. Preparación y Mitigación en Puerto Rico: Un Análisis Organizacional. *Desastres y Sociedad*, **2**, 23-40.

Rodríguez, H., Colón-Rivera, C.R., Gutiérrez-Sánchez, J. and Rodríguez-Rivera, V., 1993. Natural Disasters in Puerto Rico: An Analysis of Attitudes, Knowledge, Preparedness and Response. Applied Social Research Center, University of Puerto Rico-Mayagüez.

U.S. Bureau of the Census, 1993. Census of Population and Housing, 1990: Summary Tape File 3 (Puerto Rico) [machine-readable files]. Prepared by the Bureau of the Census. The Bureau of the Census, Washington, D.C.

Section C

VULNERABILITY AND POLICY ISSUES

8 Vulnerability to Hurricanes Along the U.S. Atlantic and Gulf Coasts: Considerations of the Use of Long-Term Forecasts

Roger A. Pielke, Jr.[1] and Roger A. Pielke, Sr.[2]

[1]Environmental and Societal Impacts Group
National Center for Atmospheric Research
P.O. Box 3000
Boulder, CO 80307-3000, U.S.A.

[2]Department of Atmospheric Sciences
Colorado State University
Fort Collins, CO 80523, U.S.A.

Abstract

The scientific community has demonstrated some skill in long-term forecasts (i.e., inter-annual/decadal) of Atlantic hurricane activity. Long-term forecasts of hurricane incidence are a component of broader political and social processes. Recognition of this broader context of a forecast is central to its utilization for societal benefits. We seek to use the concept of vulnerability to hurricanes along the U.S. Atlantic and Gulf coasts to provide a framework to assess the use/value of long-term forecasts of hurricane activity. Vulnerability provides a common goal that integrates the science and social science of hurricane impacts. Experience suggests that improved long-term forecasts, by themselves, will not reduce society's vulnerability unless accompanied by efforts to use them in processes of preparedness. To the extent that improved forecasts contribute to reduced vulnerability we will have made less serious the threat of hurricanes to coastal communities, supplied a basis of experience for actions in response to other types of extreme weather phenomena, and provided a practical demonstration of a successful connection of scientific research in pursuit of societal objectives.

8.1 Introduction: Defining the Problem

A better understanding of societal vulnerability to hurricanes helps to illuminate at least three complexes of problems facing the United States. First, hurricanes pose a serious threat to the U.S. Atlantic and Gulf coasts. Second, hurricanes are a subset of a broader class of extreme weather events that threaten the nation. And third, societal and political responses to the hurricane threat have potential to contribute to ongoing debate over U.S. science policy concerning the efficacy of research supported with federal funds. In the context of these problems, this paper seeks to (1) expand the problem definitions, (2) define and clarify vulnerability to hurricanes along the U.S. Atlantic and Gulf coasts, focusing on intense hurricanes, and (3) provide general guidance as to how long-term (i.e., interannual/decadal) forecasts of hurricane activity might contribute to reduced societal vulnerability. The paper is targeted at both users (actual and potential) and producers of hurricane information, as well as those interested in a better understanding the role of long-term forecasts in reducing vulnerability to hurricanes.

8.1.1 Intense Hurricanes

About a year after Hurricane Andrew devastated south Florida and parts of rural Louisiana, Director of the National Hurricane Center, Dr. Robert Sheets (whose home suffered extensive damage due to the storm), testified before Congress that "We were lucky" (Sheets 1993, 41). In spite of the $25 billion in damage caused by Andrew, Floridians and the U.S. were lucky because "had Hurricane Andrew been displaced only 20 miles north of its track over South Florida, two different studies show that losses would have exceeded $60 billion in South Florida alone." Indeed, over the past several decades the U.S. Atlantic and Gulf coasts as a whole may be considered fortunate because hurricane activity during this period has been below the climatological average (Gray and Landsea 1992; Landsea et al. 1996; Gray et al., Chap. 2 this volume). Coch (1994), on the other hand, argues that in 1985 residents along the northeast Atlantic coast "were just lucky" to avoid a greater impact from Hurricane Gloria. The trend of below-average hurricane activity is apparent in Figs. 9.1a, b which show the tracks of intense hurricanes for two periods: (a) 1947 to 1969, and (b) 1970 to 1987. Although Hurricane Andrew was not as costly as it might have been, it was the most costly weather-related disaster in U.S. history, and perhaps a warning that our luck with hurricanes may be running out.

Hurricanes, which are a type of tropical cyclone, are classified by their damage potential according to a scale developed in the 1970s by Robert Simpson, a meteorologist and director of the National Hurricane Center, and Herbert Saffir, a consulting engineer in Dade County, Florida (Simpson and Riehl 1981). The Saffir/Simpson scale is now in wide use and was developed by the National Weather Service to give public officials usable information on the magnitude of a storm in progress. The scale has five categories, with category 1 the least intense hurricane and category 5 the most intense. Storms are named when they reach tropical storm strength. Tropical storms are tropical cyclones below category 1 strength and have wind speeds of 39-74 mph. Table 8.1 shows the Saffir/Simpson scale and the corresponding criteria for classification. This paper focuses on intense (or "great") hurricanes, i.e., those classified in categories 3, 4, or 5.

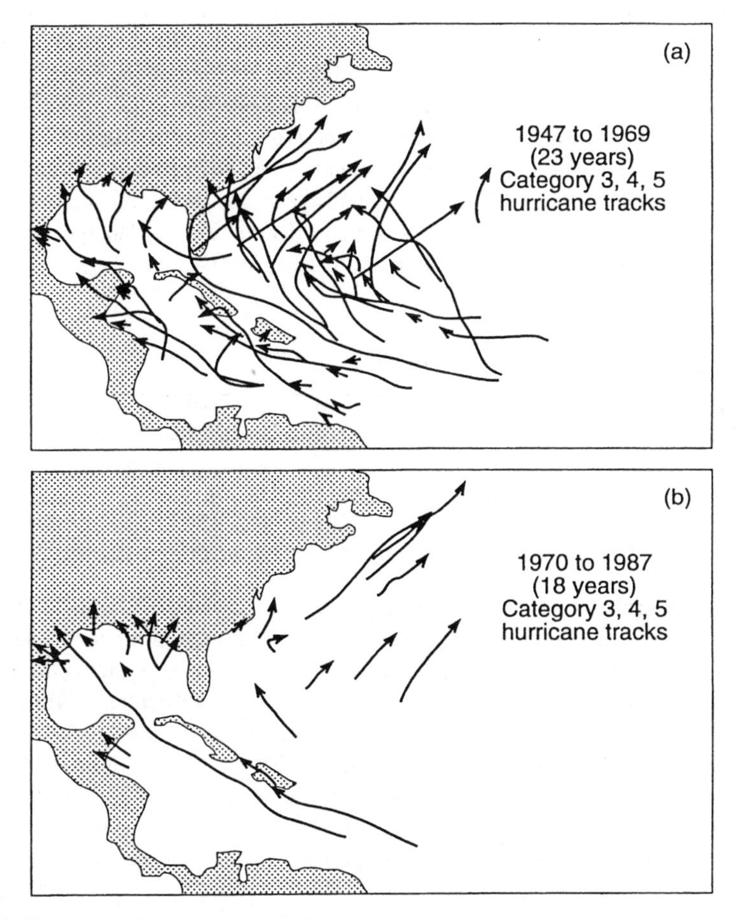

Fig. 8.1. Comparison of intense (Saffir/Simpson categories 3, 4, and 5) hurricane tracks for (a) 1947 to 1969, and (b) 1970 to 1987. Since 1989 there have been two landfalling intense hurricanes, Hugo (1989) and Andrew (1992). Graphic provided by W. Gray.

Hurricanes pose a number of threats to coastal communities including storm surge (e.g., Anthes 1982), high winds (e.g., Golden and Snow 1991), excessive rainfall (e.g., Dunn and Miller 1960), and tornadoes (e.g., Novlan and Gray 1974).[1] A storm

---

[1]Snowfall has occasionally been associated with the inland portion of hurricane circulation. for instance, in 1963 Hurricane Ginny caused more than a foot (30 cm) of snow in northern Maine (Pielke 1990)

surge is a dome-shaped area of water caused by strong hurricane winds and is related to the low surface pressure of a storm. Past disasters have focused attention on hurricane storm surge. For example, in 1900, more than 6000 deaths occurred in Galveston, Texas, primarily as a result of a hurricane storm surge. In 1957 the storm surge of over 20 feet associated with Hurricane Audrey extended as far inland as 25 statute miles and was the major cause of 390 deaths in Louisiana. Winds have also been responsible for significant loss of life. In September 1928, the waters of Lake Okeechobee, driven by hurricane wind, overflowed the banks of the lake and were the main cause of 1836 deaths associated with the storm. Loss of life due to excessive rainfall has primarily been an inland threat, such as that associated with Hurricane Camille (1969), although rainfall also provides economic benefits (Sugg 1968). Hurricane Andrew (1992) has increased attention to damage caused by strong hurricane winds (e.g., Wakimoto and Black 1993).

Table 8.1. Saffir/Simpson Hurricane Scale (after Herbert et al. 1993)

Category	Central (mb)	Pressure (inches)	Winds (mph)	Surge (feet)	Damage
1	980	≥28.94	74-95	4-5	minimal
2	965-979	28.50- 28.91	96-110	6-8	moderate
3	945-964	27.91-28.47	111-130	9-12	extensive
4	920-944	27.27-27.88	131-155	13-18	extreme
5	<920	<27.17	>155	>18	catastrophic

When they strike the U.S. coast, intense hurricanes cost lives and dollars, and disrupt communities. Figures 8.2a-j show U.S. intense landfalling hurricane tracks for each decade this century. Due largely to better warning systems, hurricane-related loss of life has decreased dramatically in the 20th century (NRC 1989). However, the economic and social costs of hurricanes is large and rising. A rough calculation shows that annual losses to hurricanes has been in the billions of dollars (cf. Sugg 1967). Landsea (1991, 62) documents $74 billion (1990 $) in hurricane related damage for the period 1949-1989 (cf. Gray and Landsea 1992, 1356). With the addition of the approximately $25 billion for Hurricanes Andrew (1992) and Bob (1991), the total is about $100 billion over 45 years, or well over $2 billion annually. Approximately 80% of the costs are attributed to intense hurricanes (Gray and Landsea 1992). However, as Landsea (1991), Sheets (1993), and others are quick to observe, this type of calculation is likely to significantly underestimate the magnitude of the actual hurricane threat because data based on past storm damage represents costs incurred prior to most coastal development. Were the same storms to landfall today, in most cases, damage would be significantly higher. Therefore, the $2 billion figure should be considered a lower bound on the annual costs of the hurricane threat facing the united states. Sheets (1993) estimates the annual monetary costs of hurricanes to be on the order of $3 billion.

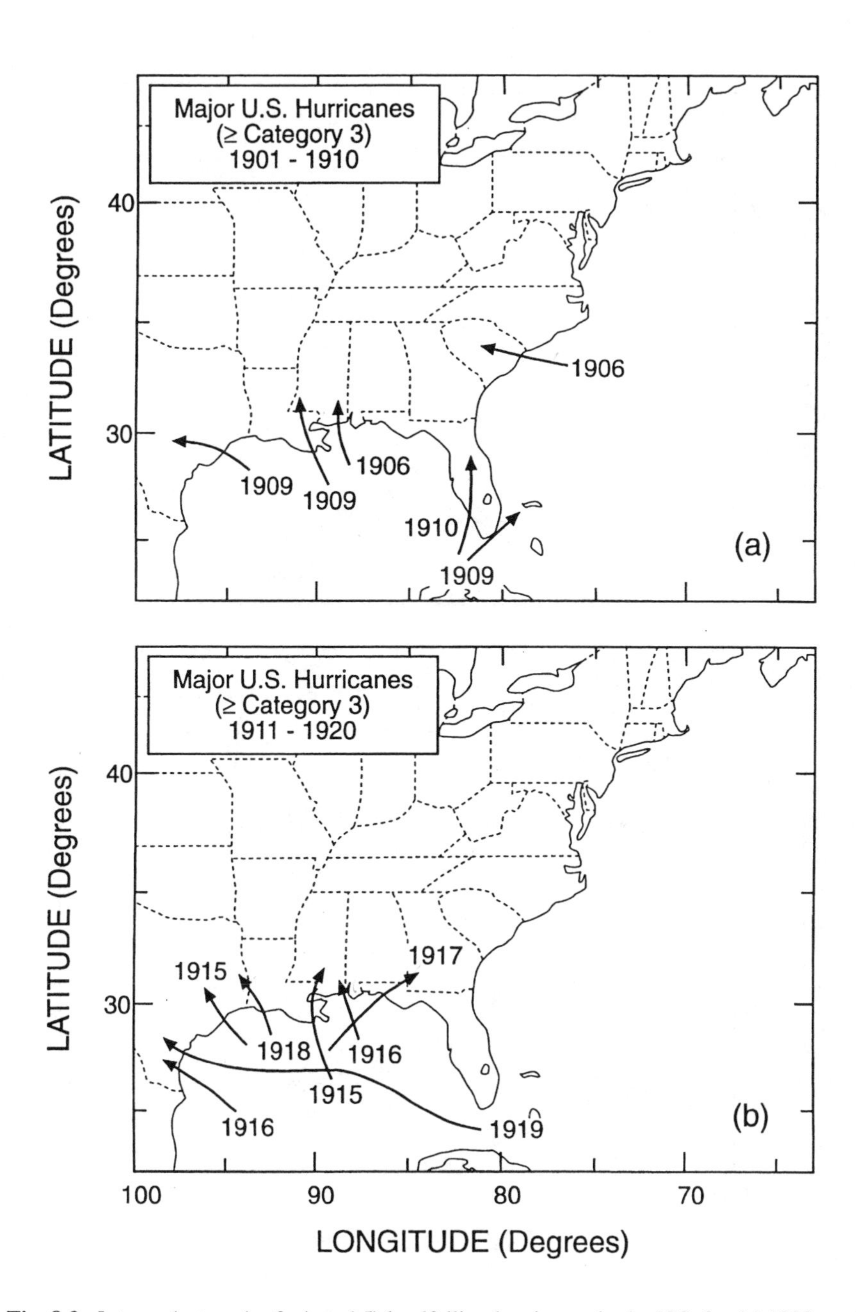

Fig. 8.2. Intense (categories 3, 4, and 5) landfalling hurricanes in the U.S. for (a) 1900s and (b) 1910s. Source: Herbert et al. (1993).

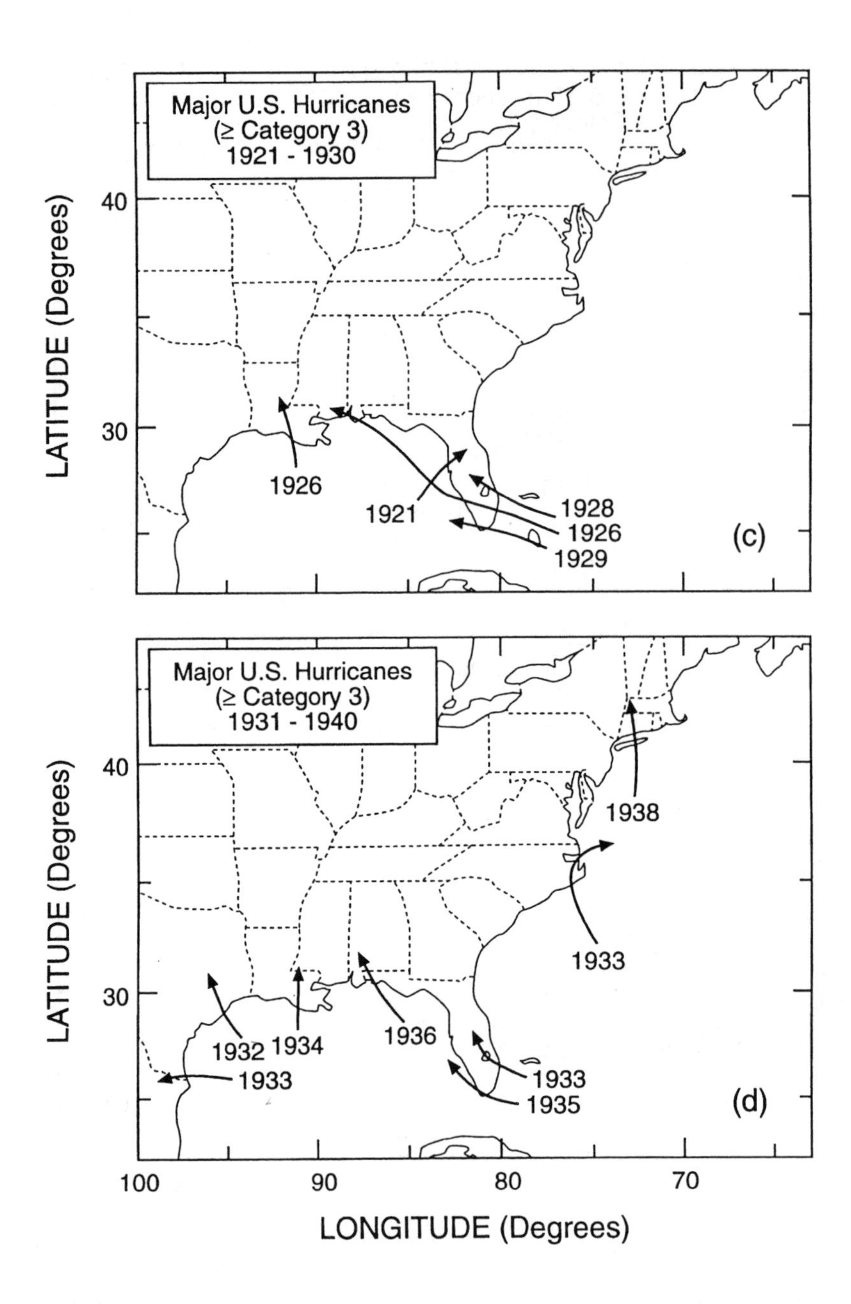

Fig. 8.2. Intense (categories 3, 4, and 5) landfalling hurricanes in the U.S. for (c) 1920s and (d) 1930s. Source Hebert et al. (1993).

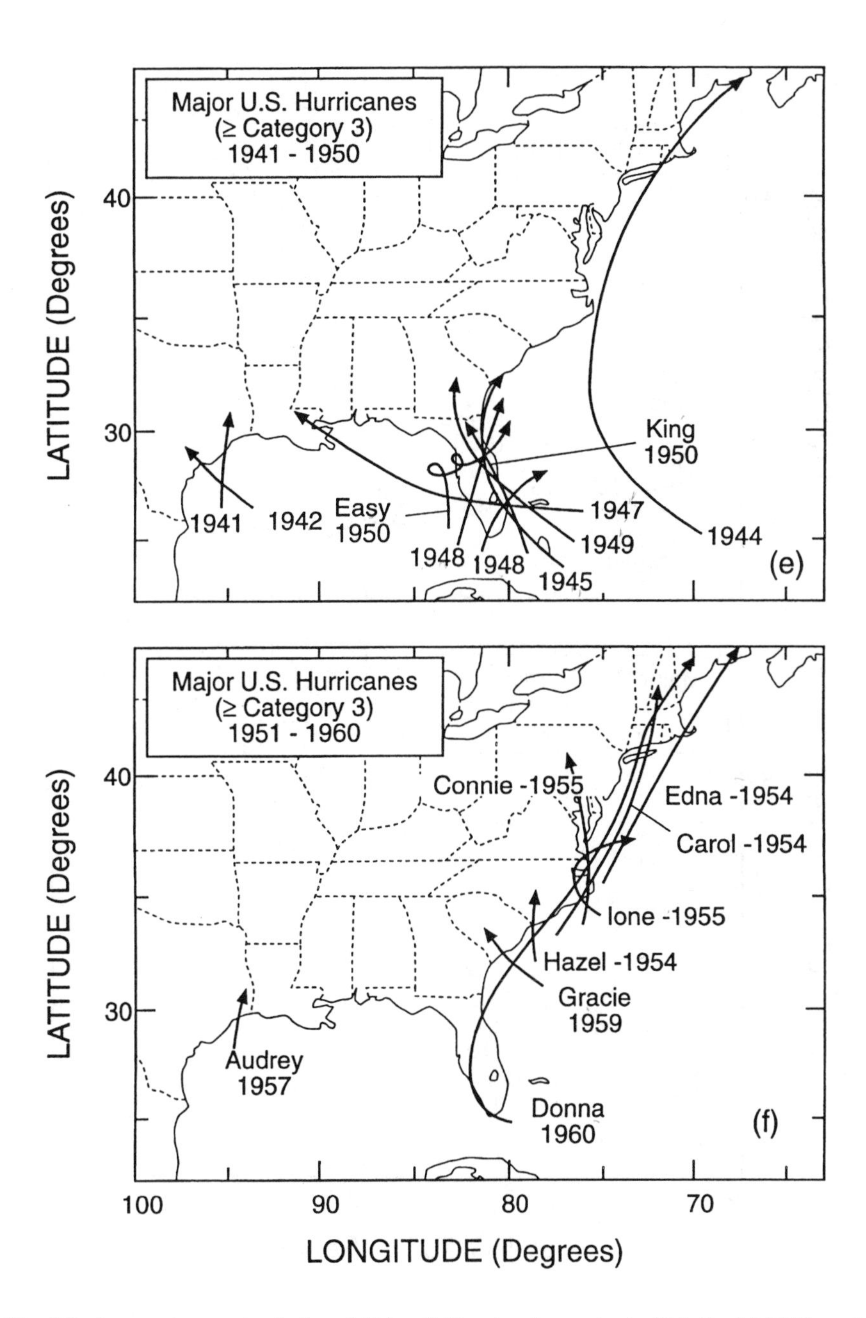

Fig. 8.2. Intense (categories 3, 4, and 5) landfalling hurricanes in the U.S. for (e) 1940s and (f) 1950s. Source: Hebert et al. (1993).

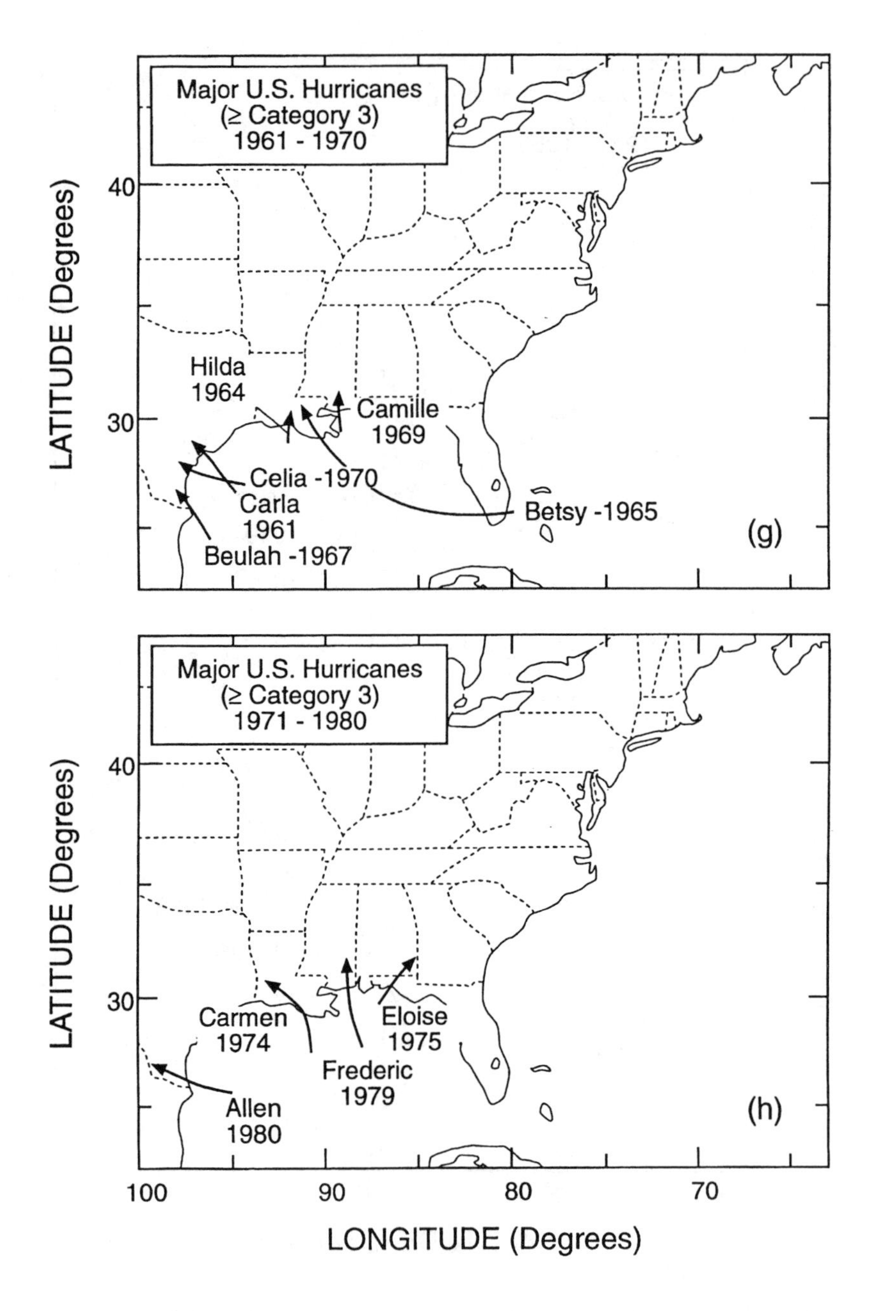

Fig. 8.2. Intense (categories 3, 4, and 5) landfalling hurricanes in the U.S. for (g) 1960s and (h) 1970s. Source: Hebert et al. (1993).

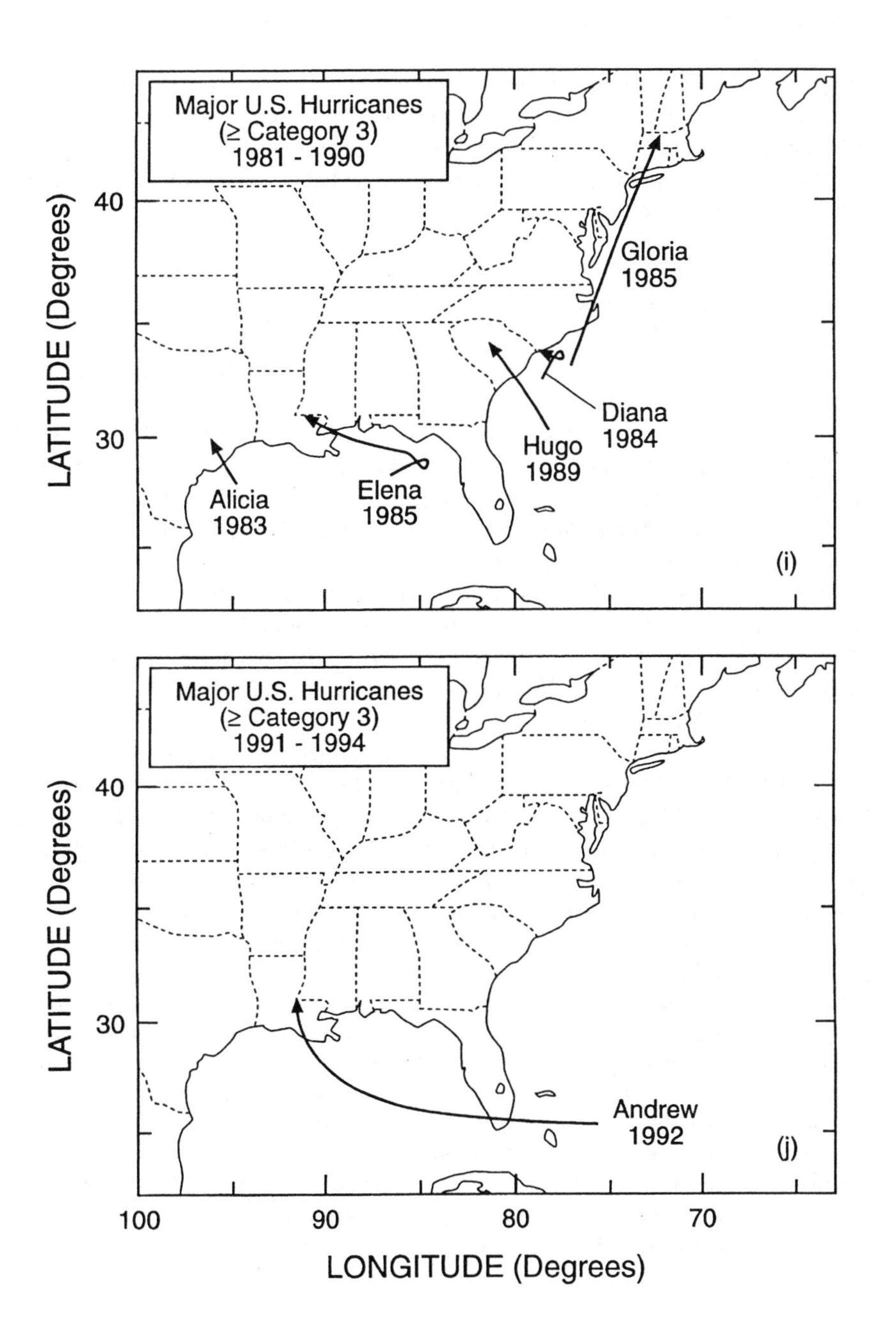

Fig. 8.2. Intense (categories 3, 4, and 5) landfalling hurricanes in the U.S. for (i) 1980s and (j) 1990s. Source: Hebert et al. (1993).

While the hurricane threat to the U.S. Atlantic and Gulf coasts has been widely recognized, only in recent years, following Hurricane Andrew, have many public and private decisionmakers sought to better understand the magnitude of the threat.[2] The U.S. will never escape the hurricane threat, although it is possible that with a better understanding of societal vulnerability to hurricanes, public and private decisionmakers may recognize ways to change behavior in ways that would significantly reduce societal vulnerability.

8.1.2 Extreme Weather Events

The hurricane is part of a broader class of extreme weather phenomena that threaten the United States. Other phenomena include winter storms (e.g., snow, sleet, freezing rain, and freezes), thunderstorms (e.g., tornadoes, heavy rains/floods, lightning, wind, and hail), and windstorms. Changnon and Changnon (1992) report insurance claims due to extreme weather events for the period 1950 to 1989. They find $66.2 billion (1991 $) in insured losses during the 40-year period due to extreme weather events, with about half due to hurricanes. For comparison, Landsea (1991) found *total* monetary losses over the same period due to hurricanes to be about twice the *insured* losses reported by Changnon and Changnon (1992). The difference is attributed to uninsured losses, damage to public infrastructure, federal disaster assistance payments, private contributions, and other costs (Pielke 1995).

In the face of large losses due to weather phenomena in recent years, the U.S. Congress established the U.S. Weather Research Program in Public Law 102-567. The program is justified on the basis that it will save the country "hundreds of lives, thousands of injuries, and billions of dollars" (USWRP 1994, xvi). While implementation of the program awaits congressional appropriations, a 1980 National Academy of Sciences assessment of the National Weather Service gives reason for caution in structuring a program to leverage societal benefits from scientific research:

> For many years the National Weather Service and its predecessor organization apparently operated on the assumption that if they produced a good product someone would come to get it and use it... Users are currently left largely to their own devices in determining what is available and how to use it; many are unaware of the information available. As a result, the potential benefits of the excellent information currently available are not being fully realized (NAS 1980).

While the hurricane threat to the U.S. Atlantic and Gulf coasts remains significant, past successes in reducing societal vulnerability to hurricanes (particularly the threat to human life) provide a significant base of experience to

---

[2] We use the term "decisionmaker" to refer to anyone faced with a decision. The term "policymaker" is reserved for elected officials and administrators at various levels of government.

serve as a prototype of how scientific information can play a more generally useful role in preparation, mitigation, and response efforts to extreme events.

8.1.3 Science for Society

In broader context, a significant aspect of societal responses to extreme weather events is the role of scientific research, particularly forecasts, in the decision processes of public and private individuals and groups. In recent years, in the U.S. as well as in other countries, the institution of science has faced close scrutiny with respect to its ability to contribute to societal problems (e.g., *Nature* 1995; Rensberger 1995). For instance, some members of Congress, both Republicans and Democrats, have called upon science to demonstrate the societal benefits that are often promised in efforts to secure federal funding (Byerly 1995). In light of changes in the environment of U.S. science policy, it is likely that sustained federal support of scientific research, including weather research, will be a function of a particular program's performance with respect to justifications made to Congress by its supporters (Brunner and Ascher 1992).

The case of the hurricane threat to the U.S. Atlantic and Gulf coasts provides the scientific community with an opportunity to demonstrate tangible societal benefits that are directly related to scientific research. Yet, demonstration of benefits is often difficult to achieve in practice as "the path between scientific research and societal benefits is neither certain, nor straight" (Brown 1992). As Glantz (1978) has noted, "adverse weather events by themselves can be devastating for society, but their effects are often exacerbated by economic, political, and social decisions made, in many instances, long before those events take place." Thus, while research holds much potential to contribute results useful to reducing societal vulnerabilities to hurricanes, if potential is to be realized, then care must be taken to understand such results in their broader political and social contexts.

8.1.4 The Challenge

The general challenge facing U.S. science policy is to connect scientific research more directly to societal benefits. In the context of the hurricane threat facing the U.S. Atlantic and Gulf coasts, the challenge is to improve public and private decisionmaking with the aid of science. With the modernization of the National Weather Service, a multi-billion dollar U.S. Global Change Research Program, and a congressionally authorized U.S. Weather Research Program, weather and climate forecasts on various time scales appear to be a growth industry. Yet, as Glantz (1986) has cautioned, "forecasts are the answer, but what was the question?"

The remainder of this paper has two purposes. First, it uses the concept of societal vulnerability to integrate the physical and societal dimensions of hurricanes in order to provide a sense of the broader context of the hurricane threat facing the continental United States. Second, the paper discusses ways in which interannual/decadal forecasts might contribute to reducing societal vulnerabilities to hurricanes, and recognizes both the opportunities and the limitations.

8.2 Societal Vulnerability to Hurricanes

Societal vulnerability (Pielke 1995) to hurricanes is a function of *exposure* and *incidence*. Clearly, if people were not exposed to hurricanes or if hurricanes did not occur (i.e., no incidence), then society would be *in*vulnerable to hurricanes. Exposure refers to the number of people and amount of property threatened by hurricanes. The gross number of exposed people and property can be reduced through preparedness efforts such as evacuation and building fortification. Incidence refers to the climatology of hurricanes -- how many, how strong, and where. Societal vulnerability, then, is determined through the societal and climatic aspects of the hurricane phenomenon.

8.2.1 Exposure to Hurricanes

Exposure is a function of (a) population at risk, (b) property at risk, and (c) preparedness (cf. Brinkmann 1975, 11-20). This section presents data on hurricane exposure at the coastal county level for 168 coastal counties that lie adjacent to the Gulf of Mexico and the Atlantic Ocean from the Mexico-Texas border to the Maine-Canada border. Figure 8.3 shows the coastal counties used in this study. Of course, societal vulnerability to hurricanes extends well inland, beyond the coastal counties. For instance, following Hurricane Andrew's landfall in Louisiana, 29 inland parishes were declared disaster areas in addition to all 11 of Louisiana's coastal parishes (USDOC 1993). The remnants of Hurricane Camille (1969) killed 109 in central Virginia as a result of up to 30 inches (0.76 m) of rain within 6 hours. Coastal counties, however, are a primary component of societal vulnerability to hurricanes.

Population at Risk. Figure 8.4 shows U.S. coastal county population by state for each of the 168 coastal counties from Texas to Maine for the years 1930 and 1990 (U.S. Census). The most readily apparent trend is the growth in population along the Gulf and southern Atlantic counties through North Carolina. For instance, in 1990 the combined population of Dade County, Florida, and its two neighbors to the north, Broward and Palm Beach counties, was more than that of 29 states. About the same number of people now live in Dade and Broward counties, Florida, as lived in *all* the 109 coastal counties from Texas through Virginia in 1930. A second trend is the very low level of growth in the coastal counties of the U.S.

northeast. Some counties north of New York City have actually experienced population decreases in recent decades. However, the population of the Atlantic coast from Baltimore to Boston remains very large.

Fig. 8.3. The 168 coastal counties from Texas to Maine used in this study.

Rising population has not only created record densities for areas such as south Florida, but also "filled in" formerly low-population areas. In the 95 coastal counties from Texas through North Carolina the number of counties with populations of more than 250,000 tripled from 1950 to 1970, from 3 (3%) to 9 (10%), and doubled again from 9 to 18 (19%) by 1990. A quarter of a million residents is about the population of Charleston County, South Carolina, where Hurricane Hugo made landfall in 1989 and caused $8.2 billion (1993 $) in damage (Hebert et al. 1993, 10). (Hurricane Hugo, however, made landfall in a relatively unpopulated stretch of South Carolina coast (Baker 1994). Had Hugo directly hit a more populated section, casualties and damage would likely have been significantly higher.) The number of counties with more than 100,000 residents went from 15 (16%) in 1950 to 21 (22%) in 1970, to 36 (38%) in 1990. Hurricane Frederic made landfall in a county of about 100,000 near Gulf Shores, Alabama, in 1979 and caused $3.8 billion (1993 $) in damage (Hebert et al. 1993, 10). Another way to look at population growth is in terms of the dwindling number of counties with very few residents. From Texas through North Carolina, the number of counties with less than 50,000 residents decreased from 75 (79%) in 1950 to 54 (57%) in 1970, and to 38 (40%) by 1990. Hurricane Andrew made landfall across two such relatively low-population counties in Louisiana in 1992 and caused about $1 billion (1993 $) in damage (Cochran and Levitan 1994). According to the U.S. Census Bureau, population growth can be expected to continue in most coastal counties (Campbell 1994).

In aggregate, the 168 coastal counties (out of more than 3,000 nationwide) are home to approximately 40 million people, or about 16% of the total U.S. population. Although the numbers are large, census populations may actually underestimate the number of people in coastal counties during hurricane season. Because the hurricane season overlaps the tourist season in many of these coastal counties, many more people in addition to permanent residents may actually be in the path of an approaching hurricane (cf. Sheets 1993).

Property at Risk. Figure 8.5 shows insured property values in each of the 168 coastal counties from Texas to Maine for the years 1988 and 1993 (IRC/IIPLR 1995). Figure 8.6 shows the increase from 1988 to 1993 as a percentage of the 1988 total. An increase of 100% represents a doubling in value, 200% a tripling, etc. Inflation accounts for 19.5% of the aggregate growth during that 5-year period (Council of Economic Advisors 1994). The remainder of growth can be attributed to expanded insurance coverage and real increases in property values.

The total amounts of insured property are staggering. Over $3.1 *trillion* worth of property was insured in 1993, an increase of 69% (50% excluding inflation) over the 1988 total of about $1.9 trillion. The 1988 total represented an increase of 64% (35% after inflation) over the 1980 total of about $1.1 trillion (Sheets 1993, 47). For comparison, the coastal counties represented about 15% of the total insured property in the United States in 1993, which was about $21.4 trillion. For 1988 the figures are approximately 14% and $13 trillion, respectively.

Compared to rates of population growth, growth in insured property in the last 5 years is startling. After adjusting for the effects of inflation, the aggregate growth in insured property for all coastal counties is 46%. Table 8.2 shows a summary by state of insured property for the coastal counties. Except for Louisiana, Florida had the slowest rate of growth in coastal insured property. Despite its lower rate of growth, the total insured coastal property in Florida exceeds the combined coastal insured property from Texas to Delaware (excluding Florida). Within states, local variations are large. Only one county, Laforche Parish, Louisiana, had property values decrease over the period (compensating for inflation). Over the 5 year period, 24 counties experienced more than a doubling in insured property. Although the amount of insured property is large, it is worth repeating a point made in the introduction that insured property represents only a portion of the total losses due to hurricanes. In addition, uninsured property and public infrastructure makes up a substantial portion of damage due to hurricanes. It is also important to recognize that the "costs" of hurricanes go well beyond those which can be easily expressed in dollars (e.g., Mauro 1992).

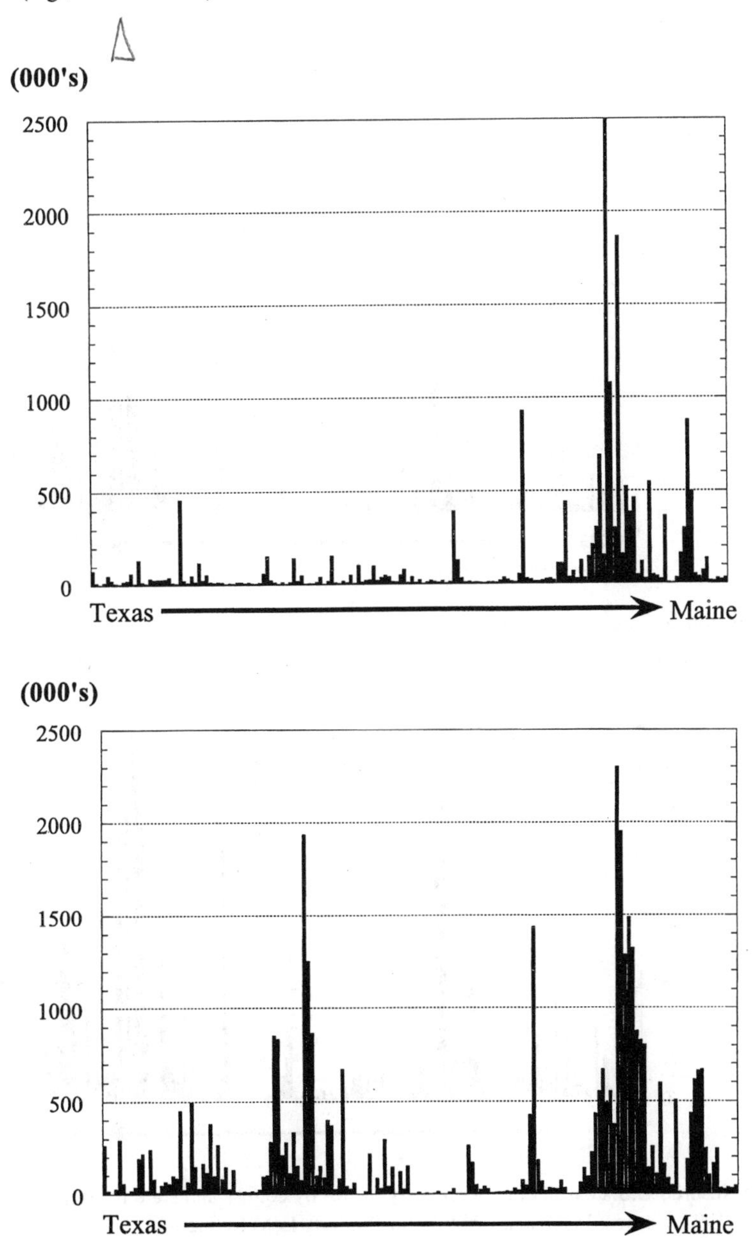

Fig. 8.4. U.S. Coastal county population by county and state for (top) 1930 and (bottom) 1990. Source: U.S. Census Bureau.

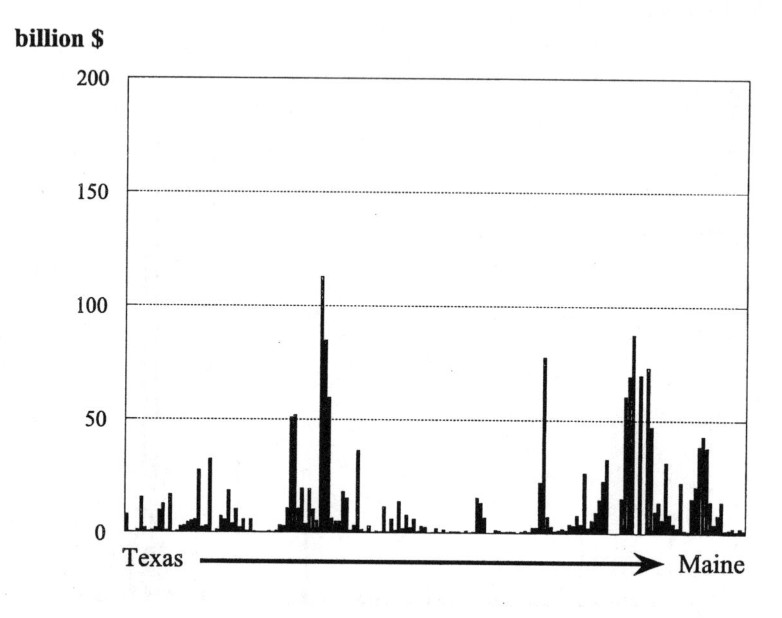

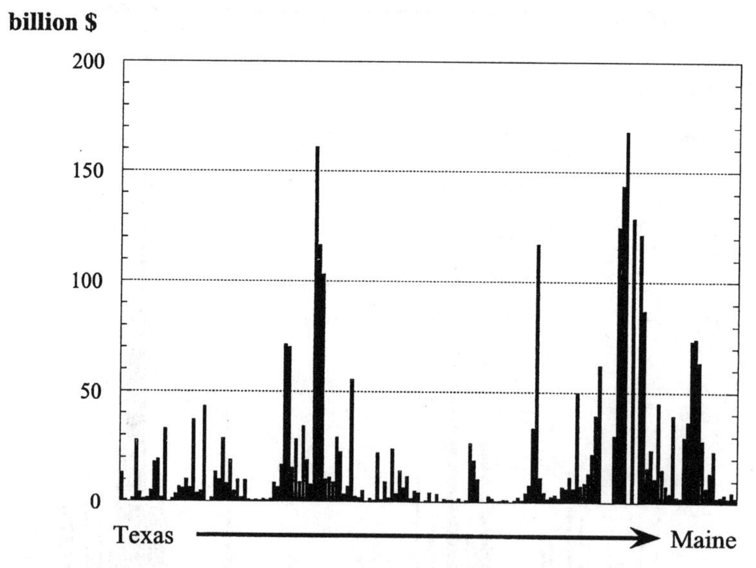

Fig. 8.5. Insured U.S. coastal county property values by county and state for (top) 1988 and (bottom) 1993. Source: Insurance Institute for Property Loss Reduction.

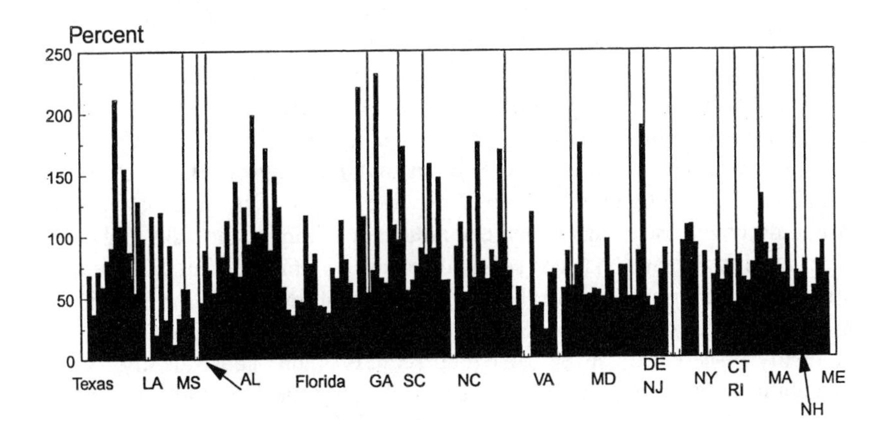

Fig. 8.6. Increase in insured U.S. coastal county property values by county and state from 1988 to 1993 as a percentage of the 1988 total. Source: Insurance Institute for Property Loss Reduction.

Table 8.2. Summary of Coastal County Insured Property by State.

State	1988 Total coastal Insured Property Values (Current billions of $)	1993 Total Coastal Insured Property Values (Current billions of $)	1988-1993 Increase as a Percent of 1988 Value
Texas	70.1	128.6	83
Louisiana	87.5	123.5	41
Mississippi	14.1	25.5	80
Alabama	22.8	36.9	61
Florida	565.8	871.7	54
Georgia	16.5	32.5	96
South Carolina	31.2	54.7	75
North Carolina	22.7	45.0	97
Virginia	42.5	67.8	59
Maryland	129.2	202.6	56
Delaware	38.7	67.7	74
New Jersey	88.5	152.8	72
New York	301.7	595.6	97
Connecticut	143.3	248.1	73
Rhode Island	52.9	83.1	57
Massachusetts	179.8	321.6	78
New Hampshire	18.5	34.9	88
Maine	32.3	54.5	68
Coastal Total	1858.1	3147.0	69
U.S. Total	12967.1	21422.0	65

Source: Insurance Institute for Property Loss Reduction.

Preparedness. Preparedness, as used in this paper, refers to all of the various efforts at various levels of public and private decisionmaking to reduce vulnerability to hurricanes. We use the term "preparedness" in a general sense to refer to the full range of emergency management activities (e.g., planning, mitigation, response, restoration) in full recognition of the significant differences between the various phases of preparedness. Preparedness can be broken down into component phases such as planning, mitigation, response, restoration, etc., and has been widely studied by the natural hazards community (e.g., hazard research centers exist at the Universities of Colorado, Delaware, and Texas, see Burton et al. [1993] for a review of the field). Salmon and Henningson (1987, 2) refer to preparedness as "mitigation planning" and they argue that "it makes excellent sense to do now those things which can reduce or minimize the risks and costs of future hurricanes, and hasten sensible recovery practices after the storm." Mitigation planning has technical, practical, and political aspects which, in large part, are often determined by the idiosyncrasies of and resources available to each community. Therefore, levels of preparedness (and consequently, societal vulnerability more broadly) vary a great deal along the U.S. Gulf and Atlantic coasts.

In general, preparedness activities have short- and long-term components (Salmon and Henningson 1987). Short-term responses focus on a particular approaching storm. Long-term responses focus on the hurricane threat more generally. Many short-term responses related to protection of the exposed population are based upon long-term studies of expected storm surge due to a landfalling hurricane (Sheets 1990; Baker 1991). For example, expected coastal flooding is calculated using the SLOSH (Sea, Lake, and Overland Surges from Hurricanes) storm-surge model (Jarvinen and Lawrence 1985).[3] Evacuation studies are then based upon the areas identified to be at risk from the output of the SLOSH modeling process. Such studies include a behavioral component that seeks to identify "realistic assumptions of how the public will behave when advised or ordered to evacuate" and a transportation analysis which seeks to identify the capacity of routes of escape, points of congestion (Baker 1993; Carter 1993), and places of "last resort refuge" (Sheets 1992). Short-term response is focused on the hurricane forecast. The National Weather Service, National Hurricane Center, local and state officials, and the media coordinate hurricane watches and warnings based upon the forecast tracks of specific approaching hurricanes (Sheets 1990).

The following isolated incident, which occurred in South Carolina during Hurricane Hugo (1989), illustrates the stakes involved with long-term planning for short-term response.

> In the village of McClellanville, the Lincoln High School was used as an evacuation shelter. The evacuation plan listed the base elevation of the school as 20.53 feet National Geodetic Vertical Datum (NGVD). Many of the residents took shelter in this school.

---

[3]The SLOSH model is run for 22 "SLOSH basins" along the U.S. coast from Texas to Maine (Jarvinen and Lawrence 1985). For each basin the model simulates from 250 to 500 different hypothetical storms. The results are maps that depict the "maximum envelope of water" for a family of storms (Sheets 1990). Errors in the forecasts of specific landfalling storms are compensated for because each map represents a composite of maximum storm surges for a range of landfall points, storm movement, and intensity.

During the height of the storm, water rose outside the school and eventually broke through one of the doors. Water rushed in and continued to rise inside the school reaching a depth of 6 feet within the building. A resident with a videocassette recorder documented people climbing on tables and bleachers to escape the rising water. As the water reached its maximum height, children were lifted onto the school's rafters. Fortunately, everyone survived the event although not without considerable anxiety.

Later examination revealed that the base elevation of the school was 10 feet, not the 20.53 feet listed on the evacuation plan. This school should not have been used as a shelter for any storm greater than a category 1 hurricane (USDOC 1990, 12).

Warning and response to Hurricane Hugo based upon the SLOSH model process has been generally judged successful (e.g., USDOC 1990; Baker 1994; Coch 1994), yet had the evacuees at Lincoln High been less fortunate such judgments would likely have been very different. The incident demonstrates the fine line between success and failure in long-term planning to reduce the hurricane threat. Another example is from Louisiana during Hurricane Andrew, when a number of emergency management officials had difficulty interpreting updated storm surge maps, and consequently relied on older, potentially dated information. Other officials in Louisiana did not have relevant FEMA software available to aid in the evacuation decision process (USACE/FEMA 1993).

The protection of property at risk also has short-term and long-term components. Designing structures to withstand hurricane-force winds is an important factor in reducing property damage (e.g., Mehta et al. 1992). An important aspect in the reduction of property damage is the establishment and enforcement of building codes commensurate with the expected risk. According to John Mulady, an insurance industry official, "a 1989 study of the damage done by hurricanes Alicia and Diana found nearly 70% of damage done to homes was the result of poor building code enforcement. However, in North Carolina, where codes were effectively enforced, only 3% of the homes suffered major structural damage" (Mulady 1994, 4). One insurance official claimed that poor compliance with building codes accounted for about 25%, or about $4 billion, of the insured losses in south Florida due to Hurricane Andrew (Noonan 1993). Other estimates range upwards to 40%, or close to $6.5 billion.

Complacency is the enemy of preparedness. The *New York Times* reported in 1993 that "of 34 coastal areas identified as needing evacuation studies, less than half have been completed, and only $900,000 a year is available for commissioning new ones" (Applebome 1993). A FEMA official complained that the lack of "funding is inhibiting an aggressive and comprehensive approach to hurricane preparedness programming" (Applebome 1993). Complacency led to Dade County's unpreparedness for Hurricane Andrew (Leen et al. 1993). For instance, in 1988 Dade county employed 16 building inspectors to serve a population of well over 1 million. On many occasions in the years preceding Andrew inspectors reported conducting more than 70 inspections per day, a rate of one every six minutes, not counting driving time (Getter 1993). Such anecdotes beg for systematic assessments of hurricane preparedness in the broader context of hurricane vulnerability *before a hurricane strikes*. There is sufficient evidence of complacency along the U.S.

Atlantic and Gulf coasts that future hurricane disasters should not come as surprises when they occur (cf. Sheets 1992, 1993).

With population-at-risk and property-at-risk rising rapidly in many coastal communities, and the rate of growth increasing as well, the key to reduced hurricane exposure lies in improved preparedness. While past responses to reduce the threat to human life have been extremely successful, demographic changes mean that the nature of the hurricane threat is ever changing. The recent experience of Hurricane Andrew in Dade County, previously believed to be among the best prepared locales, suggests that many coastal areas may not be as prepared for hurricane impact as was once thought (Pielke 1995). The worst disasters may lie ahead.

Intensity. The intensity of an Atlantic or Gulf of Mexico hurricane at landfall is directly related to its central pressure and to its speed of movement: The lower the central pressure, the higher the wind velocity (Simpson and Riehl 1981; Anthes 1982; Elsberry et al. 1988). The potential minimum central pressure is limited by sea surface temperature (Merrill 1985, 1987), a warmer sea means lower pressure (and higher winds). For example, based on analysis in Merrill (1985), a potential minimum central pressure of 964 mb (which corresponds to the minimal category 3 storm) requires a sea surface temperature of at least 26.5°C). Hurricane Gilbert (1988), a category 5 hurricane, is the most intense Atlantic hurricane on record, with a central pressure of 888 mb. Since 1899, only two category 5 hurricanes have made landfall on the U.S. coast, Camille in 1969 and an unnamed Labor Day storm that hit the Florida Keys in 1935 (Hebert et al. 1993). Hurricane Andrew (1992) was a category 4 storm.

The speed of a storm is added onto the wind speeds determined by the central pressure. For instance, the Great New England Hurricane of 1938 was travelling over 70 knots at landfall (cf. Coch 1994). Thus, winds parallel to the storm's direction of motion were increased by that amount. The 1938 storm is an example of hurricanes that accelerate out of the tropics ahead of a trough in the westerlies. Hurricanes Hazel (1954) and Carol (1954) are two other examples of such rapidly moving storms. Damage potential due to strong winds is directly related to a storm's central pressure and speed of forward movement.

On average, storms only reach about 55% of their maximum potential intensity (DeMaria and Kaplan 1994). In addition to (a) warm sea surface temperatures, favorable conditions for intensification include (b) weak vertical shear of the horizontal wind (less than about 15 knots between the upper and lower troposphere within a radius of about 4 degrees of latitude from the center), (c) an environment favorable for deep cumulonimbus convection, and (d) an upper tropospheric large-scale anticyclone over the surface low so as to evacuate mass from the region of the hurricane, thereby permitting surface pressure to continue to fall (Pielke 1990).

Under the conditions favorable for hurricane intensification, there are generally two types of maximum intensity hurricanes that threaten the U.S. Atlantic and Gulf coasts. The first type is directly related to the sea surface temperature and is most appropriate for slow moving storms such as usually occur in the Gulf of Mexico and over Florida and the Southeast coasts. The slow movement of these storms means that winds in parallel and in the same direction as the storm's motion will increase on the order of only about 10%. The second type of maximum intensity storm is one that has intensified over a warm sea surface and then is rapidly ejected to higher latitudes. In this scenario the rapid transit over cooler ocean waters does not permit much weakening of the storm prior to landfall. These storms threaten the Northeast Atlantic coast with the storm's high speed resulting in a large magnification of the winds in the storm's right front quadrant (where wind direction is in the storm's direction of motion). Consequently, the area of maximum winds remains offshore in storms that move parallel to the Atlantic coast (with the eye remaining offshore).

Occurrence. Landfalling hurricanes of categories 3, 4, and 5 are irregular events within a decadal and longer time scale, and for particular locations on the coast. Thus, when they do occur, they are intensively studied, e.g., Hurricane Frederic (Powell 1982); Hurricane Hugo (Golden 1990); Hurricane Andrew (Wakimoto and Black 1994).

On the annual time scale, Gray (1994) summarizes a statistical Atlantic Seasonal Hurricane Variability technique based on a number of climatic indices, including factors such as the state of the El Niño/Southern Oscillation (ENSO), the Quasi-Biennial Oscillation (QBO) of 30 mb and 50 mb stratospheric winds, the Caribbean Basin-Gulf of Mexico sea-level pressure anomaly in spring and early summer, the lower latitude Caribbean Basin 200 mb zonal wind anomaly in early summer, the rainfall in Africa's Western Sahel region, and a parameter expressing the trend in west to east surface pressure and surface temperature gradients in February through May in West Africa. Among the forecasts made using this method are Hurricane Destruction Potential, Intense Hurricanes (reaching at least category 3 at some time in a storm's evolution), and Intense Hurricane Days (see Gray et al., Chap. 2, this volume). Because of their record of success, the seasonal hurricane forecasts have received wide attention in user communities (e.g. Morgenthaler 1994).

It has been suggested that time periods longer than a year might be amenable to some level of forecast skill. Landsea et al. (1994) speculate to a connection between decadal variations in the "ocean conveyor belt" and sea surface temperatures, Sahel rainfall, ENSO events, and Atlantic hurricanes. Gray (1992) notes that in only one instance has an intense hurricane been observed to strike the Atlantic coast of the U.S. (Andrew in 1992) in years with simultaneous occurrences of a warm ENSO event and dry Western Sahel conditions. Gray and Landsea (1992) found that category 3, 4, and 5 hurricanes along the U.S. Atlantic coast have a strong correlation with Sahel rainfall (cf. Gray 1990). Lare and Nicholson (1994) discuss dry conditions in the western Sahel and attribute intra-year persistence of anomalous wet or dry periods to land-atmosphere feedbacks.

The track record of seasonal hurricane forecasts appears to demonstrate that some statistical skill in predicting the level of hurricane activity in the Atlantic ocean, at least on an annual basis, and with some suggestion that longer period skill may be achievable (cf. Livezey 1990). However, none of these techniques offers a tool for predicting with skill specific landfall locations at annual or longer time scales (Hess and Elsner 1994). Thus, climatology remains the best tool available to estimate coastal vulnerability to hurricane occurrence for specific locations.

Landfall frequency. Ninety-five percent of category 3, 4, and 5 hurricanes in the Atlantic Basin occur during August to October (Landsea 1993). Over the last several decades there has been an observed decrease in major landfalling hurricanes (Landsea et al. 1995, Landsea 1993). Hurricanes Hugo and Andrew and the active 1995 hurricane season have led many to suggest that annual hurricane activity may be increasing. According to Hebert et al. (1993) a category 4 storm strikes the U.S., on average, every 6 years. Hurricanes Andrew and Hugo are the only category 4 storms to make landfall on the U.S. coast since 1969.

Hebert et al. (1993) also note that on the average two hurricanes strike the U.S. coast each year, with *two intense hurricanes striking the U.S. coast every 3 years.* Climatology suggests that the average damage due to an intense hurricane would only need to be \$4.5 billion for annual U.S. exposure to be at least \$3.0 billion. Hurricanes Hugo and Andrew suggest, however, that these estimates may be low. If the average damage due to an intense hurricane is \$7.5 billion, for example, then annual exposure would be at least \$5.0 billion. Such estimates exclude the costs of landfalling tropical storms and category 1 and 2 hurricanes which are capable of extensive economic damage (cf. Landsea 1991). For example, tropical storm Gordon (1994) resulted in more than \$200 million in agriculture-related damage in Florida (NYT 1994).

Other climatological information related to landfall frequency includes (a) 35% of all hurricanes have hit Florida, (b) more than 70% of category 4 or 5 storms have hit Florida and Texas, and (c) along the middle Gulf coast, southern Florida, and southern New England half of all landfalling hurricanes have been category 3 and higher.

Jarrell et al. (1992) compile coastal county hurricane landfalls since 1900 and provide the data necessary to compute landfall probabilities by coastal county. One storm which could arguably be added to their compilation include 17-26 August 1933 storm that made landfall at the North Carolina Outer Banks with a central pressure of 960 mb. and moved north along the western side of the Chesapeake Bay (Cobb 1991). Figure 8.7 shows the observed annual probability of a direct hit by an intense hurricane expressed as a percentage for each coastal county from Texas to Maine. The highest probability is for Monroe County, Florida (0.095), immediately to the south of Dade County where Hurricane Andrew (1992) made landfall. The Gulf coast and the Atlantic coast of Florida are particularly at risk to hurricane landfall. In contrast, several counties along the Gulf coast and many along the Atlantic coast have never experienced a direct hit from intense hurricane during the period 1900-1994. Most of New England sees few direct hits from hurricanes because most storms accelerate northward or northeast on a track that typically places the storm either inland or parallel to the coast but offshore. The absence of direct hits to the northern Florida and Georgia coasts, in contrast, is partly a consequence of their orientation with respect to the more typical hurricane track as the storms begin to recurve around the subtropical Bermuda high pressure ridge. Undoubtedly, the absence of direct hits is also partly due to good fortune (cf. Kocin and Keller 1991). A landfall of Hurricane Hugo (1989) just slightly farther south would have altered these statistics. Indeed, Ludlam (1963) reports that a number of hurricanes struck the coast of northern Florida and Georgia in the 18th and 19th centuries. For instance, the Atlantic coast between St. Augustine, Florida, and Savannah, Georgia, was hit by very strong storms in 1824 and 1837.

Table 8.3 shows annual landfall probabilities (percentages) for intense hurricanes, average return period in years, and year of last direct hit by an intense hurricane for selected coastal counties and major metropolitan areas for each state from Texas to Maine through 1994. Table 8.3 also displays for each coastal county insured property values and estimated 1993 population. The return period is simply the inverse of the annual probability and represents the average number of years between direct hits for each location. An observed annual landfall probability of 0.053 for Miami, Mobile, and Galveston corresponds to a greater than 50% chance of at least one intense hurricane within the next 13 years.[4] Similarly, an observed annual landfall probability of 0.021, such as observed in Charleston, South Carolina, Wilmington, Delaware, and Providence, Rhode Island, corresponds to a greater than 50% chance of at least one intense hurricane within the next 33 years.[5] Climatology represents a baseline against which efforts to improve long-term forecasts ought to be measured. Reliable forecasts that improve upon climatology may have value to those decisionmakers who can alter their behavior accordingly.The probabilities shown in Fig. 8.7 for intense hurricane are similar to the findings of Ho et al. (1987) for all types of landfalling hurricanes. Ho et al. (1987, 77) find that:

> Highest frequency of landfalling tropical cyclones on the Atlantic coast is in southern Florida, and a comparatively high frequency appears to the south of Cape Hatteras, North Carolina. The frequency of entries drops off rapidly from Miami to Daytona Beach, Florida and from Cape Hatteras northward to Maine, except around Long Island.

Simpson and Lawrence (1971) report observed landfall probability for 50-mile segments of coast. Differences between Simpson and Lawrence, Ho et al., and the analysis of landfall probability in this paper are primarily due to different levels of analysis -- that is landfall probability percentages are given in this paper for each coastal county, whereas Simpson and Lawrence and Ho et al. give landfall probabilities for equal segments of coastline. Each method has strengths: equal segments of coastline allow for ready comparison between segments, whereas data

---

[4]The calculation which gives the period N, where N equals the number of years from now within which there exists a greater than 50% chance of at least one intense hurricane is, $(1 - P)^N \geq 0.50$, where P is the annual landfall probability based on the historical record and is assumed to be constant, and events are assumed to be independent from year to year.

[5]For Monroe County, Florida immediately to the south of Dade County (Miami) climatology suggests that there is a greater than 50% chance of at least one intense hurricane in the next 7 years.

by county may be more meaningful to decisionmakers at state and local levels. Figure 8.8 shows analyses of landfall probabilities from Simpson and Lawrence and Ho et al. for purposes of comparison. It has been suggested that hurricane intensity, occurrence, and landfall frequency may be affected by anthropogenic global warming. One hypothesis is that the oceans would warm, thereby creating more intense hurricanes over a greater geographic area (Emanuel 1987). But Gray (1990) argues that variability in hurricane incidence is a function of random climatic variability. See also Emanuel, Chap. 3, this volume. On the basis of such scientific hypotheses, a number of groups have argued that the recent impacts of Hurricanes Andrew and Hugo are evidence for global warming (e.g., Leggett 1994). In addition to faulty logic (i.e., the converse of a true proposition is not necessarily true) such arguments, no matter how well intentioned, serve to direct attention away from the documented hurricane threat. In the context of the hurricane threat facing the U.S., such arguments focus attention on responses to global warming (e.g., reducing carbon emissions) rather that on the need for increased hurricane preparedness in individual communities to reduce vulnerability. Increased effort in preparedness is necessary under *any* climate change scenario.

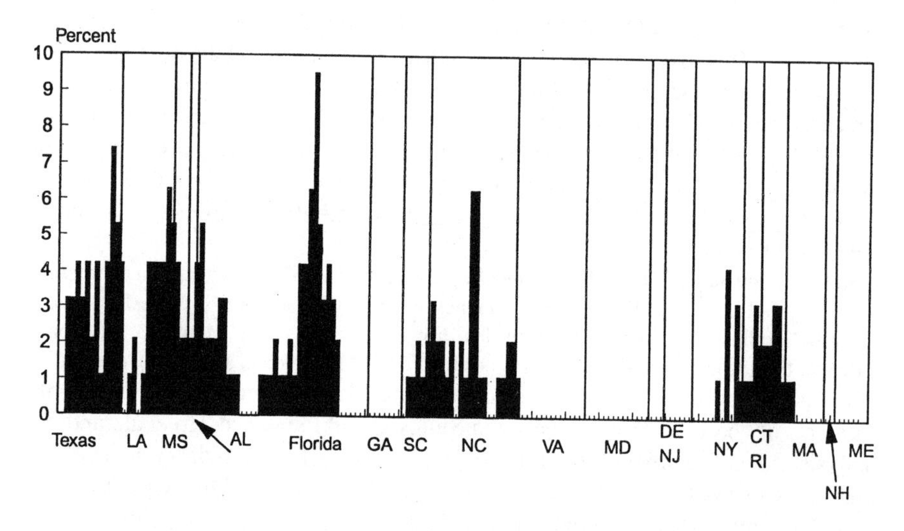

Fig. 8.7. Observed (1900-1994) annual probability of a direct hit by an intense hurricane for U.S. coastal counties from Texas to Maine.

Furthermore, from the standpoint of societal vulnerability to hurricanes, the cause of any increased hurricane occurrence, intensity, and landfall frequency is arguably less important for purposes of action because response efforts necessarily must focus on preparedness and not hurricane prevention. Consider that coastal population has increased by more than 20% over the period 1970 to 1990 and insured property has increased by more than 180% over the period 1980 to 1993. Meanwhile, over the period 1970 to 1990 hurricane incidence was well below the observed climatological average (Landsea et al. 1995). Efforts to reduce vulnerability make sense no matter what the cause of any increase in hurricane incidence. In other words, global warming is largely irrelevant to the need for actions to reduce our vulnerability to hurricanes. History alone dictates that such actions are sorely needed.

Table 8.3. Summary of hurricane statistics for selected locales in Atlantic and Gulf coastal states.

State	City	County	Population 1993 (est)[a] (000s)	Insured Property 1993[b] (billions $)	Observed Intense Landfall Probability[c]	Return Period (yr)	Most Recent Intense Direct Hit[d] (Year, Category, Storm)
TX	Brownsville	Cameron	270	13.182	3.2	31.7	1980 (3) Allen
	Galveston	Galveston	217	19.309	5.3	19.0	1983 (3) Alicia
LA	New Orleans	Orleans	491	43.340	4.2	23.7	1965 (3) Betsy
MS	Gulfport	Harrision	166	13.470	2.1	47.5	1985 (3) Elena
AL	Mobile	Mobile	384	28.606	5.3	19.0	1985 (3) Elena
FL	Tampa/	Hillsbor-	847	69.968	1.1	95.0	1921 (3)
	St. Petersburg	ough?	859	71.283	1.1	95.0	1921 (3)
	Miami	Pinellas	1979	160.844	5.3	19.0	1992 (4) Andrew
	Jacksonville	Dade	689	55.527	0.0	-	1854
		Duval					
GA	Savannah	Chatham	219	22.386	0.0	-	1893
SC	Charleston	Charleston	304	24.118	2.1	47.5	1989 (4) Hugo
NC	Wilmington	New Hanover	125	11.814	2.1	47.5	1960 (3) Donna
VA	Norfolk	Norfolk	261	18.912	0.0	-	1856
MD	Ocean City	Worcester	35	6.269	0.0	-	1933 (3)[f]
	Baltimore	Baltimore[e]	1438	117.128	0.0	-	1850
DE	Wilmingon	New Castle	450	49.794	010	-	1861
NJ	Asbury Park	Monmouth	227	62.038	0.0	-	1861
NY	New York City	Kings	2289	124.887	0.0	-	1821
CT	New Haven	New Haven	803	87.142	1.1	95.0	1938 (3)
RI	Providence	Providence	596	45.305	2.1	47.5	1954 (3) Carol
MA	Cape Cod	Barnstable	187	29.572	1.1	95.0	1954 (3) Edna
	Boston	Suffolk	649	74.299	0.0	-	1869
NH	Portsmouth	Rockinham	243	28.010	0.0	-	1788
ME	Portland	Cumberland	245	23.341	010	-	1830

[a] Source: US Census Bureau.

[b] Source: Insurance Institute for Property Loss Reduction.

[c] Source: Jarrell et al. (1992). Period of record is 1900-1994.

[d] Data for years since 1900 from Herbert et al. (1993), Table 11, 24-25. Estimates for years prior to 1900 are based on descriptions in Ludlam (1963) and Tannehill (1952).

[e] Includes both city and county.

[f] According to data in Cobb (1992) the 1933 storm was a category 3 hurricane.

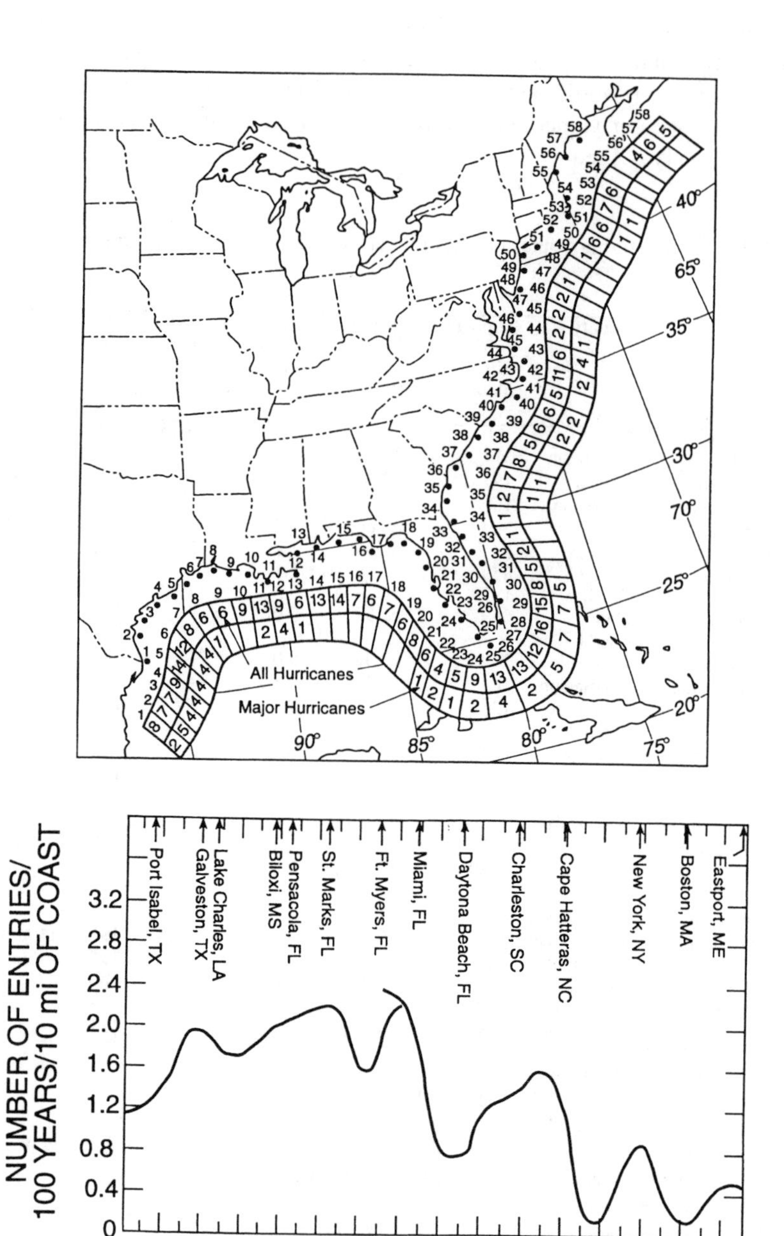

Fig. 8.8. Analysis of landfall probabilities from (top) Simpson and Lawrence (1971) and (bottom) Ho et al., (1987) for purposes of comparison.

8.3 Long-Term Forecasts and Reducing Societal Vulnerabilities to Hurricanes: Opportunities and Limitations

If we assume that for the foreseeable future little can be done to control hurricane incidence (i.e., intensity, occurrence, and frequency), then efforts to reduce societal vulnerability to hurricanes must focus on reducing exposure (i.e., population and property at risk, improving preparedness). Consequently, the value of long-term forecast might be determined through answers to the following question:[6]

> How might long-term (interannual/decadal) forecasts of hurricane incidence be used (directly/indirectly) to reduce exposure (i.e., population at risk, property at risk, unpreparedness) and thus also reduce the societal impacts of hurricanes?

To answer this question a series of assessments might be structured using the methodology applied by Glantz (1976, 1977, 1979). One such study sought to determine:

> How much flexibility decisionmakers in Saskatchewan and in Canada might have, given perfect information about their climate one year in advance. Although such a forecast will never be available, an assessment of the impact of a perfect forecast can be useful in determining the value of a less-than perfect, but feasible, long-range forecast for the Prairie Provinces. It will also enable us to examine options that various types of decisionmakers, from Provincial and Federal government officials to individual farmers, might have to minimize the impact of weather anomalies on agricultural output (Glantz 1976, 1-2).

Glantz (1982) considered the problem in another way and assessed the social costs of an inaccurate forecast. Glantz (1986, 93) concluded from the studies that "the formulation, promulgation, and implementation of a forecast must be carefully assessed, almost on a case by case basis, in order to determine its true value to society." Such assessments of use, misuse, and nonuse of forecasts could illuminate the strong sensitivity of various decisionmaking processes to improved meteorological products, and to define the upper and lower limits on the value of improved long-term forecasts.

To our knowledge, no such assessment of the value of a long-term hurricane forecast has been conducted. The observation of Gray et al. (1991, 1881) in connection with aircraft reconnaissance and tropical cyclone forecasting has general relevance for understanding the lack of assessments of potential long-term hurricane forecast use and benefits:

> While relying on reconnaissance for more than 40 years, most American [tropical cyclone] forecasters and researchers have not felt the need to make quantitative studies of just how beneficial aircraft reconnaissance has been in order to justify its

[6]Clearly, much work needs to be done to reduce vulnerability to hurricanes independent of our ability to make long-term forecasts. However, recent advances in seasonal and interannual forecasts of hurricane incidence may, with proper use, enhance efforts to reduce vulnerability.

continuation. Research is now belatedly beginning to focus on this subject.

The same claim might be said of hurricane research more generally, and indeed much of scientific research as well. Assessments of the potential and actual use and value of long-term forecasts of hurricane activity would be a valuable contribution to understanding the opportunities for and limitations on actions focused on reducing societal vulnerability to hurricanes. Furthermore, in an era where science is increasingly called upon to demonstrate societal benefits, such assessments have potential to explore how to best leverage investments in research for practical ends. The following sections present a number of heuristics for structuring and conducting an assessment of the use and value of a long-term hurricane forecast.

The phrase "integrated assessment" is a "buzz word" out of the global change community, and is generally associated with future scenarios generated by economic models for planning purposes (e.g., Dowlatabadi and Morgan 1993). As we use it, integrated assessment refers to the *integration* of knowledge of climate-related and societal processes in order to *assess* alternative courses of action under the goal of reducing societal vulnerability to hurricanes. Such an integrated assessment of the hurricane problem would focus on processes of decisions and the role of forecast information therein by public and private individuals and groups.

A focus on process is central to realistic determination of the opportunities for and limitations on use of long-term forecasts to reduce societal vulnerabilities to hurricanes (cf. Pielke 1994). To focus on process is to focus on the formulation, promulgation, and execution of particular decisions (Lasswell 1971). Often, both scientists and policymakers alike behave as if the development of scientific information (such as long-term forecasts) is sufficient to lead to better decisions. From a decision maker's perspective, a call for better information from scientists can forestall the need to make difficult decisions while placing the burden of problem solving upon the scientists (Clark and Majone 1985). From a scientist's perspective, a focus on information allows for relative autonomy from the "politics" of decisionmaking and a justification for continued funding. However, it is often the case that scientific information is misused or not used at all because of rigidities in and practicalities of decisionmaking processes (e.g., Feldman and March 1981). Therefore, an integrated assessment "addresses environmental questions and issues which reflect political, societal, regulatory, and management values and expectations, as much as, if not more than, scientific and technical information" (Davis et al. 1994, 1047-1048).

Specific decisions are made in the context of a set of alternative courses of action. For example, in order to better prepare for the hurricane threat, citizens of New Orleans might desire a range of alternative responses to the following questions: How strictly shall we enforce our buildings codes? At what point in time before an approaching hurricane shall evacuation become mandatory? How much, if any, increase in insurance should citizens be required to carry in the face of a long-term forecast of increased hurricane incidence? Alternative actions in response to each question are embedded in a broader context of values, feasibility, and efficacy. For instance, building code enforcement cannot be separated from issues such as the costs to the resident of building a reinforced structure and the tax revenues necessary to hire a sufficient number of building inspectors. In many respects, decisions to which long-term forecasts may be relevant are decisions about how a community wishes to move into the future (cf. Salmon and Henningson 1987).

8.3.1 Quantification of Benefit Requires Attention to Actual and Potential Forecast Use

Demonstration of use or value of long-term hurricane forecasts is a challenging analytical task. Glantz (1986, 93) notes that "one could effectively argue that the value of climate-related forecasts will in most instances be at least as much a function of the political, economic, and social settings in which they are issued than of the soundness of information in the forecast itself." Put another way, the solution to the hurricane problem is potentially very different in Dade County, Florida, from the solution in Worcester County, Maryland, and both of those may be significantly different from the solution in Nueces County, Texas as a result of economic, political, and civic differences between the various communities.

Apart from demonstrating value of improved long-term forecasts, accurate assessment of societal vulnerabilities to hurricanes is a very challenging task. An example of the difficulties in defining the extent and magnitude of the hurricane threat is provided by the response of the insurance industry to Hurricane Andrew. The 1992 event served as a "wake-up call" to the insurance industry. Prior to Andrew the insurance industry largely ignored hurricane climatology and instead kept records of hurricane-related deaths and economic damage, according to Russell Mulder, director of risk engineering at the Zurich-American Insurance group (Wamsted 1993). The insurance industry's records were accurate measures of their losses, but not of hurricanes: they neglected storms that did not make landfall and underestimated the potential impact of storms that made landfall in relatively unpopulated areas. Since Hurricane Andrew, the insurance industry has paid closer attention to the hurricane threat (e.g. Banham 1993; Noonan 1993; Wilson 1994). One would expect the insurance industry to be among the most sensitive to societal vulnerability to hurricanes, however, Hurricane Andrew demonstrated that even when concern exists, accurate definition of the hurricane problem is difficult (see chapters by Roth, Chap. 13, and Clark, Chap. 14, this volume).

8.3.2 Consider Forecast Use and Value in Context

Murphy (1993, 286) states that "forecasts possess no intrinsic value. They acquire value through their ability to influence the decisions made by the users of the forecasts." Yet, because numerous factors contribute to any particular decision "assessing the economic value of forecasts is not a straightforward task" (Murphy 1994, 64). That is, a forecast is, at best, only one of a multitude of factors which influence a particular (potential) user (cf. Torgerson 1985). It is often difficult to identify the signal of the forecast in the noise of the decisionmaking process. Factors external to the forecast may hinder its use.

Murphy (1994) identifies two complementary approaches to assessment of forecast value which can be summarized as use-in-theory and use-in-practice (cf. Camerer and Kunreuther 1989). Murphy (1994) calls these prescriptive and descriptive assessments of forecast value. Glantz (1976) uses the terminology of "what ought to be" and "what is." Use-in-theory refers to efforts to estimate the "value of forecasts under the assumption that the decisionmaker follows an optimal strategy" (Stewart 1995, 40). Generally, economists, statisticians, and decision theorists share expertise in assessment of use-in-theory (e.g., Winkler and Murphy 1985). Use-in-practice refers to efforts, including case studies, to understand how decisions are actually made in the real world and the value of forecast information therein (e.g., McNew et al. 1991). Political scientists, sociologists, and psychologists are examples of those with expertise in assessment of use-in-practice.

It is likely that as long-range forecasts of hurricane incidence demonstrate increased skill that the value of such forecasts will not be self-evident to most users. Hence, it may be worthwhile for producers of long-range forecasts to conduct an ongoing parallel research effort targeted at actual and potential users. Such a parallel program could focus on integrated assessments of use-in-theory and use-in-practice in order to identify opportunities for and constraints on improved and proper use of long range hurricane forecasts. Counties, states, and SLOSH basins would be appropriate levels of analysis for an assessment. Such assessments may find that in some cases a particular decision process may constrain effective use of a long-term forecast. Other assessments may find clear opportunities to leverage forecast information for reduced vulnerability. If long-term forecasts of hurricane activity are justified in terms of their value added to social processes, then the susceptibility of support for such research may depend in large part upon demonstration of actual use or value.

8.3.3 Beware of Overselling the Science

It is generally accepted that modification of hurricane incidence will remain impractical for the foreseeable future, in spite of the mid-century optimism following an intensive series of efforts to "tame" hurricanes in the 1950s and 1960s (Gentry 1974). Experience with hurricane modification does provide one very important lesson: Care must be taken not to "over-promise" expected benefits deriving from research (cf. Tennekes 1990; Namias 1980).

Consider the following statement made in the late 1940s in a talk given by Nobel Laureate Irving Langmuir at the dawn of optimism about hurricane modification: "The stakes are large and with increased knowledge, *I think that we should be able to abolish the evil effects of these hurricanes*" (quoted in Byers 1974, 15, emphasis added). On one level such claims reflect the eternal optimism of science and technology. But at another level, such claims are publicly irresponsible and potentially damaging to the institution of science (Changnon 1975). One can easily imagine a congressional appropriator, excited by the possibilities of Langmuir's claim, making an argument that "preparedness plans for hurricane would no longer be necessary because in weather modification scientists had discovered a magic bullet." Of course, taking the thought a step further, had a hurricane then hit a poorly prepared community blame would have been laid at the feet of the scientist, and not the policymaker. In the context of long-term forecasts of hurricane activity credibility with the public will be difficult to gain, and easy to lose (Slovic 1993).

Weather modification is perhaps an extreme example of the risks involved with overselling science. However, in an era when science is increasingly called upon to contribute to the resolution of many difficult societal problems, demonstration of benefits may become central to sustained federal support of research.

8.4 Conclusion

The hurricane threat facing the U.S. Atlantic and Gulf coasts is constantly changing. Along these coasts population and property vulnerable to hurricanes have increased dramatically over the last several decades. During that period hurricane incidence was well below the observed climatology. Communities that have suffered the force of recent hurricanes, particularly Dade County, Florida, with Hurricane Andrew, give evidence that the U.S. may be more vulnerable to hurricanes than has been recently thought. Furthermore, anecdotal evidence suggests that complacency is the norm in many communities. Such experiences and scattered evidence begs for systematic assessment of societal vulnerability to the hurricane threat in particular coastal communities. Such assessments could establish vulnerability with enough detail so as to provide a baseline against which one might measure the potential use and value of improved forecast capabilities (short- and long-term).

While the hurricane threat to the U.S. Atlantic and Gulf coasts is ever changing, past efforts to reduce societal vulnerability have had many successes (e.g., the relatively few casualties in Hugo [1989] and Andrew [1992]). These successes form a body of practical experience from which lessons might be distilled and used to help guide efforts to reduce vulnerability to extreme weather events more generally. Such efforts will become increasingly important as demographic changes place more people and property in the path of extreme events (themselves part of a constantly changing climate). Future successes in reducing vulnerability to hurricanes have potential to add to the base of experience available to guide action in response to weather threats. However, the lessons of past and future experience depend upon rigorous assessments of the processes of decisionmaking. Absent such assessments, policy failure will likely become apparent only in the aftermath of an extreme event. In the future, anecdotal evidence of the importance of hurricane forecasts will not suffice; for lesson-drawing there is no substitute for rigorous demonstration of use and value of the role of research in practical settings.

In the broader context of U.S. science policy, the use and value of long-term hurricane forecast, and hurricane research more generally, has not found a broad audience. There are at least three reasons why the successes of hurricane research have not reached the broader science policy community. First, as Gray (1991) argues, hurricane researchers in the past have had little incentive to conduct assessments of the use and value of their research. Second, members of the hurricane research community are poorly placed to argue for the worth of their research, as some policymakers may view such arguments as self-serving, no matter how meritorious. Finally, the question of the use and value of hurricane research is a difficult analytical question that involves assessment of the process of decisionmaking well beyond the contribution of scientific research. For these reasons, there exists an opportunity for assessment of the role of forecasts in reducing societal vulnerability to hurricanes, to contribute to recent debate on the efficacy of federally funded research.

Societal vulnerability to hurricanes -- exposure and incidence -- provides a basis for understanding the potential use and value of hurricane research, particularly forecasts, to resolution of the problems of hurricanes, extreme weather, and U.S. science policy. Reducing vulnerability provides a common goal that integrates the science and social science of hurricane impacts. Experience suggests that improved long-term forecasts, by themselves, will not reduce societal vulnerability unless accompanied by efforts to use them in processes of preparedness. To the extent that improved forecasts contribute to reduced vulnerability we will have made less serious the threat of hurricanes to coastal communities, supplied a basis of experience for actions in response to other types of extreme weather phenomena, and provided a practical demonstration of the relationship of scientific research in pursuit of societal objectives. Such progress will not come easy, but with the close interaction of social and physical scientists with actual and potential users of science, meeting these goals is a surmountable challenge.

Acknowledgments

We would like to thank the following individuals for useful comments and suggestions as well as for assistance in the preparation of this manuscript: Radford Byerly, Leighton Cochran, Mark DeMaria, Steve Dickson, Mary Downton, Leslie Forehand, Karen Gahagan, Kevin Gallagher, Michael Glantz, William Gray, William Hooke, Jan Hopper, Richard Katz, Chris Landsea, Steve Nelson, Steve Rhodes, Robert Serafin, Robert Simpson, Ken Van Sickle, and Thomas Vonderhaar. NCAR Graphics and Justin Kitsutaka expertly aided with the figures.

We would also like to thank Henry Diaz and Roger Pulwarty for providing the opportunity to prepare the manuscript and for assistance with the revision. The research in the manuscript was partially supported by a grant from the National Science Foundation. Partial support to attend the workshop was provided by the Cooperative Institute for Research in the Atmosphere at Colorado State University. The views in the paper are those of the authors and should not be construed as representing any of the individuals or institutions with whom either author is associated.

References

Anthes, R.A., 1992. *Tropical Cyclones: Their Evolution, Structure and Effects.* American Meteorological Society, Boston, MA.

Applebome, P., 1993. In the Hurricane Belt, a New, Wary Respect. *New York Times* August 18, p. A12.

Baker, E.J., 1994. Warning and response. In: *Hurricane Hugo: Puerto Rico, The Virgin Islands, and Charleston, South Carolina, September 17-22, 1989*, Natural Disaster Studies Volume 6, National Academy of Sciences, Washington, DC, 202-210.

Baker, E.J., 1993. Empirical studies of public response to tornado and hurricane warnings in the United States. In: Nemec, J., Nigg, J.M., and Siccardi, F. (eds.), *Prediction and Perception of Natural Hazards*, Kluwer Academic Publishers, Dordrecht, The Netherlands, 65-73.

Baker, E.J., 1991. Hurricane evacuation behavior. *International Journal of Mass Emergencies and Disasters,* **9**, 287-310.

Banham, R., 1993. Reinsurers seek relief in computer predictions. *Risk Management,* **8**, 14-19.

Brinkmann, W.A.R., 1975. *Hurricane Hazard in the United States: A Research Assessment*, Monograph NSF-RA-E-75-007, Program on Technology, Environment and Man, Institute of Behavioral Science, University of Colorado.

Brown, G., 1992. The objectivity crisis. *American Journal of Physics,* **60**, 779-781.

Brunner, R.D., and Ascher, W., 1992. Science and social accountability. *Policy Sciences,* **25**, 295-331.

Burton, I.R., Kates R.W., and White, G.F., 1993. *The Environment as a Hazard*, 2nd. edition, Guildford Press.

Byerly, R., 1995. U.S. science in a changing context: A perspective. In: Pielke, R.A. (ed.) *U.S. National Report to the International Union of Geodesy and Geophysics 1991-1994.*

Byers, H.R., 1974. "History of weather modification." In: Hess, W.N. (ed.), *Weather and Climate Modification*, John Wiley & Sons, New York, 3-44.

Camerer, C.F., and Kunreuther, H., 1989. Decision process for low probability events: policy implications. *Journal of Policy Analysis and Management,* **8**, 565-592.

Campbell, P.R., 1994. Population projections for states, by age, sex, race, and Hispanic origin: 1993 to 2020. Current Population reports, P25-1111, US GPO, Washington, DC.

Carter, T.M., 1993. The role of technical hazard and forecast information in preparedness for and response to the hurricane hazard in the United States. In: Nemec, J., Nigg M., and Siccardi, F., (eds.), *Prediction and Perception of Natural Hazards*, Kluwer Academic Publishers, Dordrecht, The Netherlands, 75-81.

Changnon, S.A., 1975. The paradox of planned weather modification. *Bulletin of the American Meteorological Society,* **56**, 1, 27-37.

Changnon, S.A. and Changnon, J.M., 1992. Temporal fluctuations in weather disasters: 1950-1989. *Climatic Change,* **22**, 191-208.

Clark W.C. and Majone, G., 1985. The critical appraisal of scientific inquiries with political implications. *Science, Technology, and Human Values,* **10**, 3, 6-19.

Cobb, H.D. III., 1991. The Chesapeake-Potomac hurricane of 1933. *Weatherwise,* **44**, 24-29.

Coch, N.K., 1994. Geologic effects of hurricanes. *Geomorphology,* **10**, 37-63.

Cochran, L. and Levitan, M., 1994. Lessons from Hurricane Andrew. *Architectural Science Review,* **37**, 3, 115-121.

Council of Economic Advisors, 1994. *Economic Report of the President*, US GPO, Washington, DC, February.

Davis, W. P., Thornton, K.W. and Levinson B., 1994. Framework for assessing effects of global climate change on mangrove ecosystems. *Bulletin of Marine Science,* **54**, 1045-1058.

DeMaria, M. and Kaplan, J., 1994. Sea surface temperature and the maximum intensity of Atlantic tropical cyclones. *Journal of Climate,* **7**, 1324-1334.

Dowlatabadi, H. and Morgan, M.G., 1993. Integrated assessment of climate change. *Science*, **259**, 1813, 1932.

Dunn, G.E. and Miller, B.I., 1960. *Atlantic Hurricanes*, Louisiana State University Press, Baton Rouge, LA.

Elsberry, R. L., Frank, W.M., Holland, G.J., Jarrell, J.D., and Southern, R.L., 1988. *A Global View of Tropical Cyclones*, Naval Postgraduate School, Monterey, CA.

Emanuel, K.A., 1987. The dependence of hurricane intensity on climate. *Nature,* **326**, 483-485.

Feldman M.S. and March, J.G., 1981. Information in organizations as signal and sign. *Administrative Science Quarterly,* **26**, 171-186.

Gentry, R.C., 1974. Hurricane modification. In: Hess, W.N. (ed)., *Weather and Climate Modification*, John Wiley and Sons, New York, 497-521.

Getter, L. 1993. Inspections: A breakdown in the system. In: Tait, L.S. (ed.) *Lessons of Hurricane Andrew*, Excerpts from the 15th Annual National Hurricane Conference, FEMW, Washington, DC. April, 33-36.

Glantz, M.H., 1976. Value of a reliable long range climate forecast for the Sahel: A preliminary assessment. International Federation of Institutes for Advanced Study, May 15.

Glantz, M.H., 1977. The value of a long-range weather forecast for the West African Sahel. *Bulletin of the American Meteorological Society,* **38**, 150-158.

Glantz, M.H., 1978. Render Unto Weather... -- An Editorial. *Climatic Change,* **1**, 305-306.

Glantz, M.H., 1979. Saskatchewan spring wheat production 1974: A preliminary assessment of a reliable long-range forecast. *Climatological Studies No. 33*, Canadian Government Publishing Centre, Hull, Quebec.

Glantz, M.H., 1982. Consequences and responsibilities in drought forecasting: The case of Yakima, 1977. *Water Resources Research,* **18**, 3-13.

Glantz, M.H., 1986. Politics, forecasts, and forecasting: forecasts are the answer, but what was the question? In: Krasnow, R. (ed.), *Policy Aspects of Climate Forecasting*, Resources for the Future, Washington, DC.

Golden, J.H., 1990. Meteorological data from Hurricane Hugo. *Proceedings of the 22nd Joint UJNR Panel Meetings* -- Winds and Seismic Effects, May 14-18.

Golden, J.H. and Snow, J.T., 1991. Mitigation against extreme windstorms. *Reviews of Geophysics,* **29**, 4, 477-504.

Gray, W.M., 1994. Extended range forecast of Atlantic seasonal hurricane activity for 1995. Department of Atmospheric Science, Colorado State University, Fort Collins, CO.

Gray, W.M., 1992. Summary of 1992 Atlantic tropical cyclone activity and verification of author's forecast. Department of Atmospheric Science, Colorado State University, Fort Collins, CO.

Gray, W.M., 1990. Strong association between West African rainfall and U.S. landfall of intense hurricanes. *Science,* **249**, 1251-1256.

Gray, W.M. and Landsea, C.W., 1992. African rainfall as a precursor of hurricane-related destruction on the U.S. East Coast. *Bulletin of the American Meteorological Society,* **73**, 9, 1352-1364.

Gray, W.M., Neumann, C., and Tsui, T.L., 1991. Assessment of the role of aircraft reconnaissance on tropical cyclone analysis and forecasting. *Bulletin of the American Meteorological Society,* **72**, 1867-1883.

Hebert, P.J., Jarrell, J.D., and Mayfield, M., 1993. The deadliest, costliest, and most intense United States hurricanes of this century (and other frequently requested hurricane facts). *NOAA Technical Memorandum NWS NHC-31.*

Hess, J.C. and Elsner, J.B., 1994. Historical developments leading to current forecast models of annual Atlantic hurricane activity. *Bulletin of the American Meteorological Society,* **75**, 1611-1621.

Ho, F.P., Su, J.C., Hanevich, K.L., Smith, R.J., and Richards F.P., 1987. Hurricane climatology for the Atlantic and Gulf coasts of the United States. *NOAA Technical Memorandum NWS-38.*

Jarrell, J.D., Hebert, P.J., and Mayfield, M., 1992. Hurricane experience levels of coastal county populations from Texas to Maine. *NOAA Technical Memorandum NWS NHC-46.*

Jarvinen, B.R. and Lawrence, M.B., 1985. An evaluation of the SLOSH storm-surge model. *Bulletin of the American Meteorological Society,* **66**, 1408-1411.

Kocin, P.J. and Keller, J.H., 1991. A 100-year climatology of tropical cyclones for the northeast United States. Paper Prepared for the 19th Conference on Hurricanes and Tropical Meteorology, American Meteorological Society, Miami, Florida, May 6-10.

Landsea, C.W., Nicholls, N., Gray, W.M., and Avila, L.A., 1996. Downward trends in the frequency of intense Atlantic hurricanes during the past five decades. *Geophysical Research Letters,* (in press).

Landsea, C.W. 1993. A climatology of intense (or major) atlantic hurricanes. *Monthly Weather Review,* **121**, 1703-1713.

Landsea, C.W. 1991. *West African Monsoonal Rainfall and Intense Hurricane Associations*, Colorado State University Department of Atmospheric Science, Paper No. 484, October 7, Fort Collins, CO.

Landsea, C.W., Gray, W.M., Mielke, P.W. Jr., and Berry, K.J., 1994. Seasonal forecasting of Atlantic hurricane activity. *Weather,* **49**, 273-284.

Lare, A.R. and Nicholson, S.E., 1994. Contrasting conditions of surface water balance in wet years and dry years as a possible land surface-atmosphere feedback mechanism in the West African Sahel. *Journal of Climate*, **7**, 653-668.

Lasswell, H.D., 1971. *A Pre-View of Policy Sciences*, American Elsevier, New York.

Leen, J., Doig, S.K., Getter, L., Soto, L.F., and Linefrock, D., 1993. Failure of design and discipline, In: Tait, L.S. (ed.) *Lessons of Hurricane Andrew*, Excerpts from the 15th Annual National Hurricane Conference, FEMA, Washington, DC. (April), 39-44.

Leggett, J. (ed.), 1994. *The Climate Time Bomb*, Greenpeace International, Amsterdam, The Netherlands.

Livezey, R.E., 1990. Variability of skill of long-range forecasts and implications for their use and value. *Bulletin of the American Meteorological Society*, **71**, 300-309.

Ludlam, D.M., 1963. *Early American Hurricanes*: 1492-1870, American Meteorological Society, Boston, MA.

Mauro, J., 1992. Hurricane Andrew's other legacy. *Psychology Today* **93**, 42-45.

Mayfield, M. and Avila, L., 1993. Atlantic hurricanes. *Weatherwise*, **46**, 18-24.

McNew, K.P., Mapp, H.P., Duchon, C.E., and Merritt, E.S., 1991. Sources and uses of weather information for agricultural decision makers. *Bulletin of the American Meteorological Society*, **72**, 491-498.

Mehta, K.C., Cheshire, R.H., and McDonald, J.R., 1992. Wind resistance categorization of buildings for insurance. *Journal of Wind Engineering and Industrial Aerodynamics*, **43**, 2617-2628.

Merrill, R.T, 1985. *Environmental Influences on Hurricane Intensification*, Colorado State University Department of Atmospheric Science, Paper No. 394, Fort Collins, CO.

Merrill, R.T, 1987. An experiment in statistical prediction of tropical cyclone intensity change, *NOAA Technical Memorandum NWS NHC-34*.

Morgenthaler, E. 1994. He's No Blowhard: Dr. Gray can Predict Atlantic Hurricanes. *The Wall Street Journal* August 8, p. A1.

Mulady, J.J., 1994. Building codes: They're not just hot air. *Natural Hazards Observer*, **18**, 3, 4-5.

Murphy, A.H., 1994. Assessing the economic value of weather forecasts: An overview of methods, results, and issues. *Meteorological Applications*, **1**, 69-73.

Murphy, A.H., 1993. What is a good forecast? An essay on the nature of goodness in weather forecasting. *Weather and Forecasting*, **8**, 281-293.

Namias, J., 1980. The art and science and long-range forecasting. *EOS: Transactions of the American Geophysical Society*, **61**, 449-450.

National Academy of Sciences, 1980. *Technological and Scientific Opportunities for Improved Weather and Hydrological Services in the Coming Decade*. National Academy Press, Washington, DC.

National Research Council, 1989. *Opportunities to Improve Marine Forecasting*. National Academy Press, Washington, DC.

Nature, 1995. The Year of Utilitarian Science? **373**, 1-2.

New York Times, 1994. In Florida Storm Hits Crops Hard. *The New York Times*, November 20, p. A10.

Noonan, B., 1993. Catastrophes: The New Math. *Best's Review* **83**, 41-44.

Novlan, D.J. and Gray, W.M., 1974. Hurricane-spawned tornados. *Monthly Weather Review*, **102**, 476-488.

Pielke, R.A., 1990. *The Hurricane*, Routledge, London.

Pielke, R.A. Jr., 1995. Hurricane Andrew: Mesoscale Weather and Societal Responses, Environmental and Societal Impacts Group, National Center for Atmospheric Research, June 1.

Pielke, R.A. Jr., 1994. Scientific information and global change policymaking. *Climatic Change,* **28**, 315-319.

Powell, M.D., 1982. The transition of the Hurricane Frederic boundary-layer wind field from the open Gulf of Mexico to landfall. *Monthly Weather Review,* **110**, 1912-1932.

Rensberger, B., 1995. The Cost of Curiosity. *The Washington Post Weekly Edition*, January 16-22, 11-12.

Salmon, J. and Henningson, D., 1987. Prior Planning for post-hurricane reconstruction. Report Number 88, Florida Sea Grant College, January.

Shapiro, L.J., 1989. The relationship of the quasi-biennial oscillation to Atlantic tropical storm activity. *Monthly Weather Review,* **117**, 1545-1552.

Sheets, R., 1990. "The National Hurricane Center -- past, present, and future." *Weather and Forecasting,* **5**, 185-232.

Sheets, R., 1992. Statement before Hearing of the Senate Committee on Banking, Housing, and Urban Affairs, 1992. The National Flood Insurance Reform Act of 1992, before the Subcommittee on Housing and Urban Affairs, 27 July, 102-913.

Sheets, R., 1993. Statement before Hearing of the House Committee on Science, Space, and Technology, 1993. NOAA'S Response to Weather Hazards -- Has Nature Gone Mad? Hearing before the Subcommittee on Space, 14 September, 103-55.

Simpson, R.H. and Lawrence, M.B., 1971. Atlantic hurricane frequencies along the U.S. coastline. *NOAA Technical Memorandum NWS SR-58.*

Simpson, R.H. and Riehl, H., 1981. *The Hurricane and its Impact.* LSU Press, Baton Rouge, LA.

Slovic, P., 1993. Perceived risk, trust, and democracy. *Risk Analysis,* **13**, 675-682.

Stewart, T.R., 1995. Forecast value: descriptive studies. In: Katz, R.W. and Murphy, A.H. (eds.) *Value of Weather and Climate Forecasts.* Cambridge University Press, Cambridge.

Sugg, A.L., 1967. Economic aspects of hurricanes. *Monthly Weather Review,* **95**, 143-146.

Sugg, A.L., 1968. Beneficial aspects of the tropical cyclones. *Journal of Applied Meteorology* **7,** 39-45.

Tennekes, H., 1990. A sideways look at climate research. *Weather,* **45**, 67-68.

Torgerson, D., 1985. Contextual orientation in policy analysis: The contribution of Harold D. Lasswell. *Policy Sciences,* **18**, 241-261.

U.S. Army Corps of Engineers/Federal Emergency Management Agency, 1993. Hurricane Andrew Assessment: Louisiana Version, Review of Hurricane Evacuation Studies Utilization and Information Dissemination, (July).

U.S. Department of Commerce, 1990. Hurricane Hugo September 10-22, 1989, Natural Disaster Survey Report, (May).

U.S. Department of Commerce, 1993. Hurricane Andrew: South Florida and Louisiana August 23-26, 1992, Natural Disaster Survey Report, (November).

USWRP, 1994. United States Weather Research Program: Implementation Plan, Congressional Submission, Committee on Earth and Environmental Sciences, (January).

Wakimoto, R.M. and Black, P.G., 1994. Damage survey of Hurricane Andrew and its relationship to the eyewall. *Bulletin of the American Meteorological Society,* **75**, 189-202.

Wamsted, D., 1993. Insurance industry wakes up to threat of global climate change. *Environment Week,* **6**, 39.

Wilson, N.C., 1994. Surge of hurricanes and floods perturbs insurance industry. *Journal of Meteorology,* **19**, 3-7.

Winkler, R.L. and Murphy, A.H., 1985. Decision analysis. In: Murphy, A.H. and Katz, R.W. (eds.) *Probability, Statistics, and Decisionmaking in the Atmospheric Sciences*, Westview Press, Boulder, Colorado, 493-524.

9 The Political Ecology of Vulnerability to Hurricane-Related Hazards

Roger S. Pulwarty[1] and William E. Riebsame[2]

[1]Cooperative Institute for Research in Environmental Sciences
Campus Box 449
University of Colorado, Boulder, CO 80309, U.S.A.

[2]Department of Geography
Campus Box 260
University of Colorado, Boulder, CO 80302

Abstract

Differences in research emphases into the impacts and management of natural hazards have, for the most part, resulted from two fundamentally different views environment-society relationships. In one view, a self-correcting, homeostatic process operates, in which society and environmental hazards are inexorably adjusted, toward some acceptable equilibrium. This view infers that people and institutions are committed to removing known risks from life and fail to do so only where the risk is highly uncertain. Proponents of a second view argue that vulnerability is constructed from an open-ended development process that determines the ways in which a hazard is likely to constitute a disaster. The fact that aggregate economic losses to hurricane impacts are increasing is well documented. Cases, taken from the hurricane-hazard and recent development experience in the Greater Caribbean Basin, illustrate that the progression of vulnerability cannot be addressed outside of the context of social and economic trends that play significant roles in determining "Who is vulnerable and why". An appeal is made here, for a more evenhanded (or objective) consideration of the factors that give rise to vulnerability. Political ecology provides a framework for understanding and integrating the social construction of vulnerability and climatic risk. It reflects the confluence of political economy and human ecology and refers to ways in which vulnerability is rooted in people, values, institutions, and the environment. We review the definition and study of vulnerability from these perspectives, drawing on insights and lessons developed by researchers over the past twenty-five years. These lessons are especially immediate as programs for hazard reduction and sustainable development, such as the International Decade for Natural Disaster Reduction, are being designed, implemented, and evaluated.

9.1. Introduction

Natural hazards are those elements in the physical environment which are harmful to humans and caused by forces extraneous to them (Burton and Kates 1964). Hazards (Blaikie et al. 1994) refer to the extreme natural events and sequences of events which may affect different places singly or in combination (e.g. coastlines, hillsides) at different times depending on season, varying return times, and duration. Over the past 25 years several researchers have argued that there may be an overemphasis on earth and atmospheric systems as the immediate initiating agents in the study of disasters, at the expense of examining the sociocultural (including political and economic) conditioning factors. From the point of view of actual research and funding, Morren (1983) has noted that only superficial attention is paid to the multitude of basic human problems that may accompany, interact with, or be initiated by an event, such as loss of livelihood, and reserves. He argues that the problems are recognized, but (a) are not studied with the same degree of effort as prediction, (b) little attention is paid to the ways in which ordinary people either as individuals or members of small groups undertake to cope with these and related problems, and (c) emphasis is commonly placed on technological solutions, resulting in a seemingly misleading picture of the costs, benefits, appropriateness, and effectiveness of big responses that may disrupt networks of interpersonal bonds.

Differences in research emphases have for the most part resulted from a split between two basic views of hazard-society relationships. In one view, a self-correcting, homeostatic process operates in which society and environmental hazards are inexorably adjusted toward some acceptable equilibrium (Burton et al. 1993). Other hazard researchers feel that the cumulative process of social development inevitably leads to greater vulnerability and growing potential for catastrophe; a view encapsulated in White's (1985) "perilously changing world", and Perrow's (1984) "normal accidents." Analysts subscribing to this view find little evidence of dampening feedbacks or an emerging equilibrium between social development and hazards. Rather, they envision an open-ended process leading to progressive, rather than lessened, hazardousness.

The two perspectives reflect very different views of the world and of the relationships between society, technology, and environment. Dichotomies exist between the technocentric school, to whom the solution to the disaster problem lies in the application of measuring and monitoring techniques and sophisticated structural management strategies, and the anthropological and sociological approaches which point out that vulnerability and impacts should be considered in terms of patterns of human behavior and the effects of disasters upon community functions and organizations, marginalization etc. (see for example, Maskrey, 1989). This results in a basic dichotomy between the appropriateness of structural vs. nonstructural responses to hazards. The former involve engineering methods and architectural design, while the latter includes land-use planning, risk assessment and management, and insurance.

Hewitt (1983a) has argued that vulnerability and human social organization are the critical determinants of both risk and impacts. The former approach usually assumes that as prediction of nature improves, society automatically follows with adjustments that remove risks to that level. It also infers that persons and institutions are uniformly and unambiguously committed to removing known, manageable risks from life and fail to do so only where the risk is highly uncertain (Hewitt 1983b). The restoration of productivity and re-imposing of "normal" relations become the main prescriptions of crisis relief and reconstruction. Susman et al. (1983) argue that hazardousness and vulnerability (are) increasing due to human changes, and "in the largest fraction of the world's inhabited area these changes are clearly bound up with 'development' or its failure" (p. 267).

The issue, considered in this chapter, is not simply one of reducing risk to its "positivist" or to its "cultural" constructions. Instead an appeal is made for a more evenhanded (or objective) consideration of the conditioning factors involved and their attendant uncertainties and interactions, when evaluating evidence and procedures for implementing hazard management policies (see Shrader-Frechette 1991). Political ecology as used in the title of this paper reflects the confluence of political economy and human ecology and refers to ways in which "relations of production shape the management of resources and the resiliency of groups to shocks and perturbations" (Bohle et al. 1994). These "relations of production" are located within the nexus of technical, social, political, cultural, ecological, and academic relations (Yapa 1996). In this chapter we explore the implications of this perspective for investigating local and global vulnerability to hazards, as well as for the evolution of descriptive models of environment-society interaction. The case of hurricane-related hazards in the Greater Caribbean region is used as an illustration.

The chapter is structured in three major parts. First, a review of the study of vulnerability from the human ecology perspective is provided. This traces the development and use of the human ecology model in hazards research. We argue that the advances brought by this perspective have been "short-circuited" in hazards management by a focus on structural and short-term adjustment strategies. We go on to argue that the adjustments or feedbacks presumed in the formulation of this equilibrium-centered approach are not occurring. The second component illustrates the shortcomings of applying the preparedness, response, recovery frame as a comprehensive explanatory model for vulnerability and risk management. Examples are drawn from the interactions of hurricane-hazard and recent development experience in the Greater Caribbean Basin. Lessons from these examples illustrate that the nature and progression of vulnerability cannot be understood without reference to historical, geographical and value contexts. Finally, recent advances for understanding, integrating and studying the social construction of vulnerability and the nature of climatic risk from the perspective of political ecology are discussed. The chapter is intended to synthesize insights on vulnerability and its assessment over the past 25 years by Vayda and McCay (1975),

Hewitt (1983), Ascher and Healy (1990), Berke et al. (1993), Blaikie et al. (1994), Bohle et al. (1994), Bender (1994), and White (1994), among others. It is particularly addressed to those researchers whose fields of study may not put them in direct contact with this literature. We discuss how well vulnerability trends have been and can be determined, and explore prospects for understanding and reducing vulnerability to hazards in late 20th century. The development planning field can potentially make an important contribution by suggesting factors that foster institutional capacity for undertaking adaptive learning before disaster occurs (Berke et al. 1993; Gunderson et al. 1995). It would be unfortunate if risk and vulnerability, which in the real world have co-evolutionary, complex links with the overall development process, are not placed within a larger theoretical context. A major goal here is to synthesize information that would be useful in designing external programs aimed at fostering, reinforcing, and supplementing individual and local initiatives in an effective way. The question: "Is hazardousness lessening or worsening?" is addressed in the context of physical, social, and economic trends that play significant roles in determining "Who is vulnerable and why."

9.2 Understanding Vulnerability: The Human Ecology Perspective

Differing views as to what constitutes "vulnerability" have had profound influence on the study of the impacts of natural hazards and the choice of response strategies. The "causes" of vulnerability can be categorized into three streams with specific histories of study and responses. These are summarized from Anderson (1995) as follows:

(1) Nature as cause: In this context, vulnerability is understood as the relationship between hazards (natural events, including strength, magnitude, and duration) and risk (exposure to the events measured essentially in terms of proximity). Thus, scientists, technologists, and engineers have responded through attempts to predict natural hazard events and to develop technologies that enable human structures and systems to withstand their impacts. This approach has brought significant success over time through predicting and tracking storms, mapping hazard proneness, etc.

(2) Costs as cause: Economically rational criteria are used to decide how much vulnerability reduction is rational, and then which vulnerability reduction technologies can and should be used and under what circumstances. It is presumed that if the elimination of vulnerability were free, then societies would reduce all risks to zero and that an accurate assessment of vulnerability can be made.

(3) Humans as cause: This study context is usually pursued by social scientists, policy reformers, advocates for the poor, and environmentalists. This area is the least understood of the three and will be expanded on below.

Research emphases have been primarily placed on the first two of these "causes." Although our knowledge of physical causal mechanisms is incomplete, we propose that in evaluating disaster risk, the social production of vulnerability needs to be considered with at least the same degree of importance that is devoted to understanding and addressing the physical component of natural hazards. Knowledge of the physical environment alone may be of limited usefulness for calculating the actual level of society-at-risk or understanding the nature of uncertainty in planning (Hewitt 1983a)

An origin of the human ecology perspective can be found in the ideas of John Dewey (1899). Dewey proposed that humanity exists in a hazardous natural world which results in human insecurity. This has led to an appreciation of the fact that the impacts of environmentally initiated hazards are not completely independent of society, since these perils are defined, reshaped, and redirected by human action (Mileti et al. 1995). Human ecology (Barrows 1923; Kates 1971, 1985) was seen as a way of framing how society through organizations, households, resource allocation decisions experiences the environment in terms of risks and threats, e.g. hurricane variability, flood, heat waves. The human ecology paradigm grew out of attempts by geographers to explain the continuing failure of U.S. flood control policies in the 1930s to curb losses despite heavy investments. Flood probabilities and structural engineering were regarded as the principal factors affecting a hazard while the attributes of human exposure and vulnerability were neglected (Mileti et al. 1995).

Early natural hazards research was founded on a systems model of hazard-society interaction laid out by Kates (1971). The model (Fig 9.1) describes an equilibrium-seeking process in which vulnerability and impacts are continually dampened via feedbacks called "adjustments". Mechanisms of these adjustments include technologies and social innovations, such as drought-resistant crop varieties, land-use changes, and risk-spreading and loss-sharing programs such as insurance and disaster relief. Essentially, the model is an optimistic, normative statement of nature-society-technology relationships. Societal adjustment, from this perspective, was envisioned in a cyclical framework made up of three major stages: preparedness, response, and recovery. Haas et al. (1977) for instance, envisioned an ordered progression of re-development after a natural hazard: (1) emergency responses; (2) restoration of public services, (3) replacement or reconstruction of capital stock to pre-disaster levels, (4) initiation of developmental reconstruction involving economic growth and local development (see Berke et al. 1993b).

In its original formulation, the human-ecological model's main feedback, or adjustment lines, were drawn to illustrate feedbacks to both the social system and natural system "boxes." Although many adjustments are aimed at reducing the physical threat itself (such as cloud seeding to fight droughts), hazards research as practiced by social geographers and other social scientists has focused on human adjustments that reduce underlying social vulnerability, such as land use planning and improved building codes. This adjustment process has examples in floodplain management programs and in the implementation of seismic design in earthquake-prone regions.

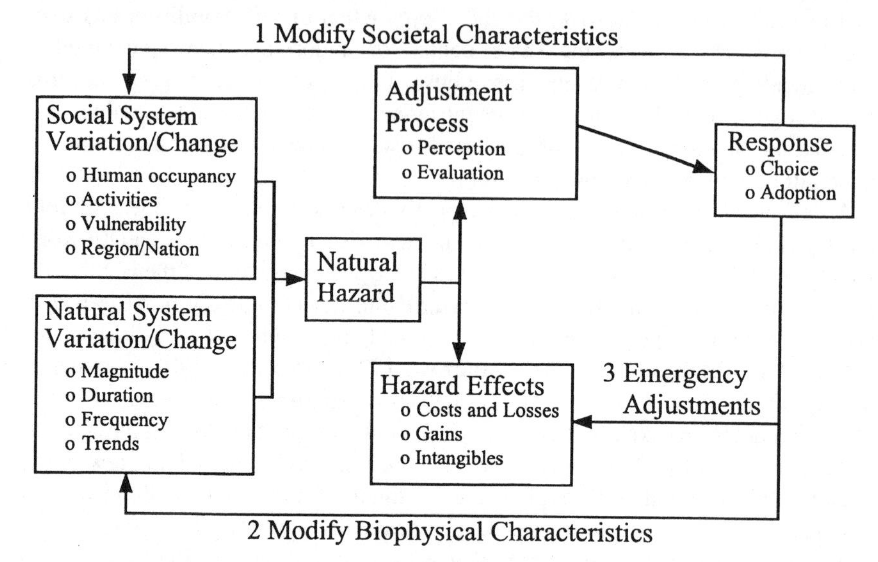

Fig. 9.1. Human-Ecology Interaction model stressing feedback to underlying physical and social processes and structures. (Adapted from Kates 1971, 1985)

9.2.1 A Short-circuiting of the Human Ecology Model

It can be argued that the social adjustment process envisioned in the human-ecological model has been "short-circuited". The great emphasis of natural and technological hazard management is not on adjustments that feed back to underlying social and natural systems, but rather on those that intervene between hazards and their first-order impacts with warnings, evacuation, emergency management, and relief (see, for example, Drabek 1986). This adjustment process is illustrated in the "emergency response" loop in Kates's model (Fig. 9.1), a process described in detail by Drabek (1987). Though Kates may have relegated emergency response to a relatively minor place in his human ecological schematic, it actually has attracted the greatest effort by hazard managers over the past two decades. Thus, hazard managers have given more attention to sophisticated evacuation planning than to land-use restrictions, to carefully maintained insurance premium pools instead of resources for hazard-proofing existing structures, and sophisticated detection and prediction systems rather than to public awareness and self-help campaigns.

While the actions of hazards managers, and broader social developments, may be short-circuiting hazards ecology, recent developments of hazard theory suggest, perhaps unwittingly, an unraveling of the ecological framework itself. This occurred as researchers reformulated the hazard-society model while shifting their focus to technological hazards (Hohenemser et al. 1983). Rather than simply appropriating the human-ecological model with its homeostatic structure, they constructed a new model based on a "causal sequence" of unidirectional, cumulative social development in which human needs and desires lead to new or worsened hazards (Fig. 9.2). While we do not suggest that the authors of the causal-sequence model explicitly rejected a homeostatic dampening of hazards, the model's structure implies that dampening feedbacks, in the form of risk assessment and management interventions, are addenda necessitated by an underlying process, not part of the process of social development itself. Indeed, Vayda and McCay (1975) have argued that the "equilibrium centered" point of view relies on "negative" feedback processes to maintain a balance between human populations and their environments and ignores "unbalanced" relationships and system disruptions resulting from society-environment interactions. In the short-circuited hazards ecology, some impacts are reduced--by moving people out of the way at the last minute and by widely sharing the economic burden of relief and recovery--but little is done to affect the primary cause of hazard losses: the juxtaposition of human development and hazard zone (Petak and Atkisson 1982; Myers and White 1993). The potential for great losses and successful recovery from especially severe events is not reduced. Indeed, the blocking of immediate effects unaccompanied by reduced underlying vulnerability, may lead to larger impacts if a false sense of sustainability develops in hazard zones. There is a direct relationship between development (including social change and modernization) and hazard vulnerability. This exposes the need to assess more closely the adaptations of people at the individual and community levels and to emphasize socially constructed components of hazardousness.

Adaptation, in varying degrees, is key to the human ecology of natural disasters. Four forms of adaptation have been described in the hazards literature (Alexander 1991): (1) persistent occupation of the hazard zone despite the risk involved: (a) with comprehensive measures for risk mitigation and hazard abatement, (b) with only warning and evacuation measures, or (c) without any protection measures (the state of maximum vulnerability); (2) cohabitation with the damage caused by past disasters; (3) abandoning damaged or destroyed structures but relocating within the risk zone; (4) migration to safer zones (a) planned or (b) unplanned. While these have been identified for a number of years (see Burton et al. 1993), we are still bound by the question raised by White et al. (1958): "Why are certain adjustments to the risk of floods preferred over others, and why, despite investment in those adjustments are societal losses increasing?" (see also Mileti et al. 1995).

The interpretive difference, described above, reflects fundamental beliefs about human development and the effectiveness of social response to environmental challenges. It balances on the question of whether and how risks are multiplied as society and technology develop, or whether increasingly sophisticated societal mechanisms lessen hazards or at least maintain an equilibrium between hazardousness and development. There is often an exaggerated optimism that good planning and design will mitigate problems without regard for other concerns.

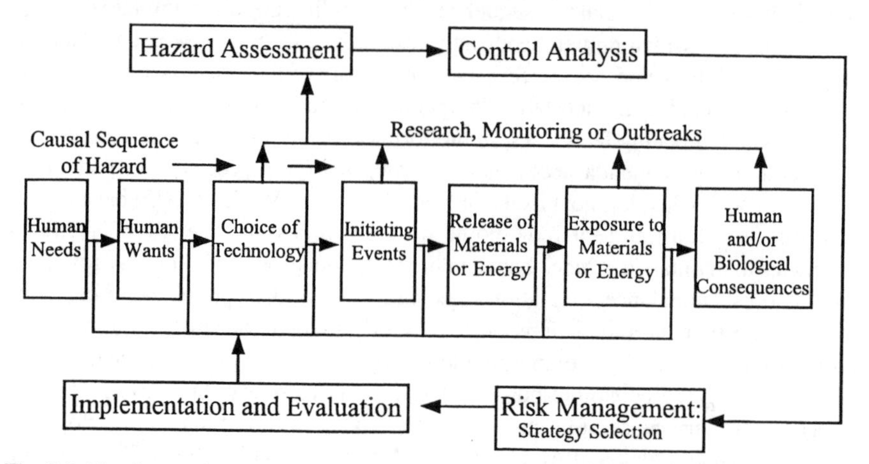

Fig. 9.2. The Causal-Chain model of technological hazards (adapted from Hohenemser et al. 1983).

9.3 The Progression of Vulnerability and the Management of Risk

Vulnerability is usefully described as the characteristics of a person or group in terms of their capacity to anticipate, cope with, resist, and recover from the impact of a natural hazard (Blaikie et al. 1994). It is "an aggregate measure of human welfare that integrates environmental, social, economic, and political exposure to a range of potential harmful perturbations" (Bohle et al. 1994). According to Bohle et al. (1994) this definition suggests three basic coordinates of vulnerability:

a) The risk of exposure to crises, stress, and shocks;

b) The risk of inadequate capacities to cope with stress, crises, and shocks;

c) The risk of severe consequences, and the attendant risk, of slow or limited recovery from surprise.

The prescriptive and normative response to vulnerability is to reduce exposure, enhance coping capacity, strengthen recovery potentiality, and bolster damage control (minimize destructive consequences) via private and public means. Social, economic and political processes are themselves often modified by a disaster in ways that make some people more vulnerable to extreme or cumulative events in the future. In addition to income there are many other factors that conjunctively influence whether an individual or group will continue to be vulnerable. For instance, infrequency of catastrophic events can result in the attitude that "it won't happen to me", decreasing the willingness to expend resources on mitigation measures (M.F. Myers, pers. comm.). Usually, the most vulnerable groups are those that find it hardest to reconstruct their livelihoods following disaster and as a result are more vulnerable to the effects of subsequent or more frequent hazard events. Livelihood, as described by Bohle and Watts (1987), indicates the command an individual, family, or other social groups has over an income and/or resources that can be used or exchanged to satisfy its needs. It includes mixes of food sources derived from production, labor or market exchanges, donations or relief (Sen 1981). This may involve information, cultural knowledge, social networks, legal rights as well as tools, land, or other physical resources, i.e. access to opportunities. For example, the urban poor use their location as the base around which activities are organized (street-trading, crafts, etc.). If the structure of urban land-ownership and rent means that access to economic opportunities requires living in hillside slums, people will locate there almost independent of landslide risk (Hardoy and Satterthwaite 1989; Bender, 1993). In this situation neither the notion of "voluntary choice" nor of "bounded rationality" is applicable.

Defining a disaster as the impact of a natural hazard upon vulnerable people, Blaikie et al. (1994) and Cannon (1994) construct a valuable model illustrating the relationship "pressures" between hazard and vulnerability in the causation of disaster and correspondingly identify areas of emphasis for study (Table 9.1). The framework demonstrates the interrelationships among economic production, natural resources, the environment, and the distribution of income and wealth. It provides a means of developing profiles that highlight population and social infrastructure vunerability to hazards. It also illustrates that changes in the frequency of events do not have to occur for vulnerability to increase.

This model has direct links to the concepts of vertical and horizontal integration of institutional networks discussed by Berke et al. (1993). Bureaucracies are designed for vertical communication yet the coordination of activities of several agencies obviously calls for horizontal communication. Examples include the networks between which households can define and communicate their needs and the processes by which organizations make decisions on aid distribution. These issues are not as well understood as shorter term economic impacts. Here the components of self-protection and social protection include building quality, location of home, work, disruption of networks, and the capacity to act (Cannon 1994).

In the next section the recent history of hurricane-society interactions in the Caribbean is discussed. Examples are chosen to illustrate how human-induced factors can coincide with extreme natural events to result in disaster.

Table 9.1. The relationships between hazard and vulnerability in the causation of disasters as rooted in environment, people, values, and institutions. (adapted from Blaikie et al. 1994 and Cannon 1994).

Hazard	Disaster	Vulnerability (measure of groups level of)	Socio-Economic and Political Economy	National and International Policy Economy
Hurricane:		PREPAREDNESS	CLASS	Generation and allocation of surplus
Return period			Income	
Duration		Self protection	Distribution	Social power and control
Magnitude	D	(location, building quality)	Livelihood	
			Opportunity	Civil security (war)
Flood	I	Social protection		Debt crises
		(building regulations	GENDER	Environmental degradation
Drought	S	level of scientific knowledge,	Household	
			Security	Floodplain occupance
Earthquakes	A	technical interventions	Nutrition	Deforestation
		and ability)		Landuse
Volcanic activity	S			
		RESILIENCE	ETHNICITY	STATE
	T		Income	
		Strength of livelihood	Assets	Institutional support
	E	Assets, income, savings	Discrimination	Regional/local
Human modification	R	Recovery of livelihood		Biases; Training
		HEALTH		
		Social precautions		
		Individual robustness		

9.4 Society and Hurricane-Hazard Interactions in the Greater Caribbean Region

From 1960 to 1989, hurricanes in the Greater Caribbean Basin resulted in the deaths of 28,000 people, disrupted the lives of 6 million people and destroyed property worth U.S. $16 billion (OAS, 1991). In 1930, three people were affected by hurricanes for each person killed. By 1989 the ratio was 100,000:1 (OFDA, 1990). Property costs rose from $5000 per person killed to $20 million per person killed. Reduced death rates, where they occur, are due almost entirely to improved warning systems and preparation. However, the results of hurricane impacts, including the ability to recover, can be said to be "pre-figured" by developments in resource-use and local to regional political economy (Hewitt 1983a; Dove and Khan 1995). Griggs and Gilchrist (1983) in reviewing high risk situations on the Gulf and Atlantic coasts of the U.S. concluded that "people want to live in the sun and be able to look at the ocean; realtors and developers want to make money, and local governments want more tax dollars." Indeed, demographic and economic pressures have transformed the sparsely developed eastern shorelines of the U.S. of the 1940s and 1950s into "cities on the beach" (Platt et al. 1991). In addition, as Platt et al. observe, "For every "volunteer" resident in a high risk coastal location ("sun and surf") there are thousands who have no alternatives because their livelihoods are tied to jobs in oil refineries, export enclaves, tourism, fishing boats, coastal farms and plantations."

Three factors are usually identified for study in assessments of the hurricane hazard: (1) weather generators specifying the frequency, magnitude, and other characteristics such as storm surge and windspeed; (2) the population-at-risk (exposure) specifying the number and geographic distribution of people and buildings in areas prone to hurricane impact; and (3) vulnerability, commonly specified as the susceptibility of the population-at-risk to injury or damage when an event of a given severity occurs (Friedman 1977, among others). The number of people in a particular place indicates the degree of exposure, and the likelihood of a particular event in particular location represents probability of occurrence. Exposure, as defined above, was identified as early as 1964 (Burton and Kates 1964) and in numerous publications by Gilbert White, and others (see Jarrell et al. 1994; Berke et al. 1993; Pielke and Pielke 1996, Chap. 8, this volume). Although there is little evidence to suggest an acceleration in the recurrence of major geophysical triggering mechanisms (earthquakes, volcanic eruptions, hurricanes) in Latin America, empirical data indicate an increase in recent decades in the number and spatial coverage of less dramatic events related in particular to flooding, landslides, and drought (Lavell 1994). Lavell notes that the lack of an adequate system for registering the economic impact of the multiplicity of smaller scale "disasters" in Central America prohibits a calculation of their cumulative impact on conventional indicators of (growth of GNP, international debt, etc.) of the types usually attempted. It has been suggested that the medium- and long-term effects of these lower range events could be equivalent to "one or more large-scale disasters" (Abril Ojeda 1982). We now turn to three examples, with a detailed focus on Jamaica, to illustrate differing conjunctive events where hurricanes have exacerbated already problematic situations.

9.4.1 Honduras: Impact of Hurricane Fifi 1974

In 1974, Hurricane Fifi struck Honduras killing 8000 people, leaving over 100,000 homeless, and resulting in losses of $540 million in production and infrastructure (OAS, 1991). Fiscal deficits reached 79%, with a 66% decline in exports and an attendant 61% increase in imports (ECLAC, 1974). As Susman et al. (1983) show, the most important factor in the destruction caused by Hurricane Fifi was the deforestation beginning in earnest in the mid-1960s. Around 1956, banana companies changed from labor-intensive to capital intensive production. United State-owned banana companies extensively developed the rich fertile valleys in northern Honduras around the San Pedro de Sula Valley, and increasingly cleared forest for banana plantations. Land on the valley bottoms were either owned by banana companies, such as United Brands and Standard Fruit, or used for large-scale irrigation, resulting in subsistence and extractive living on hillslopes. Campesinos moved to this cheaper land where forest was cleared to grow maize, increasing soil erosion and siltation of rivers. One result was that, in the town of Choloma, 2300 people were killed when a dam created by landslides into a nearby river gave way.

Susman et al. (1983) contrasted the impact of this event with a storm of similar magnitude, Hurricane Tracy (1974), which killed 49 people in Darwin, Australia. Differences in the degree of impact appears to have resulted from the pattern of rural land ownership in each place. In Honduras, where 63% of the farmers had access to only 6% of the arable land, large-scale beef ranches and banana plantations had displaced peasants over several decades into isolated valleys and steep hillsides. Here they received little warning and were at risk from landslips, due to heavy rainfall. The Susman et al. study shows that it is possible for two hurricanes of the same duration and intensity to strike areas with similar population densities, and for one to be a disaster and the other to be less of a disruption to livelihoods and future well-being (with fewer deaths, etc.).

9.4.2 Dominica: Impact of Hurricanes David (1979) and Hugo (1989)

In 1979 Hurricane David stripped much of the cover of the small eastern-Caribbean island of Dominica (PASB, 1980). After a leaf-blight epidemic in 1980, international organizations chose to spend almost U.S. $2 million to prop up and privatize Dominica's banana industry (McAfee 1991). Banana production more than doubled from 1983 to 1987, then accounting for 71% of Dominica's export earnings. In 1987, a past Minister of Agriculture was quoted as saying (McAfee 1991):

"When you take into account the debts to 'revitalize' the industry, the fees paid to foreign consultants, the cost of chemical purchases from abroad, we suspect that the cost to us of growing these bananas has been much higher than what we've earned... We've become a much more fragile and vulnerable economy before the so-called banana revitalization, as the next strong gust of wind will demonstrate."

Indeed, by 1989 the country's debt had doubled to almost 60% of GDP. In September 1989, Hurricane Hugo struck Dominica, destroying 75% of the island's banana plantations. Replanting was again undertaken. In 1995, Dominica lost 90% of its banana crop to Hurricane Luis and the remaining 10% to Hurricane Marilyn, one week later. Once again, the danger of the country's dependence on bananas for the bulk of its export earnings was demonstrated.

9.4.3 Jamaica: Before, During and After Hurricane Gilbert (1988)

Hurricane Gilbert (1988), the most powerful storm recorded in the Western Hemisphere (888 mb) traveled directly over Kingston, Jamaica, and across the full length of that island. Prior to this, Jamaica had been highly active in promoting evacuation and sheltering and hence there were only 49 fatalities including 11 people shot while presumed looting. Damage was estimated at U.S. $1 billion (Barker and Miller 1990). Jamaica lost 30% of its sugar hectarage, 54% of its coffee, and more than 90% of its bananas and cocoa. Up to 75% of all buildings sustained some damage (Eyre 1989), with one house in four and 500 of 580 schools being significantly affected. Half a million people were evacuated for a matter of hours but up to 50,000 were displaced from weeks to months (Eyre 1989). Hurricane Gilbert deprived Jamaica of more than U.S. $27 million in foreign exports in 1988-89 alone. The hurricane, with central pressure of 960 mb in Jamaica and 31 km/h forward speed, however, generated relatively minimal storm surge and little serious flooding but released strong rains at high elevation (OAS 1990). This can be contrasted with subsequent conditions induced by Gilbert in Yucatan and northern Mexico where 5- to 6-m surges occurred with sustained winds up to 290 km/h.

In terms of foreign exchange earnings, the most serious losses occurred in the banana and coffee industries. Over 70% of the coffee grown in the famous Blue Mountains was destroyed. In the previous five years several new large-scale coffee projects utilizing steep slopes, with extensive removal of tree cover, had been established (Barker and Miller 1990). Rainfall-triggered land sliding occurred on an unprecedented scale in the days and weeks following the hurricane, chiefly in the form of rapid shallow translational slides and debris flows in weathered materials. One underlying cause of disaster was failure of structures. Direct damage to roofs alone were estimated to be $2 million (Jam.). Estimates of external reinsurance transfers to Jamaica were about $370 million, but as an indicator of costs does not include the 70% of private buildings not covered by any insurance nor the devastated public sector infrastructure (Eyre 1989). Building codes existed but the parameters specified were not realistic in engineering terms and with respect to the practical cost and organizational difficulty of implementing them (Berke et al. 1993a). The major impact has however been in the area of recovery. The nature and degree of impact is tied intimately to the direct hurricane event, to national development decisions, the structure of community power and influence, and to the unfulfilled or misdirected promises of the Caribbean Basin Initiative earlier in the decade.

Polanyi-Leavitt (1991) in a detailed study of the impact of structural adjustment programs in Jamaica, describes the social effects of adjustment as related to poverty, malnutrition, equity, impact on women and children, health and education sectors, housing, transportation and the erosion of traditional values and social norms. The government's own programs for preparedness and mitigation were reduced. It was clear that Hurricane Gilbert made already existing critical problems disastrous. On a per capita basis, GDP in Jamaica had dropped 14% and private consumption by 3% in the 1979-1985 period. In 1985, in the context of low market prices for bauxite and alumina, Kaiser/Reynolds and Alcoa shut down operations in Jamaica. By 1986 the severe adjustments of the mid-1980s had reduced overall public sector deficit from previous levels of 15 to 20% of the GDP to 5.6% and economic growth had resumed at 1.8% (Polanyi-Leavitt 1991). However, the costs of purchasing the least-cost basket of minimum food requirements for a five person household, as determined by the Jamaican Ministry of Health, increased by 429% over the six-year period June 1979 to July 1985. Capital expenditure on housing in 1985/86 had fallen to 11% of that of 1982/83, and real expenditure in social services was cut by 41% over the ten-year period 1975/76 to 1985/86 (Girvan et al. 1990). From 1981 to 1985 total government per capita spending on health declined by 33%. In 1987 Jamaica registered real (aggregate) growth of 5.7% and the national economy appeared, briefly, to have begun stabilization (Robinson 1994). However, in early 1988, the then World Bank Senior Economist for Jamaica stated that "Jamaica's social and economic infrastructure is worse than it was in the 1970s" (McAfee 1991). By U.S. Agency USAID's accounting, Jamaica had "a crippling debt burden with an economic output far below the production level of 1972. Distribution of wealth and income was highly unequal, with shortages of key medical and technical personnel, severe deficits in infrastructure and housing, and where physical decay and social violence deter investment" (USAID 1989, cited in McAfee 1991). This assessment was made six months before Hurricane Gilbert made landfall.

Ironically, this reduction in resilience and capacity was developing at the very time that Jamaica was receiving large inputs of foreign aid through the United States Caribbean Basin Initiative (CBI). In particular, Jamaica received extensive U.S. foreign assistance during the Seaga administration, which came into power in 1980 and was designated the "showpiece" of the CBI (Hewan 1994). From 1980 to 1982 aid from the U.S. increased from about U.S. $14 million to $140 million. By 1987 Jamaica ranked 17th out of 115 recipients and within the top twenty recipients of U.S. aid during the administration's seven years in power. Military assistance went from U.S. $128,000 in 1980 to $2.2 million in 1987 (U.S. Dept. of Defense 1989). At the same time sugar import quotas to the U.S. were cut by 46% (Hewan 1994). In an address to Congress the then President of the United States noted that "a key principle of the program is to encourage a more productive, competitive and dynamic private sector, and therefore provide jobs, goods and services which the people of the Basin need for a better life for themselves and their children. All the elements of this program are designed to help establish conditions under which a free competitive private sector can flourish" (Office of the Press Secretary 1982). In one of the earliest analyses of the CBI, Zorn and Mayerson (1983) pointed out that the private sector which the CBI in fact supported was that based in the United States.

Jamaica has been bound for a decade and a half by IMF/World Bank Programs and has met its debt obligations to the multilaterals at the expense of social services and the erosion of future capacity of self-reliant development (Polanyi-Leavitt 1991). The fiscal squeeze is shown by the decline in education and health services expenditures from around 25% during the 1970s of public expenditures to 17% in 1987/88. Household deteriorations from the adjustments of 1983-1985 (removal of food subsidies, devaluations, increased taxes) resulted in the Seaga government introducing a Food Security Programme supplied by imports from USAID, EEC, and the World Food Programme. This still amounted to only 10% of what was actually required to service the number of people living in poverty. Interestingly, part of the foreign debt incurred to launch structural adjustment was due to loans to pay for previous hurricane damage, particularly for Hurricane Allen (1980).

Although there were portentous signs in the first half of 1988/89 with rising inflation and loss of international reserves, the hurricane of September 1988 was a watershed event (Polanyi-Leavitt 1991). It further dislocated the economy and precipitated another series of adjustment measures which are still continuing. Economic growth that year was just 1.6%. The local organizational capacity to carry out recovery initiatives was severely undermined and resulted in dependence on direct government aid which in many cases was perceived as politically partisan. Berke et al. (1993a) found that though parts of two communities in Jamaica were totally rebuilt after Hurricane Gilbert, households in other locations that had sustained similar damage were, months later, still clearing debris from the streets and had only made temporary roof repairs.

The recent developmental history of Jamaica points to both increasing vulnerability and decreasing resilience to future hurricane impacts. The annual deforestation rate is estimated to be 3.3% (Eyre 1990) with a rate of 7.25% for the 1988-93 period (World Bank 1995). Girvan (1991) notes that from 1980-1990 there was a net conversion of 1315 km^2 of forest to other uses, with 3000 km^2 seriously degraded and with 400 million tons of soil lost from surface watersheds to coastlines. Since 1950, 50 named rivers have lost perennial flow. Urbanization associated with resort areas, particularly on the north coast, places great stress on primary infrastructure of housing, potable water, sewage disposal and other services, and on the regional carrying capacity itself. On the southern end, Kingston Harbor, reputedly the sixth best natural harbor in the world, is fast becoming "a giant septic pit" (Girvan 1991). This is as a consequence of inadequately treated sewage, drainage from rivers polluted by garbage dumps, topsoil, and garbage washed down by heavy rains. The eastern portion of the harbor is now biologically dead and the rest is dying. The harbor used to be a source of pride and recreational pleasure to Kingstonians who now number over 800,000 in the Metropolitan area. Poverty-related illness are also increasing. For instance, the number of admissions for malnutrition and malnutrition-gastroenteritis of children less than five years old to Bustamante Children's Hospital has steadily increased from 1975-1985 (Polanyi-Leavitt 1991).

Girvan (1991) points to the occurrence in Jamaica of what he calls the "Haitian syndrome," a vicious cycle of deforestation-soil erosion-declining soil fertility-overcrowding-subdivisions of holdings-further deforestation. This is a direct consequence of inequality in land distribution together with neglect of land use or the undermining of planning and conservation practices at the level of the individual, the household and the state.

9.4.4 Prospects for Hurricane-Hazard Reduction in the Greater Caribbean

The overall record on mitigation of hurricane risk in the Caribbean and Latin America does not appear very encouraging. There are many cases of new investments in public or protective sectors that were exposed to significant hazard because of inappropriate design or location and even of projects that were rebuilt on the same site and in the same way after having been destroyed a first time (OAS 1991). Other cases can be cited of schools funded with bilateral aid that were built to design standards suitable for the donor country but incapable of resisting hurricane-strength winds characteristic of the recipient country. The tourism sector in the Caribbean has become notorious for its apparent disregard of the risk of hurricanes and associated hazard (OAS 1990).

Patterns of deaths and damage due to these storms and the ability of people to reconstruct their livelihoods show differences according to wealth, history and socio-political organization (Blaikie et al. 1994). Recovery following hurricane impacts in different parts of the Caribbean in 1988 and 1989 showed such contrasts. A revealing study by Berke and Wenger (1991) showed that after Hurricane Hugo (1989) struck the Caribbean island of Montserrat an international donor organization used the disaster as an opportunity to build capacity of a pre-existing community action group to undertake self-directed long term development initiatives. Nicaragua mobilized a nationwide effort to help victims of Hurricane Joan on the its Atlantic coast. In part, this offered an opportunity for the Sandinista government to build support among sections of the people who had not benefited substantially from the 1979 revolution, and in some cases had supported the opposition Contras.

By contrast, relief efforts in Jamaica following Hurricane Gilbert (1988) were rife with partisan politics and corruption, so much so that mismanagement was one of the factors that led to a change in the government in the elections that followed. In the U.S., although Rubin and Popkin (1990) identified problems in all phases of emergency management following Hurricane Hugo (1989), they observed that the major obstacle to relief and reconstruction on the South Carolina coast was bureaucratic blindness to the needs of poor (often illiterate) people who lacked insurance and other support systems. These were often African-Americans (Miller and Simile 1992). Two decades earlier when Hurricane Camille devastated the Mississippi delta, the U.S. Senate also investigated charges that relief assistance had been racially biased (Popkin 1990).

The experience with such storms suggests that planners attempting to reduce vulnerability would need different strategies than those appropriate in Jamaica or the United States. Socio-political organization can be as significant as national wealth in disaster preparedness. In wealthy countries, such as the U.S., and poorer former colonies, such as Jamaica, there are large towns and cities on hurricane-prone coasts (Blaikie et al. 1994). However competing constructions of vulnerability can result in very different impacts (Dove and Khan, 1995). Potential losses are a function of the types of urbanization (Havlick 1987). In places where people experience multiple hazards the impact of one may be less or more serious than another. More is needed than looking simply at how much money is spent on meteorological and hurricane warning services, or the surplus available to a country for hurricane-proof buildings and planned relief systems. A highly vulnerable group may be badly affected by a relatively weak event or succession of events, while a low vulnerability group may be little affected by a strong one. Recovery capacities are similarly distributed and disagreement then arises as to which factors are the most important. One promising approach, to which the Blaikie et al. (1994) frame can be applied, is that of starting with a community and asking what combination of action, research, information, planning, and program, would achieve optimal risk for the facility and people (White 1994). Clearly the assessment of vulnerability is a complex undertaking. It is also increasingly clear that a dearth exists in understanding how the structure of a society determines the way in which a hazard is likely to affect it.

9.5 Trends in the Co-evolution of Disasters and Development

"If we are not careful, we may end up where we are going." (Chinese Proverb)

A working definition of a disaster is a crisis event that surpasses the ability of an individual, community, or society to control or recover from its consequences. "Recovery" has been shown to include the psychological and physical recovery of the victims, the replacement of physical resources and the social relations required to use them (see Rodríguez, Chap. 7, this volume). Careful attention needs to be focused on whether the vulnerability approach advocated here involve irreconcilable conflicts, especially since people must live with governments, businesses and systems (national and international) that can maintain the inequities which increase vulnerability. Decisions that those in power make today determine the resources available to those who come after (Bock 1967).

"Development" is traditionally understood as a change toward the greater wealth of a community (Bender 1994). While some development activities will require sustained application in the foreseeable future, in Latin America and the Caribbean, population pressure and other demands on the regions' ecosystems have exceeded their natural capacity to provide sufficient goods and services, especially to meet the basic needs of the urban poor. Development projects assign resources (time, knowledge, money, goods, services) intended for the betterment of some segment of the population (Bender 1994). Shifting to formal exploitation of resources can remove the opportunities for the poor to have access to those resources. In addition, administrators shape outcomes through their discretion and decisions and thus create policy in a concrete sense (Ascher and Healy 1990). Vulnerability reduction is therefore an increasingly important measure of "who" is responding and "for whom" development projects are intended (Vayda and McCay 1975).

Evaluating the ways in which natural events affect a community's surroundings in terms of the attributes and hazards of the events is part of an integrated approach to environmental management. Some landscapes become more vulnerable because naturally occurring mitigation elements have been damaged or removed due to development (environmental degradation) or because development in hazardous areas have created a vulnerable social and economic infrastructure where none existed before. While the long-term economic benefits of living in areas where the physical threat has been modified (by barriers) can outweigh the local losses for some groups, distributional effects, such as opportunities lost to other locales further inland due to beachfront development, are not as readily considered (Wiener 1996).

Bruce (1994) identifies three types of global change that affect human and economic losses due to natural hazards. These are: (1) variability and change in frequency and severity of natural hazards, (2) increasing economic development, especially along coastlines, in flood plains, and other hazard-prone areas, and (3) changes in land surfaces and vegetation. There are, in addition, three types of natural hazards-related natural resource accounts to be integrated into evolving national income accounts (Bender 1992):

a) losses due to the impact of natural events. These must include losses to naturally occurring mitigation elements in the ecosystem as well as infrastructure;

b) the value of natural structures (e.g. reefs) and functions (e.g. flood water absorption capacity of wetlands) that mitigate against the impacts of hazards;

c) the value of natural resources that can be damaged or destroyed by natural events (e.g. beaches by hurricanes) and the depreciated value of economic and social infrastructure due to natural hazard risks.

The unraveling of the human-ecological hazards model concisely illustrates fundamental questions about environment-society interaction: is hazardousness lessening or worsening? and, what is the impact of "surprises" in the system on these trends? In human ecology theory, hazards are lessened as human systems interact with them (the homeostatic, human-ecological model). But the technological hazards model implies a progressive, open-ended trajectory of worsening as humans develop new technologies and seek to fill increasing needs and desires. The core of the model is clearly not a self-dampening, human-ecological system, but rather a development trajectory which creates new and often unanticipated problems that must be fixed by a risk management superstructure (Kasperson and Stallen 1991). The impression is of an open-ended process not readily diverted by mechanisms inherent to human experience and response. This of necessity complicates the analysis of vulnerability. One of the most significant of the continua is that which exists between disasters that strike abruptly and those that have long drawn-out impact and are hence known as "creeping disasters" (e.g. desertification, soil erosion, sequences of events).

Though we cannot yet determine, unambiguously, whether nature-society-technology relationships in the late 20th century are becoming more or less hazardous, we can identify broad social and environmental trends that may exacerbate or decrease the impacts of extreme events in the future (Table 9.2). The list is not intended to be exhaustive. An understanding of the nature and conditioning factors on trends in each of these areas has important implications. Among those that obviously increase risks are raw population growth, rapid coastal development, industrialization and the spread of hazardous facilities, and encroachment of agriculture into marginal lands. Often disaster educators hope to see that if people suffer once from their vulnerability they will be motivated to undertake hazard preparedness and mitigation efforts that will reduce their vulnerability in the future. However, especially among the poor and marginalized groups, if resources are destroyed, if assistance promotes dependency, if family or social systems are undermined, they will have less resilience for facing subsequent hazards (Anderson and Woodward 1989). It is also evident, that for developing areas, these trends will get worse as the regions or countries involved attempt to rapidly increase economic productivity.

Preparedness has been the area most recognizable in disaster planning, because it can relate to various technical interventions commonly seen as necessary for disaster avoidance (e.g. warning systems). But it is clear that people's ability to protect themselves depends on their livelihood strength, and their relationship to the network of social and political structures. As in the case of Jamaica, the state commonly recognizes, officially, the need to offer social protection to reduce vulnerability but is also a party to the economic and social processes that lead people to be unable to protect themselves in the first place. Depending on the intensity and duration of any given hazard event, the depth and extent of the resulting disaster will depend on the number of people who are vulnerable in terms of their lack of self-protection, social protection, and the strength of their livelihoods (and differing combinations of these components of vulnerability). Many natural hazards have amplified impact because already vulnerable people have to cope and survive by engaging in practices which increase vulnerability. The impact of hurricanes on Jamaica, Honduras, or Mexico usually has far worse effects in terms of loss of life, injury, and livelihood disruption than is the case with similar hurricanes in the U.S. The U.S. has a different level of preparedness and individual livelihoods are more robust. Reconstruction costs are more easily absorbed because large surpluses are publicly controlled. This does not mean that disaster vulnerability is absent from wealthy countries. Indeed, even in the wealthiest parts of the world class, age, gender, and ethnicity are still likely to be very significant indicators of the variable impact of hazards (see Platt et al. 1991).

Planning strategies must be prepared to acknowledge degrees of ignorance in exploring how each given project would do in different contexts (Ascher and Healy 1990). Reduction of vulnerability appears to be contingent on the way disaster planning is located within development planning. It is clear that this requires attention not only at the level of ordinary policy-making (i.e. specific projects) but at the level of systems of authority and control, as well. Economists usually tend to dismiss distributional concerns, how policies may be perceived and the extent and affecting hazardousness are as follows (see also Anderson, 1995).

Table 9.2. Social and environmental trends that condition vulnerability.

Lessening Effect		Worsening Effect
	Social Trends	
Improved building technology Better detection/warning systems Improved health care systems Community-based initiatives Alternative uses for traditional crops		Increasing hazard zone occupancy Aging population Third World urban migration More hazardous facilities and decaying social services Widening rich-poor gap Decreasing food self-sufficiency Belief in rationality through pricing alone
	Environmental Trends	
Warming climate in northern areas Increase moisture in low latitudes Reduced use of some pesticides Clean-up of toxic sites Creation of coastal reserves		Sea-level rise Mid-latitude drying Forest destruction Soil erosion Wetland loss Coral reef decline Loss of genetic diversity Use of marginal lands Effluent production

Vulnerabilities that are cumulating, tending toward irreversibility, and expanding through their interconnectedness pose far more serious problems and demand more immediate attention than those that exhibit the opposite characteristics (Bender 1994). While complexity implies the need for good information, the uneven quality of information and how this information is controlled influences the focus of attention. Access to usable information is one of the most potent connections between the technical and political aspects of the policy process. Communicating the range of causal structures, choices, and alternatives of the types discussed in the previous section, to potential victims and their allies can itself become part of the process by which society is changed to avoid and reduce vulnerability being generated. Information must be communicated not only to decision makers but to people about whom decisions are being made. Thus, alternative sources of information for the development of adaptive measures is extremely important.

The trade-off between impacts intervention and fundamental hazard mitigation deserves more careful analysis as new programs are designed for hazard reduction. As Mitchell (1988) points out, both the International Geosphere-Biosphere Program and the International Decade for Natural Hazard Disaster Reduction (IDNDR) appear to be little informed by knowledge about human behavior in the face of hazard. This does not imply that criticism of these programs be made at every step, but that simply a clearer focus on identification of incentives for cooperation is necessary. Broad societal processes that create dynamic pressures and unsafe conditions (Blaikie et al. 1994) are not easy to change, yet are fundamental to human vulnerability. The involvement of outside, supra-local groups may render permanent the effects of an otherwise short-duration problem laying bare direct links between vulnerability and the operation of regional to global economies. What then matters, may not be appeals to increases in national GDP indicators but the availability of tangible use values - food, water, health care, housing and so on (Yapa 1996). The economic and social trends must be in the direction of resilience to natural events and not in greater dependence on assistance. Indeed, the federal disaster assistance in the U.S. ($27.6 billion from 1989-1993) has been referred to as "another entitlement program" (Platt 1995). Development where large segments of the population have no choice but to live and work in high-risk environments, only perpetuates dependency. If present trends are followed, expected growth and environmental change pose grave risk of continued resource degradation and vulnerability to disasters (Bender 1994). Principles on which studies and mitigation of hazards and disasters should be based are increasingly being identified from disparate sources in the literature. These can be summarized as follows:

a) The causes, internal features, and consequences of natural disasters and recovery are not explained solely in terms of disaster events themselves, but rather in terms of the ongoing social order and processes (and individual behavior) and the everyday relationship with the environment (e.g. Varley 1994b);

b) Development can foster dependency and specialization on the part of individuals and communities, reducing their ability to respond effectively or narrowing the range of normal environmental variability with which they are able to cope on their own (e.g. Morren 1983);

c) Equilibrium-centered views should be abandoned and inquiry focused on change in relation to homeostasis and changes in context (e.g. Vayda and McCay 1975);

d) There is a need to make disaster mitigation, for all hazards (including technological), as much of a priority in planning and application as emergency preparedness, response and recovery (e.g. Quarantelli 1994).

Particular attention must be paid to the ways in which some risks are acknowledged while others go unnoticed or ignored and, to the importance of the acceptability of risk and proposed management policies by lay persons affected (Mayo and Hollander 1991). Central governments and international donor organizations should recognize a disaster as providing a window of opportunity to pursue activities that are not necessarily related to the disaster event, particularly through the use of local organization. Insurance cannot take on the role of compliance enforcement, i.e. the enforcement of building codes and land use policies. This requires that actions taken must be mutually supported, with no single group acting alone.

Social models are typically based on norms that researchers think rational people should aspire to, including transactional efficiency, profit maximization, risk minimization, and social equity. Subtle philosophical traps arise when a model built for informing policy is used to predict social behavior. Indeed it has been argued that "the very act of modeling and prediction, the basis of quantitative risk analysis, strip away the interconnections that defy known methods of forecasting" (O'Riordan and Rayner 1991). Enough is known about the processes of choice to be reasonably sure that it cannot be accurately described as a simple effort to maximize net marginal return (Slovic et al. 1974). As Quarantelli (1989) and others have pointed out, the re-development process is neither as ordered nor as predictable as indicated by linear models of development. What this implies is that a single policy should not be expected to maximize everyone's expected utility, much less so for more value-based criteria, and that such promises should not be blithely made.

9.6 Conclusion: Sustainability and the Search for Guidance

Disasters occur when physical impacts coincide with human vulnerability. Vulnerability is complex, dynamic, compounding, and cumulative (Anderson 1995). Whereas previous assessments focused on the acts of nature that come from outside human agency, later assessments have begun to acknowledge that it is largely human actions, decisions, and choices that result in people's vulnerability to natural events (Varley 1994b). This is usually represented in the literature by those who have identified the differential character of vulnerability and the central role of humans in creating vulnerability. These include roles in economic poverty, lack of access to resources, and other factors that in any given setting cause people to live in circumstances that put them at high risk through the interaction of natural hazards with market, social, political, or other changes.

Emphasis on the combination of human ecology with political economy perspectives, i.e. political ecology, underscores the ways in which the natural environment is transformed by humans (see Emel and Peet 1989). These are rooted in interactions of the climate, geomorphology, and ecological contexts with key aspects of socio-physical processes, such as economic production, which takes place within and across physical regions (Ascher and Healy 1990; Bohle et al. 1994). What is advocated here is a way to introduce the areas of ambiguity before the development objectives and mitigation goals are prioritized or ranked. Indeed if this is not done, different concerns will be weighted simply by including and excluding particular types of data, emphasizing the measurable, and considering only first-order consequences or consequences of well-understood relationships (Ascher and Healy 1990). The political ecology perspective advocated here, is a way of understanding both the risk environment which vulnerable groups confront, and the "quality" of their resource endowments (Sen 1981, Bohle et al. 1994).

It is not enough to say that given local control, provisional shelter can be transformed into acceptable safe housing or that low-income settlements can evolve into stable communities (Bender 1992). Where possible, all-hazard insurance and government re-insurance as risk-spreading mechanisms will have to be used (Kunreuther 1995). However, most of the risk-based principles of insurability (Kunreuther 1995, Roth 1996, Chap. 13, this volume) may not be met in developing countries. In fact, a perfect forecast of decadal-scale hurricane risk may lead insurers to reduce their financial exposure by not providing coverage for the most vulnerable regions (Kunreuther 1993). The issue remains one of how to better integrate natural hazard management concerns into the development planning process (Vermeiren 1990). As White (1994) points out "What is to be avoided is occupation of hazardous area that to remain viable requires uneconomic support in the way of resource degradation, subsidies and public assistance."

Mileti et al. (1995), in linking natural hazards to sustainability and global change paradigms, call for (1) broader natural hazards goals rather than local loss reduction, (2) a revised paradigm that links natural hazards to their global context, environmental sustainability and societal resiliency, and (3) hazards mitigation that is compatible with sustainability. Their thesis is that natural hazards are the consequence of unsustainable development and that the move to sustainability requires a "shift in culture". The conditions and factors which govern vulnerability and define the specific coordinates of risk exposure, coping capacity and recovery potential, although qualitative, are increasingly acceptable as causes of risk to researchers and decisionmakers. From a hazards management standpoint the definition of sustainability employed by Ascher and Healy (1990) is particularly useful. They define sustainable development as a process of development that achieves the following goals:

a) an optimal rate of use of natural resources over time (growth);
b) distributional equity;
c) environmental protection (including functioning of complex natural systems);
d) participation of all sectors of society in decisionmaking.

Within this frame, emphasis is placed on clarification of processes through which development can be defined, promoted, retarded or guided by public institutions. These institutions subsidize certain activities or products and tax others, ration access to physical resources or markets, regulate private business, and engage in state-owned investments and products (Ascher and Healy 1990). As required by Bohle et al. (1994), there is explicit recognition of the complex interrelationships of how social organizations and growth have direct implications for sustainability and how the environment is experienced in terms of risk and threats (e.g. human-induced global warming, decadal-scale variability of hurricanes). It also recognizes that policy making is a sequence of many actions by many actors, each with potentially different interests, information, roles, and perspectives (Lasswell 1971). As Wescoat (1991) points out, the conditioning factors of formal resource management have been, in the global perspective, war (including preparations), colonization (i.e. appropriation of other peoples resources), trade and finance, price shocks and restructuring and, ecological crises. Research emphases must thus be placed on the processes by which political power is mobilized and in the identification of trends and patterns that can deny resource management of the adaptiveness that is so necessary in practice. It also points to the question of who should have the "right" to manage and to what end (Emel and Peet, 1989).

In an excellent review titled "Human Dimensions of Environmental Hazards: Complexity, Disparity and the Search for Guidance," Mitchell (1990) defined the "search for guidance" as being driven by an awareness of the need to devise problem orientations for choosing among diverse problem definitions and proposed actions. Information about the future is always subject to interpretation by experts, resulting in ambiguity and different recommendations on what the appropriate actions should be (Kunreuther, 1993). Mitchell (1990) emphasized the need for better organization of the intellectual domain if there is to be more effective use of hazards research knowledge to combat mounting hazard management problems. The necessary links must be made between decision-making rules as ways of evaluating alternatives and political economy perspectives which elaborate processes that create and maintain hazardous conditions. The problem appears not to be only the lack of recognition of the social, economic, cultural, and historical contexts of vulnerability but a general failure or unwillingness to put this recognition into practice (Whyte 1986). Varley (1994a) and others indicate that the very reason that political economy perspectives as described in Hewitt (1983) have not commonly been adopted is due to its early focus on the economic poor. As is evident along the shores of the eastern U.S. seaboard, poverty is not a sufficient proxy for vulnerability (Varley 1994b). Indeed while people may be affected differently by virtue of "economic" vulnerability, the type of warning systems (if any), means of communication, and inadequacy of mitigation measures make everyone vulnerable (Cannon 1994).

In 1987, a U.S. National Research Council committee estimated that the Decade's efforts would result in a 50% reduction in losses within 10 years (NRC, 1987). There has been much activity in processing and distributing scientific findings on natural hazards' occurrence and risk in the first half of the IDNDR. However, by the middle of the decade, losses were significantly greater than in any previous five-year period (White 1994). While the IDNDR resolution stipulates that substantial knowledge exists on technical solutions to disaster reduction, less work has been done on understanding why these solutions have not been implemented (Berke et al. 1993b). National and international relief and development agencies have increasingly recognized that local organizational and individual capacity to use these solutions is inadequate (OAS 1990, OFDA 1990). An IDNDR strategy that sought to integrate hazard reduction with basic societal demands would have compelling benefits, including social and cultural acceptability, political feasibility, and economic and environmental sustainability (Oaks and Bender 1990). Researchers should expand their investigations out from conventional conceptions of hazards as damaging events along "linkages" that connect with larger or different contexts, i.e. the nexus of production relations (Mitchell 1990; Yapa 1996). Given the complexity this entails, necessary steps must be to clearly document how risks were reduced from a past disaster, barriers to learning, and how the appropriate lessons can be learned from innovations. A key component would be to insure that comprehensive post-audits of programs be undertaken (White 1988, M.F. Meyers pers. comm.). Specific attention must be paid to the requirements for including evaluation as a tool in the implementation process and to the standards of social and environmental values to be used in auditing outcomes (White 1988). Another dilemma is then posed by the observation that "there is still considerable resistance in the scientific community to adoption of an integrated view of hazards and development" (Mitchell 1990).

The methods of contextual inquiry into vulnerability, described here, are resonated in the adaptive management efforts from the ecological literature (Gunderson et al. 1995) and prototyping studies from the policy sciences arena (Lasswell 1963). These, together, provide promising ways of linking case-study and experimental planning approaches to increase practical learning at socially relevant scales, with value considerations at the center. It is too often assumed, implicitly, that programs can be designed optimally in advance (Stern 1993). What needs to be analyzed in each case is how the structures of society and individual behavior determine the ways in which a hazard and its variability of occurrence is likely to constitute a disaster. Coping with climatic variations or future "surprise" must be rooted in a full appreciation of the complex structures, uncertainties and causes of present vulnerability, and how these may evolve over the coming decades. Complexity is thus introduced not only in the non-linear interactions of the physical system but in the continual dilemma of ascertaining the acceptability of risk and solutions, how and when actions are to be taken and, for whose benefits. This underscores the need to understand the social process as rooted in people, values, institutions, and resources (Lasswell and McDougal 1992). The overarching issues concern how to deal with the basic links in environmental and social systems so that the human family may be sustained (White 1992). The primary guiding principle for any proposed solution must be, as Gilbert White has said on many occasions, "to reaffirm the dignity of human choice" in the social process.

Acknowledgments
We would like to thank Mary Fran Myers for contributing many useful comments and insights to the chapter.

References

Abril Ojeda, G., 1982. The Role of Disaster Relief for Long Term Development in LDCs with Special Reference to Guatemala after the 1976 *Earthquake. Monograph No. 6* Institute of Latin American Studies, Stockholm.

Alexander, D., 1991. Natural disasters: A framework for research and teaching. *Disasters*, **15**, 209-223.

Anderson, M.B., 1995. Vulnerability to disaster and sustainable development: A general framework. In, Munasinghe, M., and Clarke, C. (eds), *Disaster Prevention for Sustainable Development: Economic and Policy Issues*. IDNDR and World Bank, 41-60.

Ascher, W. and Healy, R., 1990. *Natural resource Policymaking in Developing Countries.* Duke University Press, Durham, N.C., 223 pp.

Bacon, P., 1989. Assessment of the Economic Impacts of Hurricane Gilbert on Coastal and Marine Resources in Jamaica. *UNEP Regional Seas Report No.110*. Kingston, Jamaica.

Barker, D. and Miller, D., 1990. Hurricane Gilbert: anthropomorphing a natural disaster. *Area*, **22**, 107-116.

Barrows, H., 1923. Geography as human ecology. *Annals, Association of American Geographers,* **13**, 1-14.

Bender, S.O., 1992. Disaster prevention and mitigation in Latin America and the Caribbean: Notes on the decade of the 1990s. In Aysan, Y., and I., Davis, *Disasters and the Small Dwelling: Perspectives for the UN IDNDR.* James and James Science Publ.London. 45-57

Bender, S.O., 1993. Urban growth in transition: Environmental change and natural hazard management in Latin America and the Caribbean. *International Seminar on Latin American Regional Development in an Era of Transition 6-8 December,* Sao Paulo Brazil.

Bender, S.O., 1994. The sustaining nature of the disaster-development linkage. *Ecodecision,* April, 50-52.

Berke, P.R., Beatley, T., and Feagin, C., 1993a. Hurricane Gilbert strikes Jamaica: Linking disaster recovery to development. *Coastal Management,* **21**, 1-23.

Berke, P.R., Kartez, J., and Wenger, D., 1993b. Recovery after disaster: Achieving sustainable development, mitigation and equity. *Disasters,* **17**, 93-109.

Berke, P. and Wenger, D., 1991. *Montserrat: Emergency Planning Response and Recovery Related to Hurricane Hugo.* Hazard Recovery Center, Texas A and M University, College Station, TX.

Blaikie, P., Cannon, T., Davis, I., and Wisner, B., 1994. *At Risk: Natural Hazards, Peoples' Vulnerability and Disasters.* Routledge, London, 284 pp.

Bock, E., 1967. "The last colonialism: Governmental problems arising from the use and abuse of the future. CAG/ASPA. Cited in Ascher, W., and R. Healy, 1990: *Natural resource Policymaking in Developing Countries.* Duke University Press, Durham, N.C. 223 pp.

Bohle, H.G., Downing, T.E., and Watts, M.J., 1994. Climate change and social vulnerability. *Global Environmental Change,* **4**, 37-48.

Bohle, H.G. and Watts, M.J., 1987. The space of vulnerability: the causal structure of hunger and famine. *Progress in Human Geography,* **13**, 43-67.

Bruce, J., 1994. Natural disaster reduction and global change. *Bulletin of the American Meteorological Society,* **75,** 1831-1835.

Burton, I. and Kates, R.W., 1964. The floodplain and the seashore. *Geographical Review,* **54**, 366-385.

Burton, I., Kates, R., and G.White, 1993. *The Environment as Hazard,* Oxford University Press, New York, 240 pp.

Cannon, T., 1994. Vulnerability analysis and the explanation of "natural" disasters. In: Varley, A., *Disasters, Development and Environment,* J. Wiley and Sons, Chichester, 13-30.

Cuny, F., 1983. *Disasters and Development,* Oxford University Press, Oxford.

Dewey, J., 1899. The School and Society. In, Boydston, J., 1981 *The Middle Works: 1899-1924. Vol. I,* Southern Illinois University Press, Carbondale Illinois.

Dove, M.R. and Khan, M.H., 1995. Competing constructions of calamity: The April 1991 Bangladesh Cyclone. *Population and Environment,* **16**, 445-471.

Drabek, T.E., 1986. *Human System Responses to Disaster.* Springer-Verlag, New York, 509 pp.

Drabek, T.E., 1987. The Professional Emergency Manager. *Monograph No. 44, Natural Hazards Research and Applications Information Center,* University of Colorado, Boulder.

ECLAC, 1974. Informe sobre los daños y repercusiones del Huracán Fifi en la economica hondureña. *Economic Commission for Latin America and the Caribbean AC.67/2/Rev. 1* Santiago, Chile.

Emel, J., and Peet, R., 1989, Resource management and natural hazards. In: Peet, R. and Thrift, N., (eds), *New Models in Geography: The political-economy perspective.* Unwin Hyman, London, 49-76

Eyre, L.A., 1989. Hurricane Gilbert: Caribbean record-breaker. *Weather*, **44**, 160-164.

Eyre. L.A., 1990. Forestry and watershed management. Paper presented at the *Consultation Workshop on the National Conservation Strategy, 25-27 April,* Jamaica.

Friedman, D.G., 1977. Assessment of the magnitude of the hurricane hazard. In: *11th Conference on Hurricanes and Tropical Meteorology, December 13-16, 1977,* Miami Beach Florida. AMS. Boston, 294-301.

Girvan, N.P., 1991. Economics and environment in the Caribbean: An overview. In, Girvan, N.P. and Simmons, D.A. (eds), *Caribbean Ecology and Economics,* Caribbean Conservation Association, St. Michaels, Barbados, 260 pp.

Girvan, N., Rodriguez, E., Sevilla, A., and Hatton, M., 1990, *The Debt Problem of Small Peripheral Economies: Case Studies from the Caribbean and Central America,* Association of Caribbean Economists, Kingston.

Griggs, G.B. and Gilchrist, J.A., 1983. *Geological Hazards, Resources, and Environmental Planning,* 2nd ed. Wadsworth, Calif.

Gunderson, S., Holling, C., and Light, S. 1995. *Barriers and Bridges to the Renewal of Ecosystems and Institutions,* New York: Columbia University Press, 593 pp.

Hardoy, J.E. and Satterthwaite, D., 1989. *Squatter Citizens: Life In the Urban Third World,* Earthscan, London.

Haas, E., Kates, R.W., and Bowden, M., 1977. *Reconstruction following Disaster,* MIT Press, Cambridge, MA.

Havlick, S., 1986. Third world cities at risk: Building for calamity. *Environment,* **28**, 6.

Hewan, C.G., 1994. *Jamaica and the United States Caribbean Basin Initiative: Showpiece or Failure?,* Peter Lang Publishing, N.Y. 153 pp.

Hewitt, K., 1983a. *Interpretations of Calamity from the Viewpoint of Human Ecology,* Allen and Unwin, Boston, 304 pp.

Hewitt, K., 1983b. The idea of calamity in a technocratic age. In, Hewitt, K., (ed), *Interpretations of Calamity from the Viewpoint of Human Ecology.* Allen and Unwin, Boston, 3-32.

Hohenemser, C., Kates, R.W., and Slovic, P., 1983. The Nature of Technological Hazards. *Science*, **220**, 378-384.

Jarrell, J., Hebert, P., and Mayfield, M., 1992. Hurricane experience levels of coastal county populations from Texas to Maine. NOAA Technical Memorandum NWS NHC 46.

Kasperson, R. and Stallen, P.J., 1991. *Communicating Risks to the Public, International Perspectives,* Kluwer Academic Publishing, London, 481 pp.

Kates, R.W., 1971. Natural hazards in human ecological perspective: Hypotheses and models. *Economic Geography,* **47**, 438-451.

Kates, R.W., 1985. The interaction of climate and society. In: Kates, R., Ausubel, J., and. Berberian, M., (eds), Climate Impact Assessment SCOPE 27. John Wiley and Sons. Chichester. 3-36.

Kunreuther, H., 1993. Comments. Panel on Benefits of Improved Decadal Hurricane Forecasting. NHRAIC Natural Hazards Workshop, July, 1993. Boulder, CO.

Kunreuther, H., 1995. The role of insurance in reducing losses from natural hazards. In: Munasinghe, M. and Clarke, C., (eds.), 1995: *Disaster Prevention for Sustainable Development: Economic and Policy Issues*, IDNDR and World Bank, 87-102.

Lasswell, H.D., 1963. Experimentation, prototyping, intervention. In: Lasswell, H., *The Future of Political Science,* Atherton Press, NY, 95-123.

Lasswell, H. and McDougal, A., 1992. *Jurisprudence for a Free Society,* Kluwer, Dordrecht 1588 pp.

Lavell, A., 1994. Prevention and mitigation of disasters in Central America: Vulnerability to disasters at the local level. In, A. Varley, *Disasters, Development and Environment.* J. Wiley and Sons, Chichester, 49-64.

Maskrey, A., 1989. *Disaster Mitigation: A Community-Approach,* OXFAM, Oxford England

Mayo, D. and Hollander R., 1991 (eds) *Acceptable Evidence: Science and Values in Risk Management,* Oxford University Press, New York, 292 pp.

McAfee, K., 1991. *Storm Signals: Structural Adjustment and Development Alternatives in the Caribbean,* Southend Press (with Oxfam America). Boston, MA.

Mileti, D., Darlington, J., Passerini, E., and M.F., Myers, 1995. Toward an integration of natural hazards and sustainability. *Environmental Professional,* **17**, 117-126.

Miller, K. and Simile, C., 1992. *They could see stars from their beds: The Plight of the Rural Poor in the Aftermath of Hurricane Hugo,* 26 March, Society for Applied Anthropology, Memphis, Tennessee.

Mitchell, J., 1988. Review: Confronting Natural Disasters: An International Decade for Natural Hazards Reduction. *Environment,* **30**, 25-29.

Mitchell, J., 1990. Human dimensions of environmental hazards. In: Kirby. A., (ed.), *Nothing to Fear: Risks and Hazards in American Society,* University of Arizona Press, Tucson. 131-175

Morren, G., 1983. A general approach to the identification of hazards and responses. In: Hewitt, K., Interpretations of Calamity. Allen and Unwin, 284-297.

Myers, M.F. and White, G.F., 1993. The challenge of the Mississippi flood. *Environment,* December, 35, p. 6.

OAS, 1990. Primer on Natural Hazards Management in Integrated Regional Development Planning, Department of Regional Development and Environment, Organization of American States.

OFDA, 1990. *FY 1989 Annual Report,* U.S. Office of Foreign Disaster Assistance. Agency of International Development, Washington, D.C.

Oaks, S. and Bender, S., 1990. Hazard reduction and everyday life: Opportunities for integration during the Decade for Natural Disaster Reduction. *Natural Hazards,* **3**, 87-89.

O'Riordan, T. and S. Rayner, 1991: Risk management for global environmental change. *Global Environmental Change,* **3,** 91-108.

Natural Hazards Research and Applications Information Center, 1992. *A Bibliography of Weather and Climate Hazards.* Topical Bibliography No. 16. University of Colorado, Boulder. 331 pp.

NRC, 1994. *Facing the Challenge: The U.S. National Report to the IDNDR World Conference on Natural Hazard Reduction,* 23-27 May, 1994, Yokohama, Japan, 78 pp.

PASB, 1980. *Disaster Reports: The Effects of Hurricane David, 1979 on the Population of Dominica.* PAHO/Pan American Sanitary Bureau, Washington, D.C., 60pp.

Perrow, C., 1984. *Normal Accidents: Living with High-Risk Technologies,* Basic Books, New York, 386 pp.

Petak, W.J. and Atkisson, A.A., 1982. *Natural Hazard Risk Assessment and Public Policy. Anticipating the Unexpected,* Springer-Verlag, New York, 489 pp.

Platt, R.H., 1995. Review: Sharing the challenge: floodplain management into the 21st century. *Environment,* **37**, 25-28.

Platt, R., Beatley, T., and Miller, H.C., 1991. The folly at Folly Beach. *Environment*, **33**, 7-32.

Polanyi-Leavitt, K., 1991. *The Origins and Consequences of Jamaica's Debt Crisis, 1970-1990*, Consortium Graduate School of Social Sciences, University of the West Indies. Mona, Jamaica. 67 pp.

Popkin, R., 1990: The history and politics of disaster management in the United States. In: Kirby, A. (ed.), *Nothing to Fear: Risks and Hazards in American Society.* University of Arizona Press, Tucson, 101-130.

Quarantelli, E.L., 1989. *A Review of the Literature in Disaster Recovery Research,* Disaster Research Center, University of Delaware, Newark, DE.

Quarantelli, E.L., 1994. *Future Disaster Trends and Policy Implications for Developing Countries,* Disaster Research Center, University of Delaware, Newark, DE.

Robinson, J.W., 1994. Lessons from the structural adjustment process in Jamaica. *Social and Economic Studies*, **43**, 94-113.

Rubin, C. and Popkin, R., 1990. Disaster Recovery after Hurricane Hugo in South Carolina. *Natural Hazards Research and Application Information Center Working Paper No. 69., University of Colorado,* Boulder, CO.

Sen, A., 1981. *Poverty and Famines: an Essay on Entitlements and Deprivation.* Clarendon press, Oxford, 257 pp.

Shrader-Frechette, K., 1991. Reductionist approaches to risk. In: Mayo, D.G. and Hollander, R., (eds), *Acceptable Evidence: Science and Values in Risk Management,* Oxford University Press. NY. 218-248

Slovic, P., Kunreuther, H., and White, G., 1974. Decision processes, rationality and adjustment to natural hazards. In, G.F. White (ed), *Natural Hazards: Local, National and Global.* New York, 187-205.

Stern, P., 1993. A second environmental science: Human-environment interactions. *Science*, **260**, 1897-1899

Susman, P., O'Keefe, P. and Wisner, B., 1983. Global disasters: a radical interpretation. In: Hewitt, K., 1983a: *Interpretations of Calamity from the Viewpoint of Human Ecology,* Allen and Unwin, Boston, 263-283.

Varley, A., 1994a. *Disasters, Development and Environment,* J. Wiley and Sons, Chichester, 167 pp.

Varley, A., 1994b. The exceptional and the everyday: Vulnerability analysis in the International Decade for Natural Disaster Reduction. In, Varley, A., *Disasters, Development and Environment.* J. Wiley and Sons, Chichester, 1-12.

Vayda, A.P. and McCay, B.J., 1975. New directions in ecology and ecological anthropology. *Annual Review of Anthropology,* **4**, 293-306.

Vermeiren, J., 1991. Natural Disasters: Linking economics and the environment with a vengeance. In, N. Girvan and D. Simmons, Caribbean Ecology and Economics. Caribbean Conservation Association, 127-142.

Wescoat, J., 1991, Resource management: the long-term global trend. *Progress in Human Geography,* **15**, 81-93

White, G.F., 1945. Cited in White et al., 1958, Changes in Urban Occupancy of Flood Plains in the United States. *Dept. of Geography Research Paper 7. Univ. of Chicago Press*, IL., 205-226.

White, G.F., 1966. Formation and role of public attitudes. In: Jarrett, M. (ed.) *Environmental Quality in a Growing Environment,* Johns Hopkins Press Baltimore, 105-127.

White, G.F., 1985. Geographers in a perilously changing world. *Annals, Association of American Geographers,* **75**, 10-15.

White, G.F., 1988. When may a post-audit teach lessons? In: Rosen, H. and Reuss, M. (eds.) The Flood Control Challenge: Past Present and Future. Public Works Historical Society, 53-65.

White, G.F., 1994. A perspective on losses from natural hazards. *Bulletin of the American Meteorological Society,* **75**, 1237-1240.

Whyte, A.V., 1986. From hazard perception to human ecology. In: Kates, R.W. and Burton, I. (eds.) *Geography, Resources and Environment: Vol. II,* Themes from the Work of Gilbert F. White. University of Chicago Press, Chicago, 240-271.

Wiener, J.,1996. Research opportunities in aid of Federal flood policy. *Policy Sciences* (submitted).

World Bank, 1990. Social Indicators of Development. Johns Hopkins Press, 350pp.

World Bank, 1995. Social Indicators of Development. Johns Hopkins Press, 412 pp.

Yapa, L., 1996: What causes poverty: A postmodern view. *Annals, Association of American. Geographers* (in press).

Zorn, J.G., and H. Mayerson, 1983. The Caribbean Basin Initiative: A windfall for the Private Sector. *Lawyer for the Americas*, **14**, 523-556

10 Incorporating Variability in the Disaster Planning Process

Charles C. Watson, Jr.[1] and Jan C. Vermeiren[2]

[1]Watson Technical Consulting
110 Deerwood Ct.
Rincon, GA 31326, U.S.A.

[2]Department of Regional Development and Environment
Organization of American States
Washington, DC 20006, U.S.A.

Abstract

Historically, disaster planning has been a fairly static process. Recent research indicating the quasi-cyclic nature of tropical cyclone formation, when combined with human-induced changes to the environment, requires disaster planners to consider the problem of how to create dynamic disaster plans. Geographic Information System technology and remote sensing can provide data bases for numerical models which can be rapidly updated. The Caribbean Disaster Mitigation Project's experiences in the Caribbean show that while the technical challenges to incorporating variability in the disaster planning process are achievable, the political and economic implications are extensive, and relatively more difficult to overcome.

10.1 Introduction

The General Secretariat of the Organization of American States (OAS), under an agreement with the Office of Foreign Disaster Assistance of the U.S. Agency for International Development (USAID), is executing a five-year Caribbean Disaster Mitigation Project (CDMP). The principal objective of the CDMP is to establish sustainable public-private sector mechanisms which measurably lessen loss of life, reduce the potential for physical and economic damage, and shorten the disaster recovery process in participating Caribbean countries.

In attempting to fulfill this ambitious objective, CDMP is examining critical aspects of the disaster planning process for opportunities to apply some of the latest advances in computer technology and in scientific development. Traditional disaster planning has been a relatively static process. The recent recognition of the natural variability of tropical cyclone formation, and of the impact of the rapid pace of anthropogenic changes on the environment, is the basis for the efforts of CDMP aimed at reducing the time needed for preparing and updating disaster plans. By making extensive use of Geographic Information System (GIS) technology and remote sensing, disaster plans can be turned into dynamic documents capable of responding to both natural and human changes to the environment.

This paper is presented in three sections. The first discusses the impact of variability on the disaster planning process from a theoretical perspective. The next section describes some practical strategies to cope with change in the disaster planning process. The third section describes how these issues are incorporated into the CDMP, and the reactions from various users in the region. In particular, we illustrate the disaster planning process with a case study for Montego Bay, Jamaica.

10.2 Theoretical Considerations

Much of the disaster planning process begins with the determination of risk. Risk determination consists of two components: what impact a particular event will have, and how often the event is likely to occur. In the case of tropical cyclones, both of these components are subject to variability from natural and anthropogenic sources. Rapid development in many coastal regions has the obvious effect of placing more people and a greater investment in harm's way, but can also increase the severity of a given event, such as by increasing flooding due to runoff from impervious surfaces. The impact of human activities on the frequency and intensity of future storms through the process of global climate change is currently the subject of intensive study and much debate.

There appear to be natural cycles to tropical cyclones (see Gray et al., Chap. 2, this volume). An analysis of historical climate data has demonstrated the existence of natural variability in frequency and intensity of past tropical cyclone activity. Projecting these patterns into the future represents a serious challenge for the scientific, engineering, and disaster planning communities. Regardless of the cause, the primary effect of variability and change in climatological phenomena on the disaster planning process is to require the periodic revision of disaster plans. Unfortunately, most disaster plans tend to be fairly static documents which are only evaluated after they are put to the test in an actual event. This is partly due to the expense and difficulty of updating the plan. Through the use of Geographic Information Systems (GIS), and GIS-based computer models, remote sensing, and the Global Positioning System (GPS), the update cycle and preparation costs for disaster plans can be reduced dramatically, and these plans can be made into dynamic, as opposed to static, documents. As more is learned about the effects of various natural and human sources of variability, this knowledge can be incorporated in the disaster planning process through the use of such techniques. The effects of possible change scenarios on critical elements of a disaster plan can be tested with GIS based computer models.

10.2.1 Natural Sources of Change in Tropical Climatology

There are two primary sources of natural variability and change which the disaster planner must consider. The first is the historical variability of both frequency and intensity of tropical cyclones. The second is the effect of natural environmental changes on the impact of a particular event.

Historical Variability of Tropical Cyclone Frequency and Intensity. Recent research indicates that there may be patterns to tropical cyclone frequency and intensity in the Atlantic Basin (see Gray et al., Chap. 2, this volume). An examination of return times and intensity for most sites in the Caribbean and U.S. Atlantic coasts reveals that, regardless of the exact mechanism involved, tropical cyclones do tend to come in groups. Table 10.1 shows an analysis of return times for Montego Bay, Jamaica. The observed return times are not easily fitted into known statistical distributions. The apparent interdecadal variability in return times presents a serious problem for the planner: should he or she plan for the worst case return times, or for the averages? How does the planner, when assessing the risk to a project, deal with the additional uncertainty of variation in risk over time? As discussed later, there are many political, social, and economic implications to these questions.

Natural Environmental Changes. Coastal regions are among the most dynamic areas on Earth. The changing wave climate causes beaches to erode and barrier islands to form and disappear. The land itself may be rising or subsiding due to the effects of plate motions or other geological phenomena. While some of these effects are slow, in many areas the changing coastal configuration is rapid enough to change the impact a given tropical event may have in the decadal time frame. At the individual structure level, changes may be rapid enough so that a structure which is protected from a Saffir/Simpson Category 2 storm by a dune one year may be exposed the next. If this structure happens to be a critical one for the community, such as a hospital, utility, or major employer, the effects of something as simple as a dune migrating a few hundred meters down the beach could be dramatic.

Some sectors of the disaster planning community in the U.S. have attempted to deal with these issues, such as NOAA periodically revising its SLOSH basins (Jelesniasksi et al. 1992). However, in many areas the data upon which risk maps such as the FEMA (U.S. Federal Emergency Management Agency) flood hazard zones are based are approaching 20 years old. While the areas at risk have been mapped in some way for most of the U.S., many communities are struggling to complete their first comprehensive disaster plan, much less face the challenge incorporating of periodic updates. In the Caribbean, most, if not all, countries are in a similar situation of not yet having a first comprehensive plan to update, and many areas have no hazard maps at all.

Table 10.1. Return time Analysis for Montego Bay, Jamaica, 18°N, 77°W, using data from 1886-1992.

Total Storms: 39
Total Years with Storms: 31
Total Years with Multiple Storms: 7
Total Years with Multiple Hurricanes: 0

Intensity	Events	Heading	Average Speed
Tropical Storm	25	296.1	10.4 knots
Category 1	3	297.5	8.6 knots
Category 2	5	299.5	17.0 knots
Category 3	5	286.3	17.1 knots
Category 4	1	35.5	8.6 knots

Overall Average Motion: 296.1 degrees, 11.6 knots

Tropical Storm Event or Greater Interval Analysis
27 intervals found
Average Interval: 3.44 years
Maximum Interval: 10
Minimum Interval: 1

Year Interval	Number of Occurrences
1	9
2	4
3	4
4	3
5	1
6	1
7	2
8	1
9	1
10	1

Category 1 Event or Greater Interval Analysis
12 intervals found
Average Interval: 7.75 years
Maximum interval: 16

Minimum Interval:1

Year Interval	Number of Occurrences
1	3
3	1
7	2
8	1
9	1
11	1
13	1
16	2

Category 2 Event or Greater Interval Analysis
9 intervals found
Average Interval: 10.33 years
Maximum interval: 29
Minimum Interval:1

Year Interval	Number of Occurrences
1	2
3	1
7	2
8	1
9	1
28	1
29	1

Category 3 Event or Greater Interval Analysis
5 intervals found
Average Interval: 18.44 years
Maximum interval: 36
Minimum Interval:7

Year Interval	Number of Occurrences
7	1
8	1
9	1
32	1
36	1

Category 4 Event or Greater Interval Analysis
Only 1 event and 0 intervals found
No Analysis Possible

Category 5 event or Greater Interval Analysis
No events found
No Analysis Possible

10.2.2 Anthropogenic Sources of Change and Tropical Climatology

Perhaps the most dramatic source of change in coastal areas is that of human development. Coastal areas have seen a tremendous increase in population in recent years, resulting in major changes to the land cover of coastal regions. These changes often have the effect of increasing the impact a given tropical cyclone will have, independent of the fact that there is more to be damaged in the first place. Engineers and disaster planners should pay careful attention to the cumulative nature of this effect. The effects of global climate change on tropical climatology are still poorly understood, which leaves disaster planners involved in the creation of long range plans in the 25 to 100 year range with little or no guidance.

Changes in Land-Cover Characteristics. Changes to local land cover as a result of development constitutes the largest single source of change for the disaster planner. In a first instance, the planner must consider how to evacuate larger numbers of people from threatened areas, and how to protect more property at risk. Furthermore, human activities can influence the magnitude of the effects of a given storm event. Examples are the destruction of reefs and the change to underwater features through dredging, each of which can result in higher waves reaching the coast. The destruction and modification of dunes and coastal vegetation can have significant implications for disaster planning. An example of this is at Hilton Head Island, South Carolina. The FEMA Flood Insurance Rate Maps for Hilton Head were created using maps which indicated a relatively intact dune system. In many areas, human activities have destroyed these protective dunes, exposing hundreds of millions of dollars of property to the potential for direct wave assault by a tropical cyclone. Finally, the increase in impervious surfaces and disruption to inland vegetation can increase the area flooded by rainfall from a storm event. Another consequence of this type of activity, especially in the Caribbean, is that steep slopes denuded of vegetation tend to suffer from soil failure and mud slides. In recent years the loss of life to mud slides from tropical systems has equaled or exceeded that caused by storm surge or winds.

Global Climate Change. The potential impact of human activities on climate has profound implications for many sectors of disaster planning. From the standpoint of planning for tropical cyclones, there are two primary impacts that can result from potential global warming scenarios. The first is the rise in sea level, while the second is the effect of global warming on the frequency and intensity of storm events.

A rise in sea level has the obvious impact of higher starting water levels to which storm effects are added. However, there are other, more complex implications. As noted in an Environmental Protection Agency report (EPA 1988), sea-level rise may result in fewer coastal wetlands. Coastal wetlands are important buffers for mitigating storm effects (see Doyle and Girod, Chap. 6, this volume). In addition, changing water depths and higher sea water temperature may have the effect of changing coral reef health and growth patterns. The higher water level will result in higher water tables, reducing inland drainage capacity and increasing flooding from stormwater runoff. Presently, most sea-level rise scenarios over the next century are within the range of uncertainty for both the data and models used for determining the impact of storm surge (Mercado et al. 1993). In the same study, the authors conclude that the indirect effects of sea-level rise, such as shoreline displacement due to increased erosion, will introduce the need to periodically update coastal maps and other input data to the storm surge modeling process.

Several Caribbean disaster planners have already suggested to the authors that sea-level rise be incorporated in their mitigation workplans. Given the wide range of values suggested in various sea-level rise scenarios, the decision has been left to the user as to which scenario they desire to use. The sea-level rise forecast is then added directly to the total water levels computed by the storm models. This method does not take in to account the secondary effects noted above, and is almost certainly inadequate, but does produce a least some margin of safety.

There is an active ongoing debate over the impact of various global warming scenarios on tropical cyclone frequency and intensity (see Emanuel, Chap. 3, this volume). While this debate has little implication for evacuation planning, it has significant implications for the insurance industry, developers, banking, and disaster planners. The natural variability in storm frequency noted above is rarely incorporated in storm frequency calculations used in the disaster planning community or in the determination of return periods for engineering design. Attempting to take into account an increase (or decrease) in future variability is not likely until the scientific community reaches some consensus regarding the impact of global warming on hurricane frequency.

10.3 Techniques to Cope with Change and Variability in Disaster Planning

Given the diverse sources of change noted above, disaster plans and hazard assessment techniques must be designed to be as flexible as possible. Modern mapping technology can play an important role in this process by making the task of keeping up with change easier. A Geographic Information System, or GIS, is a computer-based mapping system which allows data to be stored and analyzed in a spatial and temporal context. Satellite-based remote sensing allows the user to rapidly update land cover, and in some cases even update bathymetry and topography.

10.3.1 Monitoring Changes in Coastal Areas

Changes to the structure and physical parameters of the coastline and foreshore directly influence the impact of tropical weather systems. When using numerical models to assess coastal storm hazards, bathymetry, and coastal configuration data should be updated periodically in order to produce reliable estimates for storm surge and wave action. Extensive use of GIS and remote sensing can greatly facilitate this updating process. The use of satellite-based remote sensing is a well established technique for monitoring land use and development (Elachi 1986). As noted in Watson (1992a), dynamic underwater features can be mapped using simple GIS techniques and satellite imagery down to depths of 10 m. In the shallow waters which control wave refraction and breaking, and influence the magnitude of the storm surge, data from Landsat and Spot satellite imagery can produce reliable bathymetric information. Off the southeastern U.S. coast, this technique is limited to depths of around 5 m, due to the sediment content of the water. In Caribbean, with its clear waters, bathymetry of 20 m or more can be mapped (ERIM 1985). A storm hazard assessment model which uses a GIS as its data base manager can readily incorporate remotely sensed data.

10.3.2 Refining Return Period Estimates

Two serious problems exist with the determination of return periods for tropical weather events. The first problem is that, at best, just over 100 years of observations on tropical cyclone activity exist, with only 30 years of reasonably reliable data (Atlantic Basin Best Tack Documentation file provided by C. Landsea). This historical record represents too narrow a basis to project return periods much beyond 20 years with sufficient statistical reliability. Planners and engineers however need to set design standards for structures with lifetimes of 50 years and beyond. Such structures should be able to withstand forces that have only a small probability of being exceeded during their lifetime, i.e. forces related to events with return periods of 50 years or more.

The second problem is that of the natural variability of storm occurrences at a particular location. Return times seem to cluster, and reflect an inter-decadal variability (see Clark, Chap. 14, this volume). This renders most statistical methods for determining return times, such as the traditionally used Weibull distribution, unsuitable, as they assume a uniform probability of events over time.

When CDMP needed to simulate return periods for future hurricane events for storm hazard assessments it planned to carry out in the Caribbean, it was decided not to build a stochastic model in light of the two problems stated above. Instead, a technique was used similar to the one used by Young (1994) who simulates daily climate regimes based on random selections and linear transformations of historical data. Using the National Hurricane Center's Best Fit tracks, a summary of all of the historical return periods for a particular event or greater is created (as shown in

Table 10.1). Although this method can not determine return times longer than what is available in the data sets, it has the advantage of clearly conveying to the user the historical information content of the data, as well as the limitations of the data.

10.3.3 Integrating Development Trends in Damage Projections

Emergency managers tend to attach great value to historical information on damages caused by past tropical cyclones in estimating the effects of similar events in the future. In order to achieve an acceptable degree of reliability in their estimates, emergency managers need to resolve questions about accuracy of the historical data, and how development may change the impact of a given event. The determination of the secondary or collateral effects of such development can be difficult, and almost always involves some kind of numerical modeling. These effects are usually overlooked but can account for a significant portion of the damages.

One example is the contribution of run-off to coastal flooding caused by storm surge. Flood hazard simulations carried out by CDMP for Montego Bay, Jamaica, indicated that the increase in impervious terrain from roads and buildings would result in water levels of as much as 1 m higher in some locations if a storm identical to that of the November 1912 storm were to strike today. Another example are drainage works, culverts, and bridges designed for runoff associated with a specified return period and constructed over the past 3 decades in several of the Caribbean islands. Expansion of agricultural production and in particular banana cultivation has drastically changed land cover in many of the watersheds, changing run-off characteristics. Runoff levels once associated with a 30 year rainfall are now regularly exceeded by rainfalls of lesser intensity, causing localized flooding, subsidence, and damage to the infrastructure. Again, numerical models using GIS-based data sets can be used to determine the effects of events on the current and projected land cover for an area.

Traditional approaches to development planning and project formulation tend to view projects in isolation. As the above examples indicate, the risk to a project may not only depend on when, where and how it was built, but on what happens around it and how distant neighbors decide to use their land. Assessing the risk to any project needs to take into account the different ways in which the hazards that impact on the project can be affected by development decisions in the surrounding area. Hazard assessment and simulation models should be constructed to be able to incorporate these effects. A GIS-based storm model, such as the model used by CDMP, can rapidly incorporate changes to land cover. When a project is proposed, the projected changes to land cover can be made to the GIS data base, and the effects of future development estimated. Most GIS packages can incorporate the files used by engineering design software, such as AutoCAD. Unfortunately there is resistance in the development community to the use of predictive modeling, largely due to their uncertainty regarding the regulations which may result rather than any specific problem with the model themselves. The CDMP experience in Montego Bay, described below, illustrates this resistance.

10.4 The Montego Bay Case Study

One component of the Caribbean Disaster Mitigation Project is the assessment of potential hazards generated by tropical storms, in particular the potential for storm surge, coastal flooding and extreme wind. CDMP is using a numerical model to produce estimates of maximum sustained wind vectors at surface, and still water surge height and wave height at the coastline, for any coastal area in the Caribbean basin.

The model, named TAOS (The Arbiter of Storms) was developed by one of us (CW), and was originally created to analyze tropical storm risk to Hilton Head, South Carolina. TAOS is structured along the principles outlined above. The model relies on a generic data base structure, and uses U.S.G.S. digital data for deep ocean bathymetry, Defense Mapping Agency digital data for land boundaries and rough topography, satellite imagery for near-shore bathymetry and land cover, and the National Hurricane Center data for storm information. Model runs can be made for any historical storm, or for probable maximum events associated with different return periods. Model inputs and outputs can be configured for several common GIS formats.

Jamaica was selected as the site for a pilot project in coastal flood hazard assessment. Initial contacts were made with public and private sector agencies with interest in this activity, and a technical working group was established which includes the physical planning agency, several natural resource agencies, the University of the West Indies, the Association of General Insurance Agencies, and the Jamaica Institution of Engineers. With the collaboration of these agencies, coastal flood hazard maps were produced for the Montego Bay area at a scale of 1:12,500.

10.4.1 Production of Coastal Flooding Maps

In consultation with the Jamaican agencies participating in the project, it was agreed to represent the storm hazard in Montego Bay by selecting probable maximum storm events for a 100 year period and a 50 year period. Based on a statistical analysis of storms impacting Jamaica over the period 1886-1992 (see Table 10.1), a storm of category 3 was selected as the 50 year event, and a strong category 4 storm was selected as the 100 year event. While most storms passing through Jamaica maintain a distinct westerly motion, the more intense storms have moved in other directions, including a strong southwesterly component. Modeling indicated that due to the shoreline configuration of the Montego Bay area any southwesterly component increased the surge significantly. Therefore it was decided to take the above storms on a SSW track at the average forward speed indicated by the analysis of the historical data. A Maximum Envelope of Water (MEOW) map was created for the 50 and 100 year design storms by combining the maximums of each of 18 parallel runs spaced at 5 km intervals. The high resolution TAOS model was used, which incorporates estimated rainfall runoff, wave action and setup, and calculates storm surge on a 50 m cell grid.

While government agencies, and the insurance and disaster planning communities were generally pleased with the results, the local engineers, and in particular those representing large developers of coastal lands in the affected area, expressed concerns about the high surge values projected by the model. Whereas the selection of the storm-strength categories corresponding to a 50 and 100 year return period were generally accepted, the engineers felt that the particular family of tracks used in running the model was unfavorable, and would occur with much lower probabilities than those implied in the 50 or 100 year return periods. Another factor raised by the engineers in reviewing the model was that recent changes in coastline configuration through land reclamation may not have been reflected in the base maps that were digitized for use by the model. Finally, there was probably some confusion as to the incorporation of runoff estimates and wave heights to obtain the total water levels displayed on the maps.

Given the disagreement over the selection of the 100 year storm, it was decided to produce a series of maps for each Saffir/Simpson storm category, and for the three major directional groups of storms likely to impact the Montego Bay area. The result is a series of 15 MEOW maps. These maps, together with a statistical analysis of the storm history for the area, were presented to the engineers and distributed as a report to government planning agencies, insurance companies and other interested parties. Rather than CDMP selecting a 50 or 100 year event, it is now being left up to the individual user to decide the acceptable level of risk by selecting a design storm for a given project design life. The updated coastal configurations actually caused an increase, albeit small, in total water levels due to increased wave action in some locations.

10.4.2 Model Evaluation

While the TAOS model has been tested using historical storms in a number of locations throughout the Caribbean and U.S. (Watson 1995), it was hoped that some verification could be made in Jamaica itself. Attempts were made to test the reliability of the model's surge projections against historical data. The quality and quantity of surge information available in Jamaica is very limited. There are apparently no reliable tide gauge observations of tropical cyclone events. The most recent and systematic observation of surge phenomena was carried out for Hurricane Allen, which passed north of Jamaica in 1980 by the Government of

Jamaica (internal report 1980). The report consisted of a series of visual and circumstantial descriptions of high water marks. Unfortunately the observers made no consistent distinction between storm surge, still water levels, wave action, or wave run-up, causing some readers to either vastly over- or underestimate the true hazards. For example, the report mentions 40 ft (12.2 m) storm surges along sections of the north coast, immediately adjacent to areas where 3 to 4 ft (0.9 to 1.2 m) were noted. Examinations of topographic and bathymetric data, along with a model run of Hurricane Allen, revealed that the 12.2 m observations were most likely ocean waves striking steep bluffs, unprotected by reefs, which front on deep water. (Watson, internal CDMP analysis, 1995).

With regard to the statistical analysis of return times, the Jamaica Institution of Engineers presented information they felt contradicted the NHC data base with regard to the tracks and intensity of several storms occurring in the late 1800s and early 1900s. An analysis of this information revealed that the complaints brought up by the engineers were largely based on a misreading of the original historical observations, and that the corrected NHC data base was reasonable. When the engineering association was presented with this analysis they agreed to drop their technical reservations, and are now advocating that the maps be used in the building code. It is interesting to note that in the end the design storm selected by the engineers as the basis for the building code is virtually identical to the storm originally selected by CDMP for the first set of maps.

While it would be easy to dismiss the concerns expressed by land developers and their engineers as being based on economic self-interest, the underlying worry of the development community should be addressed in a constructive manner. The development community does not trust the government land-planning process to take in to account their economic concerns. Developers are willing to accept limitations on lifeline infrastructure in hazard zones, and perhaps more stringent construction requirements for other projects. They continue to be concerned however with overly restrictive regulations for general development in these zones, or with the imposition of such onerous requirements as to make development economically unfeasible. Planners thus face a daunting task of setting levels of acceptable risk for various categories of land use, by using mechanisms such as zoning, taxation, or subsidies.

10.4.3 Communicating Long-term Risk to Planners and Investors

The rapid growth of tourism over the last few decades in the Caribbean islands has attracted substantial investments in tourism facilities, economic infrastructure, and housing to coastal areas exposed to significant storm risk. For a variety of reasons the principal mechanisms for reducing losses from disasters, land use planning and the application of appropriate building standards, have been demonstrably ineffective. Both have traditionally suffered from serious deficiencies in enforcement. By involving the investment and insurance sectors in the hazard planning process, it is hoped that development can be redirected away from risk prone areas for economic reasons rather than by regulatory means.

CDMP has adopted a strategy of promoting the self-interest of parties subject to natural hazard risk as a means of overcoming the serious limitations inherent in systems relying uniquely on enforcement. The strategic focus is on producing accurate estimates of hazard risk and disseminating this information in easy to understand and readily used formats for decision makers in public and private sectors involved in development planning. The CDMP supports technical agencies in both sectors with training and technology transfer aimed at preparing these agencies to produce the required hazard risk information, and to use the information for land use zoning and development plans. It is hoped that once this information is available, insurers will be reluctant to insure in high hazard areas, and more willing to insure in lower risk zones, thus providing an incentive for investors to locate their development there.

10.5 Conclusions

Given the variability in the frequency, intensity, and impact of tropical cyclones, disaster planning should be a dynamic process. Through the use of GIS, remote sensing, and GPS, models used to produce disaster plans can be kept up to date in a timely and cost effective manner. This enables the disaster planner to keep up with rapidly changing coastal areas. While it is difficult to convey the implications of natural and anthropogenic variability to the user community, early involvement of all users of hazard data can assist in this process.

Appendix: Technical Notes on the TAOS model

10.A.1. Introduction

Numerical simulation of storm systems is an essential component of coastal zone management. SLOSH (Sea, Lake, Overland Surges from Hurricanes), the model used by the National Oceanic and Atmospheric Administration (NOAA) to predict storm surge and create evacuation zones, and WHAFIS (Wave Height Analysis for Flood Insurance Studies), the program used by the Federal Emergency Management Administration (FEMA) to create flood insurance zones, are two of the most notable examples. Unfortunately for the users of these models, they require custom data bases and long development times. For example, the recent creation of a SLOSH model basin for the northern Bahamas (New Providence and

Eluthera) took over a year (personal communication with Arthur Rolle, Bahamas Weather Service). Although SLOSH shows considerable skill in forecasting storm surges, the end products can be rather difficult to use by planners. The FEMA WHAFIS program, which calculates wave damage zones, requires storm surge data from external sources, manual compilation of topological and land cover data along single transects, manual compilation of the results, and extrapolation to nearby areas (FEMA 1989). Finally, modeling of wind effects are presently not considered by any widely available model, even though wind damage is arguably one of the most significant hazards associated with storms. These facts, along with the author's previous experience with coastal zone management at the Town of Hilton Head Island, South Carolina (Watson 1992b), led to the development of a new storm hazard model. Building on the legacy of SLOSH for storm surge modeling, and by adding wind damage estimates and explicit wave effects, The Arbiter Of Storms system (TAOS) was born. The numerical techniques for solving the complex equations of fluid flow came from the astrophysics community, who often have difficult factors to consider when modeling fluid flow in stars and gas clouds, such as electromagnetic and relativistic effects (Bowers and Wilson 1991). In addition, by paralleling two microcomputers and a Sun workstation, large basins at higher resolution may be modeled. By using the latest techniques of computer based mapping and remote sensing, basin development times are vastly reduced. For example, runs for the Bahamas area (including the SLOSH model basin mentioned above) were completed in three working days.

10.A.2. System Overview

Figure 10.A.1 is a block diagram of the TAOS system. It is quickly apparent that GIS technology and remote sensing are critical aspects of the overall TAOS system. Data management on both the input and output sides of the model are handled entirely by GIS. Creation of the data base for any physical modeling process often requires data in different formats. For TAOS, sources of data include satellite images, paper maps, and existing digital data bases. Satellite images from low earth observation satellites such as LANDSAT and SPOT are an excellent source of information. Through image processing, land cover may be determined, and water depths in the near shore area may be calculated by using the amount of light reflected from the bottom. TAOS can directly use data bases created by three of the more popular GIS packages. Idrisi, from Clark University, is the most widely used and one of the most inexpensive GIS packages available. Idrisi is the GIS package of choice by many in the research community, and is the standard package for the Caribbean Disaster Mitigation Project. GRASS, a public domain effort of the U.S. Army Corps of Engineers (USACOE; U.S. Army 1992), is also widely used. The UNIX source code for GRASS is freely available, making it an attractive option for the computer modeler. The UNIX version of TAOS uses GRASS as its data base manager. Arc/Info, a commercial GIS package from Environmental Systems

Research Institute, Inc., is widely used by U.S. Federal, State, and local governments. TAOS also supports several Arc/Info binary formats. As indicated in Fig. 10.A.1, the GIS processes the raw data in to three of the four input parameters: topography, land cover (for damage estimation purposes), and friction layer (to determine water and wind flows). The fourth data set, the meteorological parameters, comes from the U.S. National Hurricane Center (NHC) reports and data bases.

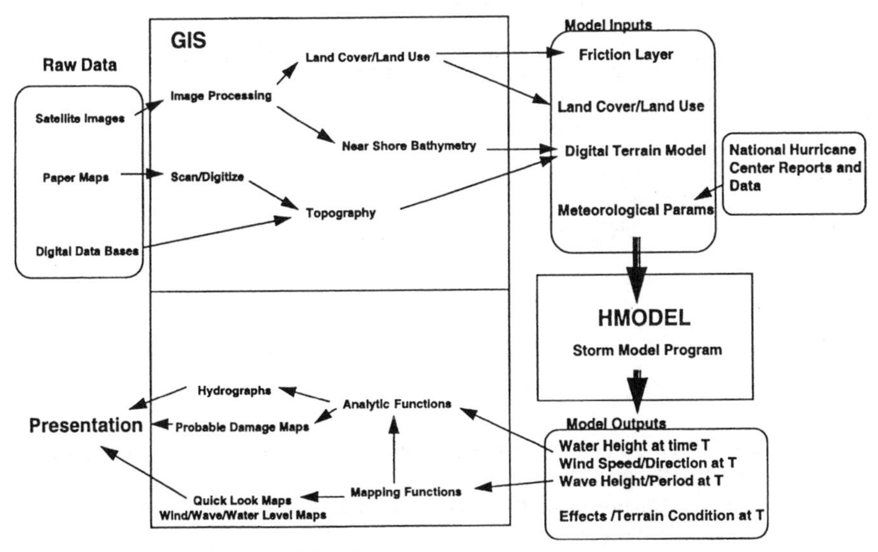

Fig. 10.A.1. TAOS Storm Modeling Process.

10.A.3. Verification and Comparisons with Existing Models

The ultimate test of any computer model is how well it compares with the real world. Also of interest is how well it performs when compared to existing techniques. A TAOS run was made for Hurricane Hugo, 1989, along its entire track using a 4 km grid cell size. Storm track and intensity data were from NHC. Deep water depths were obtained using the U.S.GS ETOPO5 data base, while topographic and other data was derived from the DMA (U.S. Defense Mapping Agency) Digital Chart of the World. Near shore depths were extracted from nautical charts, satellite images, and the ETOPO5 data. Inland friction data was obtained by using recent NOAA AVHRR data. The table below indicates the results for this run, and for two other tide gauge readings for runs of hurricanes Allen and David using the same data base. Also noted in the table are the results of two other models. The SLOSH runs are from data provided by Dr. A. Mercado of the University of Puerto Rico, while the data from the Meteo-France model was provided by the Caribbean Meteorological Institute.

The above values are in meters, including tides. TAOS runs were using the uncalibrated Regional Forecast System (RFS) run on a 4 km per grid cell data base. OB = Observed (estimated from high water marks), TG = tidal gauge readings, and are in bold type. MF = The Meteo-France model for Guadeloupe and Martinique, SLOSH = NOAA SLOSH model results from the Univ. of Puerto Rico (From A. Mercado)

Table 10.A.1. Comparisons of Peak Surge Results in the Carribean.

Storm	Location	Observed	TAOS	Other Models
Hugo	Pointe-a-Pitre	1.5 OB	1.51	1.48 (mf)
Hugo	La Puntilla	1.1 TG	1.20	0.82 (SLOSH)
Hugo (SLOSH)	Lugillo	1.3 OB	1.50	1.2
Hugo	Los Machos	2.0 OB	2.00	1.9 (SLOSH)
Allen	Le Robert	0.6 TG	0.60	0.53 (MF)
David	Pointe Fouillele	0.4 TG	0.40	0.25

References

Bowers, R. and Wilson, J., 1991. *Numerical Modeling in Applied Physics and Astrophysics*, Jones and Bartlett, Boston, 705 pp.

Elachi, C., 1986. *Introduction to the Physics and Techniques of Remote Sensing*, John Wiley and Sons, New York, 43 pp.

Environmental Research Institute of Michigan (ERIM), 1985. Evaluation of landsat thematic mapper data for shallow water bathymetry, unpublished, internal paper.

Federal Emergency Management Agency, 1989. *Guidelines and specifications for wave elevation determination and V zone mapping*, Washington, D.C., 175 pp.

Government of Jamaica, 1980.

Jelesnianskski, C., Chen, J., and Shaffer, W., 1992. *SLOSH: Sea, Lake, and Overland Surges from Hurricanes*, NOAA Technical Report NWS 48, Silver Spring, MD, 71 pp.

Mercado, A., Thompson, J., Evans, J., 1993. Requirements for modeling of future storm surge and ocean circulation, In: Maul, G.A. (ed.), *Climatic Change and the Intra-Americas Sea,* Edward Arnold, New York and London, 389 pp.

U.S. Army Coastal Engineering Research Center, 1992. *Shore Protection Manual*, Ft. Belvoir, MD, 690 pp.

U.S. Environmental Protection Agency (EPA), 1988. *Greenhouse Effect Sea Level Rise and Coastal Wetlands*. EPA Report 230-05-86-013, Washington D.C., 188 pp.

U.S. Geological Survey, 1990. *National Water Summary 1988-1989, Hydrologic Events and Floods and Droughts*, Reston, VA, 379 pp.

Watson, C., 1992a. Using satellite based remote sensing to monitor near and off shore sediment formations, *Archives of the International Society for Photogrammetry and Remote Sensing*, Washington, D.C., VII, 188-195.

Watson, C., 1992b. GIS aids in hurricane planning, *GIS World Special Issue*, **4**, 1, 46-52.

Watson, C., 1995. TAOS: A GIS based meteorological hazard model. *National Weather Digest,* **20**, 2, 2-11.
Young, K.C., 1994: A multivariate chain model for simulating climatic parameters from daily data. *Journal of Applied Meteorology,* **33**, 661-671.

11 Communicating Climate Research to Policy Makers

Arthur A. Felts[1] and David J. Smith[2]

[1]Institute for Public Affairs and Policy Studies
University of Charleston
66 George Street
Charleston, SC 29424, U.S.A.

[2]Southeast Regional Climate Center
SC Department of Natural Resources
Water Resources Division
1201 Main Street, Suite 1100
Columbia, SC 29201, U.S.A.

Abstract

This study reports on surveys of natural resource managers and state level administrators in four states in the Southeastern United States. The surveys were designed to ascertain how natural resource policy makers viewed emerging scientific research on long-term climate variability and where they acquired their information. The data suggest that little first-hand information gets to policy makers. Rather, research is filtered through various sources, notably professional journals and the mass media. These filtering processes may influence information in ways not intended by researchers and may present data in ways that actually distort. Analysis of the educational background of the respondents revealed consistent differences in approaches to inconclusive research on climate variability between those with natural science training and those with social science training. Natural scientists were quicker to form judgements and more critical of methodology while social scientists clearly exhibited a greater tolerance for ambiguity.

11.1 Introduction

Over the past two decades, researchers have focused on effective ways to communicate hazards posed by natural disasters to the general citizenry (e.g.,

Turner 1981; Quarantelli 1984; Meleti and Sorensen 1990). The aim has been to develop a general framework for risk communication which will ultimately reduce the threat that natural hazards pose to human life.

In risk communication, a great deal of attention has been paid to how individuals process hazards-related information and deal with probabilities. Existing research suggests that many tend to downplay the likelihood of disasters. One only need observe human activity before a known natural disaster, such as a hurricane, to see lack of planning in the face of many natural hazards risks. Irrational behavior and poor planning is in part a result of the frequency of occurrence of the natural hazard itself. The risk of hurricane landfall on a specific coastal areas is rare enough to allow complacency on the part of the public.

As research on natural hazards risk communication to the general citizenry has developed, attention has inevitably focused on more specifically defined audiences. This paper is intended to contribute to that literature by presenting research on how and what information natural resource policy professionals are processing related to decadal climate variability. The focus here is not on risk communication as a methodology or technique, but rather upon what risk is being communicated from an emerging body of scientific research, namely decadal climate change.

Reported here are the results of preliminary empirical investigations directed at ascertaining how natural resource policy makers are currently acquiring and processing information associated with decadal climate variability. Much of the investigation focused on ascertaining what specific types of information policy makers desired, but to get at this, it was necessary to investigate what they already knew and where they typically acquired information. By providing information about how emerging research-based information about decadal climate variation is being acquired and processed by natural resource industry professionals, the ultimate aim is to provide information which may improve communication about the risks posed by natural hazards and couple this with policy research literature.

Investigating how natural resource policy makers acquire and process information related to it offers a unique opportunity to explore how newly developing knowledge is disseminated, judged, and incorporated into cognitive maps. Clearly, research into decadal climate variability indicates an increased vulnerability of the Southeastern United States to more frequent and severe hurricanes (and conjoint phenomena such as regional changes in seasonal rainfall patterns) over the next two decades. But since the research is in its initial phases, it is subject to considerable debate on both conceptual and methodological grounds. Whatever the tentativeness of the research, the data presented here suggest that many policy makers are already in the process of acquiring information, and, in many cases, reaching conclusions. The significance of this for future efforts to communicate risk associated with increased hurricane vulnerability and long-term climate variability cannot be overstated.

An operant hypothesis guiding this research is that an individual's organizational role and background will influence their decision making. Naturally industry policy-makers will respond differently to risk-related information than the general citizenry because their decision making environments are different from those of private citizens. A popular tire ad displaying a baby, safely ringed by a tire, illustrates this. Assuming the tire is more reliable, a parent, acting as a parent, might spend considerably more - say 50% - to obtain a marginal increase - say 5% - in tire safety/reliability. But the same parent when acting as an industry policy maker might deem an additional 50% expenditure unacceptable for the marginal gain in safety/reliability when deciding on equipping company cars. Moreover, there will be differences in how information dealing with risks posed by climate variation are analyzed among different industries, and individuals will deal with the information in different ways which are influenced by factors associated with their particular areas of policy making or management of a resource.

11.2 Research Design

Under research contract from the Southeast Regional Climate Center funded by grants through the United States Environmental Protection Agency and National Oceanic and Atmospheric Administration with additional funding from the South Carolina Sea Grant Consortium, staff from the Institute for Public Affairs and Policy Studies at the College of Charleston surveyed natural resource managers in Florida, Georgia, North Carolina, and South Carolina to determine how they viewed the emerging information on long-term regional climate variability.

Four specific natural resource areas were selected: agriculture, energy, water, and forestry. The forestry industry was split off from the remainder of those engaged in agricultural activities because it was hypothesized that foresters would exhibit significant differences as a result of their longer crop cycles and well-established, regulated management practices. As well, each of these natural resource areas may be differently affected by climatic phenomena associated with long-term climate variation The selection of the four states was intentional; all are located in the Southeastern United States, and thus maximally exposed to catastrophic hurricane activity. While the risk is greatest in coastal areas, the risk from wind and flooding in inland areas is significant as well.

The research was conducted in two phases. After the initial group of policy makers was surveyed, a subsequent survey of state-level policy makers involved in natural resource policy making was conducted. Natural resource policy makers were surveyed beginning in late 1992 and continuing through the summer of 1993. Mail, telephone, and face-to-face surveys were administered, all using the same instrument. Data bases for the survey were drawn from lists of registered water users, energy producers, commodity boards (such as Peach, Tobacco, Soybean, etc.), and presidents of Farm Bureau county chapters. These lists were supplied by state regulatory agencies and commodity boards. The forestry sample included growers, consultants, professional managers, associations, researchers, and industry manufacturers. In the second phase, state natural resource administrators in South Carolina and Georgia were interviewed, using essentially the same instrument, in person in a follow-up study in 1994. But the focus was upon qualitative data collection in the second phase.

A two-part survey instrument was employed throughout the research. The first part asked for general information on educational background, industry training, prior positions, and information and personal/professional views on emerging information related to long-term climate change specifically and environmental policy generally. The second part asked industry or natural resource area specific questions relating to what types of and severity of events would influence behavior or what steps have been taken in hazards mitigation. These were intended to elicit responses that would allow insights into how natural resources were viewed within the industry as well as determine how policy makers thought long-term climate variation might affect different resource areas. Because of the complexity of the issue and the need to probe in several different areas, the questionnaire was lengthy, with a total of forty-nine questions, some of which could not be readily answered but required some thought.

The survey instrument was constructed with the assistance of an advisory board assembled with support from staff of the Southeast Regional Climate Center. The advisory board was composed of at least two representatives from each natural resource policy area, state-level policy makers, and staff from the Center itself. After construction, the instrument was circulated among the advisory board for revision and comments. After two iterations using this process, the survey was pretested, using names drawn randomly from the data base of policy makers. Pretesting occurred through telephone interviewing and telefaxing. Thirty policy makers participated in the pretesting activity. After all revisions were made based on the pretest, the survey was once again circulated among the advisory board for final comment.

Given the lack of operant hypotheses based on prior empirical research and the fact that research on long-term climate variability is relatively new, a broad research strategy was chosen. For example, presurvey interviews indicated that many policy makers would not be aware of the most recent research on climate change and those who did might not be clear on what risks were suggested by the research. The survey instrument was designed with this in mind, and was thus probed on a number of fronts, in some instances, posing hypothetical questions which were suggested by recent research.

Surveys were sent to 2242 natural resource managers in the four-state study region; 442 responded, giving the study a 19.7% response rate. In interviews with the advisory board during the survey construction and in pretesting, it quickly became apparent that a survey of this type would produce a relatively low response rate; though it should be noted that the actual response rate is not at all inconsistent with similar surveys. Lack of survey response has been noted at least since 1838 (Porter). The primary threat that it poses is that the respondents would not be representative of the population, and thus the findings cannot be generalized.

Several methods have been developed to enhance responses (Ayidiya, 1990). Some of these, such as call-backs, were employed in this survey.

There are several obvious reasons to explain the final response rate. As noted above, decadal climate variation is a relatively new area for scientific research. Thus, despite the inclusion of a detailed cover letter, a significant number of respondents likely viewed the survey as irrelevant for their activities, not seeing the saliency of the issue or research.

Pretesting of the survey indicated that decadal climate variability was conceptually connected by many policy makers with theories of climate change connected to hypotheses about the potential for global warming. Again, even though the cover letter clearly explained that hypotheses about global warming and decadal climate variability should and/or could be understood as conceptually distinct, some dismissed the latter on the basis of skepticism about the former.

Finally, the length of the questionnaire and the rate of response must be noted. Research on response rates has consistently noted the linkage between response rates and the length of the survey. The longer the survey, the lower the response rate. Given the complexity of the subject and the need to gather data on the background and perceptions of the respondents as well as data on how information was acquired and whether or not it was being used, a decision was made that fewer responses with more complete data would be more valuable than a larger number with less information.

The representativeness of the respondents to the entire survey population can of course be questioned. Over the years, numerous authors have addressed themselves to this issue (e.g., Baur 1947; Donald 1960; Armstrong and Overton 1977). Nearly all involve statistical weighting of the respondents to reflect better the demographic attributes of the population or are based on statistical adjustments made possible through additional data, for example, comparisons between pre-election surveys and the actual election outcome. Some involve weighting of extreme responses on Likert-type scaled answers. Only one means of testing for bias among the respondents in this survey was employed. Responses were stratified into four groups based on when the survey was returned; that is, the first 110 respondents were grouped into one group, the next 110 respondents into a second group, and so on. These were then compared on a range of demographic variables and response patterns to survey questions with the hypotheses being that if significant variations occurred, then the representativeness of the respondents to the population would be more suspect. No significant variations on any of the selected variables were noted among the subsamples of respondents.

11.3 Data: General Observations

Extensive analysis of the survey data revealed considerable interaction among and between several explanatory variables, including type of industry, state, natural resource area, and background. Thus, it is not possible to analyze satisfactorily the data for explanations of the observed variations without making several qualifications. For example, in some cases water resource managers from Florida answered like other water resource managers from the other states. In other cases they showed differences that appeared to be based on educational background. Finally, they also showed some similarities with other resource managers from Florida.

This directly suggests a complexity of decision environments that perhaps changes with hazard type, the proximate immediacy of the hazard, and relative threat to a natural resource or economic commodity. To give an example of this decision environment complexity, during Hurricane Hugo in South Carolina, risk was communicated to state and local government and natural resource managers through a flow of information as shown in Fig. 11.1.

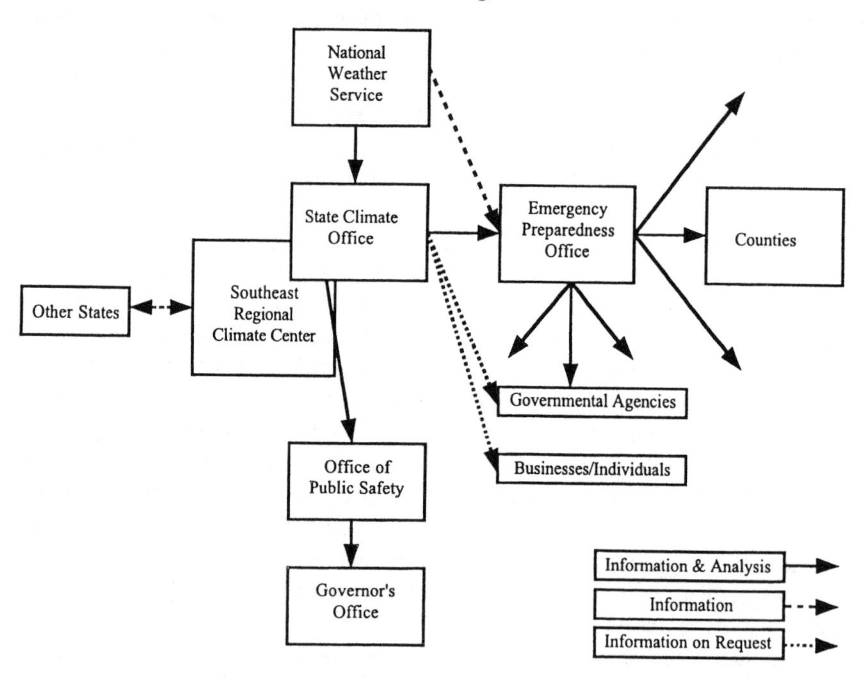

Fig. 11.1. Flow of information through South Carolina Climate Office.

This flow of information allowed climatologists and meteorologists access to pertinent information regarding the hurricane and in turn prompted a response. Throughout the process, it was unclear whether immediate responses would translate into long term decision making. Qualitative data from the interviews conducted in this study indicate that managers tend to separate planning strategies from short-term responses. This is likely due to the fact that events in the short term require response now and are generally considered to be of higher risk than potential or statistical long term climate changes.

Another broad interpretation of the findings suggests that there was minimal evidence indicating resource managers knew that research into decadal climate variability suggests that the risk of hurricanes rises over time and then falls, that is, is not broadly constant within yearly variations. Moreover, as illogical as it sounds, risk seemed to be interpreted in a relative, comparative way within resource areas. In discussions with decision makers in South Carolina, North Carolina, and Georgia, it was noted that as long as the risk for hurricanes is less than in Florida, there was little concern with changes in decadal frequency of hurricanes. This suggests that some policy makers look for cues from those they think may be exposed to greater risk; they will not act until they see these others act. There was also some indication from the survey results, particularly from private commodity groups, that risk was interpreted relative to what the competition or an economically influential group (such as the insurance industry) interpreted.

11.4 Observations

Some findings, however, can be reported more directly. For the purposes of the interest here in risk communication to natural resource policy makers, the most relevant data obtained from the study have to do with the information acquisition patterns of the policy makers surveyed.

The first item of note is whether or not risk information is being communicated to policy makers. Table 11.1 contains the frequency responses for the policy makers in response to a question which asked if they have read any information specifically addressing the potential impact of long-term climate variability on their particular industry.

Table 11.1. Are you aware of information about general climate change with respect to your policy area?

	Natural Resource Policy Makers	State-level Policy Makers
Yes	47%	78%
No	52%	22%

More than one-half the natural resource policy makers indicated they were not aware of such information. This contrasts with state policy makers among which more than three-quarters indicated some awareness. No real analysis of this observation can be made since no comparative data are available. A reasonable hypothesis would be that the number of managers aware of the potential for climate change is growing. But what factors make it grow should be investigated. One hypothesis would be that awareness ebbs and expands in connection with specific events such as Hurricanes Hugo, Andrew, and Opal (Karwan and Wallace 1984). The busy 1995 hurricane season might also raise some awareness. If, however, 1996 showed a decline, would that, in turn, lead to some complacency? In the end, this type of speculation raises the question of precisely what is necessary for policy makers to adopt mitigation strategies and enhance preparedness. As noted above, natural resource managers may view these types of decisions with a more business-like eye, but still be subjected to human proclivities to discount risk. For example, after Hurricane Hugo, a great deal of press was given to noting that the event was a "once in one hundred year" hurricane. This was clearly a statistical statement, yet many noted that they would not be alive one hundred years from now and thus did not need to worry about another Hugo-like hurricane. Inland farmers might react similarly to the intense flooding associated with Opal in 1995. Insurance may be viewed as sufficient preparation for a future event. This is where the differences between natural resource managers and a general citizenry are of note. If an individual directly experienced two or three hurricanes in a relatively short period of time, say four years, he or she might well decide to relocate to a less exposed area inland. But natural resources are fixed in place, and thus the exposure is more constant. But again, there is no evidence that the policy makers interviewed in this research clearly saw this.

Researchers investigating climate change will likely report their findings in traditional forums. This usually means professional association meetings and academic journals. In some cases, professionals may access academic journals directly, but in other cases, they will prefer information sources that have a more applied focus. This raises the question of how research data gets filtered over to policy makers.

Extensive bibliographic searches revealed no systematic studies of how technical information is disseminated. The process thus appears stochastic. A news reporter might stumble across a journal article, or hear about scientific research from a contact. An applied or industry journal editor might attend an academic meeting and sit in on a panel or plenary session. A policy maker might seek some guidance in developing a policy. In some instances, specific entities are responsible for disseminating research information to policy-makers. Most notable are the regional climate centers, state climatologists, Consortium for International Earth Science Information Network (CIESIN) in Michigan, Carbon Dioxide Information Center (CDIAC) in Tennessee, and National Center for Atmospheric Research-Environmental and Societal Impacts Group in Colorado. But no tracking has been done to see how information gets disseminated into secondary and tertiary venues.

The point to be made about patterns of information dissemination beyond academic forums is that there is no way to predict in advance exactly how it will be disseminated, nor, for that matter, in exactly what context it will be portrayed. One only need look at the vast amount of hyperbole and crude simplification entailed in presenting information about potential global warming during the mid to late 1980s to get a sense of this process. Many researchers were careful to qualify their findings and conclusions; but little of this came through.

To ascertain where they acquired information, the respondents in both studies were specifically asked where they acquired information about climate variability. The responses to this question are contained in Table 11.2.

Table 11.2. Sources of information about climate variability.

	Industry Policy Makers	State-level Policy Makers
Professional Journal	51%	
Mass Media (Newspaper/TV/Magazines	43%	38%
Professional Meetings/Conferences	27%	46%
Industry Journals	20%	17%
University Publications	15%	13%
Government Publications	15%	29%
Federal Government Publications		46%
State Government Publications		25%
State Climatologists		17%
Academic Journals		8%

Note: Columns total more than 100% because respondents indicated more than one source.

While the general diffuseness of information sources indicated in Table 11.2 is of note, it is equally as important to note the lack of reliance upon original resources. If we assume that most research on climate variation is first published in academic journals or university publications with some appearing in government documents, it is obvious that a great deal of risk information is being filtered to policy makers through professional and trade journals, and, alarmingly, the mass media. Table 11.2 indicates that there is, in short, no direct link between government and university researchers and the natural resource policy makers. The implications of this for information dissemination on issues such as climate variation on a decadal scale needs careful study.

To gain some insight into how professional journals make decisions about content, a survey was constructed and administered to journal editors. The journals selected were those specified by the natural resource policy makers as sources of information. Editors from 10 journals were interviewed for the survey. The majority of the editors (70%) classified their publication as a professional, rather than academic or research. Each publications is printed by a society or organization and the mean circulation of all the journals was 23,380. Commercial advertisements were found in 80% of the journals, while 50% of the journals contained job advertisements.

When asked, 8 of the 10 journal editors reported extensive use of college and university research as a source of information for their articles, with personal research and government reports coming in second. This indicates that scientific research is being communicated, but leaves unanswered a host of questions dealing with how it gets filtered into the cognitive maps of different professionals. With respect to issues such as climate variability, we need a greater understanding of how editors see their publications; that is, are they designed to respond to what their subscribers want, or do they have a more autonomous role with respect to decisions on specific articles. We also do not know how much input researchers have directly with these professional journals. Is the research published by journals interpreted by staff and/or free lance authors, or do they encourage researchers to report on their own findings.

However one interprets it, it is striking that the most significant percentages in Table 11.2 are the indicated reliance upon general news media as a source of information by policy makers. Although some industry-specific professionals reported a higher reliance on professional journals, the reliance on general news media was systematic with more than 50% in all segments reporting it either first or second. What may we speculate from this?

While few specific studies look at climate change per se, various studies of how information is reported in mass media invariably point to a process of accentuation and simplification. In the case of information about a complex natural science area of research such as regional climate variability, crude simplifications are promulgated.

The reliance on mass media as an information source also makes professionals look like the general citizenry with respect to information acquisition about risks associated with natural hazards. But if they process information differently in accordance with their professional roles, then the general news media may be a particularly unsatisfactory source of information. News media target general audiences and seek simplification where additional complexity may be needed in order to arrive at a professional judgment (Spencer and Triche 1994).

The above observation has provided some support in the results reported in Table 11.3. These are the response rates by the natural resource policy makers to the question of whether or not they found information about climate variability useful for their specific decision making or not.

Table 11.3. For those that did get information about climate change, did respondents find information about climate change helpful in making decision?

	Industry Policy Makers	State-Level Policy Makers
Yes	26%	60%
No	70%	35%
Unsure	4%	5%

Among public and private natural resource managers, nearly 3 out of 4 suggested that information they had acquired to that point in time was not helpful. The data for State-level policy makers was more encouraging, but still cause for concern in that more than 1 out of 3 suggested information was of little or no help as well.

The fact that the policy makers surveyed did not view the information they had acquired on climate variation to be helpful is perhaps the most critical finding of the entire survey. Extensive presurvey interviewing of natural resource managers led to an operant hypothesis that there was not a great deal of concern with climate variation as a single factor in decision making. This perhaps explains why many did not view the information they had received on climate variability to be particularly helpful. That is, unless the information on climate variation is presented in a fashion that allowed analogues, such as comparisons to previous periods like the 1930s drought in the U.S., or correlated with relevant data, such as the effect that increased rainfall had on specific crops in past periods, the policy makers tended to discount emerging information. Face-to-face interviews during the survey also revealed the need by policy makers for more certainty in the models and forecasts. This suggests that researchers are not doing an effective job of conveying information in a format that suggests the relative risks (as opposed to certain risks) posed by decadal variations in climate.

The data in Table 11.3 become all the more intriguing and noteworthy when it is observed that 22% of the respondents answered affirmatively to a question which asked whether or not they believed that a change or variation in the climate will cause problems in their lifetime. The data from this question are presented in Table 11.4. Intuitively one could assume that this group would differentiate on the basis of age; that is, older respondents, who have seen more overall climate variability might be more likely to believe variation in the climate will cause problems. But the converse can also be hypothesized; that is, because younger respondents anticipate living longer, they might believe the climate will vary over their lifetime. This latter hypothesis was supported by the data.

Table 11.4. Views on whether or not climate change will cause significant changes in lifetime.

Yes	67 (22%)
Don't Know	73 (24%)
No	164 (54%)
Total	304 (100%)

The mean age of those answering "no" in Table 11.4 was 48.11, while the mean age of those answer "don't know" was 46.57, and, finally, the mean age of those answering "yes" was 45.15. Those who reported either "don't know" or "yes" to this question were below the mean age of 47.01 for the group while those who responded "no" were above it. Though the difference in age among these is not great, the observed differences are statistically significant. Standard deviations on this data are such that all categories of response overlap, largely accounted for by the lumping of the ages of the respondents into the 40s; however, the consistent and positive association of this answer with age suggests the need for follow-up research.

11.5 Absorption of uncertainty and educational background

As mentioned, the data presented consistent problems in analysis due to interaction among and between key variables. However, after extensive analysis, one set of persistent breakouts were displayed by the data. These are exhibited when educational background is controlled for whether or not respondents had training in the natural sciences such as chemistry, biology, and forestry or the social sciences. The data showing this with respect to the same question reported in Table 11.4 are in Table 11.5.

Table 11.5. Science vs. Nonscience background views on whether or not climate change will cause significant changes in lifetime.

	Nonscience Background	Science Background
Yes	23 (33.3%)	44 (18.5%)
Don't Know	12 (18.8%)	61 (25.6%)
No	32 (46.6%)	132 (55.5%)
Total	67 (98.5%)	237 (99.6%)

Note: Total percentages are less than 100% because of missing values in the original data set.

As Table 11.5 shows, a significantly higher percentage of those without a natural science background thought problems associated with climate change to be more likely. What can account for this? Face-to-face interviewing suggested that most of the natural scientists typically felt constrained to arrive at quicker conclusions regarding the significance of scientific research because they were supposed to know about it. In addition, natural scientists indicated dissatisfaction with the quality of research used to generate predictions associated with climate change.

From this, we may hypothesize that risk communication to professionals will be influenced by whether or not they are expected to utilize their educational background in interpreting scientific information. For those with a natural science background, the concern will be less with the risk predictions and more on the models used to generate the predictions.

One could assume that the natural scientists who seem more skeptical than their nonscience counterparts of long-term climate change have developed a more "informed opinion" on climate variability; that is, they have read about it and decided that it posed no problem. This, however, does not appear to be the case. When asked if they had read any professional books, articles, or reports in the past two years that specifically addressed climate change, a higher percentage, but statistically nonsignificant number of the natural scientists indicated they had done so. Thus, there is no reason to believe it is information acquisition on climate variability that differentiates the two groups.

That the data do not support this idea of an informed opinion among scientists is further reinforced when responses to a follow-up question addressed to those who reported reading something about climate variation in a professional journal, article or report were asked about the utility of the information, that is, whether it was helpful or not. Again, there was no relation between exposure to information and judgments as to its utility and educational background.

A different perspective on this idea of informed opinion is gained when those who reported they expect a significant climate change in their lifetime are correlated with whether or not they reported exposure to literature on climate change. Table 11.6 shows the results of this, and Table 11.7 differentiates the group according to educational background.

Table 11.6. Response on exposure to literature on climate change for those who reported they expect a significant climate change in their lifetime.

	No Exposure	Exposure
Yes, Climate change	30 (18%)	46 (28%)
Don't Know	61 (36%)	28 (17%)
No, Climate Change	80 (47%)	94 (56%)
Total	171	168

Chi Square P Value = .0009

Table 11.7. Science vs. nonscience background response on exposure to literature on climate change for those who reported they expect a significant climate change in their lifetime.

	Nonscience Background: No Exposure	Nonscience Background: Exposure	Science Background: No Exposure	Science Background: Exposure
Yes, Climate change	12 (32%)	11 (38%)	14 (13%)	30 (24%)
Don't Know	6 (16%)	7 (25%)	42 (40%)	19 (15%)
No, Climate Change	20 (53%)	11 (38%)	51 (48%)	79 (62%)
Total	38	29	107	128

The two tables offer a markedly different view of the target population. Table 11.7 shows that natural scientists are more likely to form an opinion when exposed to literature, and generally, that opinion will tend to be a negative assessment of the prospects for significant climate change. This assertion is supported by the fact that a higher percentage move into the "no" cell than into the "yes" cell (from "don't know") among those reporting exposure to literature. Two radically different explanations offer some possibilities for further research.

First, an argument can be made that the above data support a hypothesis that natural scientists possess a base knowledge that allows them to judge information being presented to them. This directly suggests that policy responses to research, when either recommended or developed by professional scientists, will be better informed on the basis of research coming from climatologists and others researching climate change. Generalists, on the other hand, will not be able to absorb research data as it is currently being presented in a way that leads to an informed judgment with respect to either recommending or developing policy.

The hypothesis of informed judgment on the part of scientists as opposed to generalists is at least partially undermined by the observation of the high degree of uncertainty that currently characterizes much existing climate change research. That is, it suggests that as professionals read the literature, they are inclined to absorb uncertainties reported in the research and make a "yes/no" judgment, as opposed to a "maybe." March and Simon (1958, p. 165) report on the negative dimensions of uncertainty absorption:

> Through the process of uncertainty absorption, the recipient of a communication is severely limited in his ability to judge its correctness. Although there may be tests of apparent validity, internal consistency and consistency with other messages, the recipient must, by and large, repose his confidence in the editing process that has taken place, and, if he accepts the communication at all, accept it pretty much as it stands.

This suggests that if natural scientists are performing a gatekeeper function within their organizations, they are doing so in a fashion that absorbs the uncertainty surrounding the current research on climate variation and presents the research as not meriting serious attention.

It is reasonable to speculate that natural scientists, when compared to others in similar policy making positions, have a greater understanding of scientific research and the reporting of findings than do nonscientists. If we assume that, in reporting scientific research, natural scientists tend to absorb the uncertainty about their findings and research as they report it, then the skepticism shown by scientists toward the research of other scientists may be an effort to reinsert this uncertainty they know is present. Thus, they will tend to diminish the significance of the findings, suggesting that they are even more uncertain than they appear to be as presented. In an area of known research uncertainty such as climate change, this will incline the group to be even more skeptical.

Existing research on climate variability is filled with uncertainty, and the literature has consistently emphasized this uncertainty. However, there is some evidence that rather than tolerating this uncertainty, natural resource managers are reaching conclusions rather than reserving judgment. The implications of this are unclear. On the one hand, it may be relatively easy, if climate research takes a more conclusive turn, to change policy. On the other, the fact that opinions are being formed may be a problem because once formed, they may be more resistant to change.

Experts do not appear to be experts when they suggest ambiguity or waffle with respect to models. Indeed, expertise may be predicated upon a certainty of a base knowledge that prompts one to make conclusions, even when conclusions are unwarranted. If this is happening with respect to developing policies appropriate to deal with climate change, then clearly problems may emerge in a number of natural resource areas. If, on the other hand, the judgments of professionals are appropriate because they have been developed in the context of an informed judgment, the concerns raised here are unwarranted. We can, at best, be uncomfortable with these alternatives.

11.6 Conclusion

In the 1960s and 1970s, Frederick Mosher (1982) noted the upsurge of increasingly narrow, professional groups in government. In the public sector, the broadly based American Society for Public Administration has declined in membership throughout the 1980s and 1990s. At the same time, there has been an increase in membership in more specialized public service associations such as the Government Finance Officers Association and American Planning Association. There are good reasons for believing the same segmentation is occurring throughout society. What this means for risk communication bears serious consideration. The data presented here suggest that professional association journals are a primary means of information dissemination to natural resource policy makers. But this entails a filtering process about which little is known. Moreover, it suggests a fragmentation of the information dissemination process. Good research may get done. But whether or not it is integrated into the cognitive maps of policy makers appears more as a serendipitous process.

The emergence of professional groupings also raises the question of the utility of information. Professionals may be more inclined to expect to be treated as such; they tend to reinforce professional values and languages among themselves. Thus, they may come to expect that information should conform to their categorizations of the world, and not those of climate researchers. The fact that information reaching policy makers was judged not very useful is extremely important. In the interviews, policy makers consistently stated they needed more precise information than they were getting. They also stated they needed more *certain* information. These may be judged unreasonable demands by climate researchers, but the point is that no dialogue is occurring. Rather, information is a one-way flow to policy makers and yet policy makers show signs of wanting to engage researchers rather than be passive in the process.

The data presented here that should perhaps cause the most concern relate to whether or not policy makers have already judged research which suggests decadal climate variation as unfounded. Those trained in the natural sciences showed a greater proclivity to judge information than those coming from social science or business backgrounds. This should come as no surprise. It is unlikely that scientists are trained in organizational skills as part of their education; yet just as likely that social scientists will be, after all, organizations are a natural feature of social life. Thus scientists may be more uncertain of their organizational role, quicker to judge the information on an emerging issue and "close the books," so to speak. Yet organizational literature is rife with examples of how destructive such quick judgment in the face equivocal data can be. One need look no further than H. Wilensky's (1967) penetrating analysis of the consequences of refusing to accept the data that daylight bombing raids of Germany during World War II were not accomplishing any stated objectives, to see this. The ability to see ambiguity and not make a hasty judgment may be the key to organizational survival. The fact is now that, at worst, research on climate variation should induce a "wait and see" attitude. Rather than waiting, some policy makers have already made up their mind what they see. This, in and of itself, suggests we need to pay more attention to how risk is communicated to professional policy makers.

Acknowledgments

The authors wish to acknowledge the research assistance of Mr. Edgar Barnett, Research Associate, and Ms. Dorothy McFalls, Graduate Research Assistant, both with the Institute for Public Affairs and Policy Studies at the University of Charleston.

References

Armstrong, J.S. and Overton, T.S., 1977. Estimating nonresponse bias in mail surveys. *Journal of Marketing Research*. **14**, 396-402

Ayidiya, S.A., 1990. Response effects in mail surveys. *The Public Opinion Quarterly*, **54,** 229-247.

Barber, B., 1963. Some problems in the sociology of the professions. *Daedalus*, **92**, 4, 669-688.

Baur, E.J. 1947. Response bias in a mail survey. *Public Opinion Quarterly*, **11,** 594-600

Campbell, L.J. 1994. Grassroots organizations learn from disaster awareness survey, *Natural Hazards Observer*, **18**, 6, 6-7

Dahe, K., 1992. Myths of nature: Culture and the social construction of risk. *Journal of Social Issues*, **48**, 4, 21-37

Donald, M.N., 1960. Implications of nonresponse bias for the interpretation of mail questionnaire data. *Public Opinion Quarterly*, **34**, 99-114.

Karwan, K.R., and Wallace, W.A., 1984. Can we manage natural hazards? *Public Administration Review*, **44**, 2, 177-181.

Kearney, A.R., Understanding global change: A cognitive perspective on communicating through stories. *Climate Change*, **27**, 419-441

March, J.G. and Simon, H.A., 1958. *Organizations*. John Wiley & Sons, Inc., New York, 262 pp.

Miller, G.A., 1970. Professionals in bureaucracy: Alienation among industrial scientists and engineers. In: Grusky, O. and Miller, G.A. (eds.) *The Sociology of Organizations: Basic Studies*, Free Press, New York, 503-515.

Mosher, F.C., 1982, *Democracy and the Public Service*. Oxford University Press, New York, 251 pp.

Office of Technology Assessment, 1993. *Preparing for an Uncertain Climate*. 3 Volumes, OTA-0-567, US Government Printing Office, Washington, DC.

Pavalko, R.M., 1988. *Sociology of Occupations and Professions, 2nd Edition.* F.E. Peacock Publishers, Ithaca, 355 pp.

Porter, G.R., 1838. Agricultural queries, with returns from the country of Bedford. *Journal of the Statistical Society of London, Vol 1,* 89-96.

Quarantelli, E.L. 1984. Perceptions and reactions to emergency warnings of sudden hazards. *Ekisicts,* pp. 309.

Spencer, J.W., and Triche, E., 1994. Media construction of risk and safety: Differential framings of risk and safety, *Sociological Inquiry,* **64,** 2, 199-213

Turner, R.H., 1981. Waiting for disaster: Changing reactions to earthquake forecasts in Southern California. *International Journal of Mass Emergencies and Disasters,* **1**, 307-334.

Waters, M., 1989. Collegiality, bureaucratization, and professionalization: A Weberian analysis. *American Journal of Sociology,* **94**, 945-972.

Weber, M., 1947. *The Theory of Social and Economic Organization.* Translated by A.M. Henderson and Talcott Parsons, Talcott Parsons (ed.), Free Press, New York, 436 pp.

Weick, K.E., 1969. *The Social Psychology of Organizing.* Addison-Wesley Publishing Company, Chicago, pp. 121.

Wilensky, H., 1967. *Organizational Intelligence: Knowledge and Policy in Government and Industry.* Basic Books, Inc., New York, 226 pp.

Willigen, J.V., 1984. Truth and effectiveness: An essay on the relationship between information, policy, and action in applied anthropology. *Human Organization,* **43**, 3, 277-282.

12 Hurricane Mitigation Efforts at the U.S. Federal Emergency Management Agency

Gil Jamieson and Claire Drury
Mitigation Directorate
Hazard ID & Risk Assessment Division
Federal Emergency Management Agency
Washington, D.C. 20472, U.S.A.

Abstract

The U.S. government began funding to address the hurricane peril only in the late 1970s by providing limited resources to develop evacuation studies. These studies serve as planning and implementation tools for the emergency management community and play a major role in reducing injuries and loss of life.

During this same period, however, property damages and the social and economic costs of hurricanes have escalated. Not until recently has the Federal government begun to commit resources to reduce these losses. The National Mitigation Strategy published by the U.S. Federal Emergency Management Agency (FEMA) in 1995 calls for a reduction of all losses caused by natural hazards within 15 years. To support this goal, FEMA is developing a risk assessment/loss estimation methodology that will provide a consistent means of estimating disaster losses and will identify needs and opportunities for specific measures to reduce those losses. We anticipate that the methodology will contribute to a greater public awareness and understanding of hurricane storm losses, and that what we learn will challenge us to think more creatively about how to meet the goals of the National Mitigation Strategy.

12.1 An Historical Perspective

More than 75 million people live in counties along the Atlantic and Gulf coasts. Despite the increased population in coastal areas, U.S. hurricane fatalities have decreased significantly from the beginning of the century until now.

Credit for this decreasing trend in the U.S. death toll from tropical cyclones can be attributed to hurricane preparedness including forecasting, communications and the capability of the emergency management community to plan and carry out hurricane evacuations. Hurricane evacuation studies also play a significant role in hurricane preparedness by providing vital information for State and local emergency managers which can enhance their readiness and response to a direct hurricane threat.

Federal funding for hurricane preparedness did not begin until the late 1970's. At that time the U. S. Army Corps of Engineers (USACE) developed a hurricane preparedness methodology and initiated several seminal hurricane evacuation studies in Florida. In 1981, FEMA became involved when it provided limited funds to complete one of those studies. However, Federal resources for addressing hurricane threats have been and continue to be limited, necessitating a Federal focus on the highest priority--public safety and saving lives. From 1983 through 1993, FEMA's hurricane budget, which represents the largest share of the Federal contribution to this effort, remained static at the $896,000 level, with approximately $800,000 of this annually going to the development of hurricane evacuation studies and the remainder to training of emergency managers and public awareness activities. Since 1981, it is estimated that FEMA, the USACE and the National Oceanic and Atmospheric Administration (NOAA) have contributed approximately $17 million to fund hurricane evacuation studies.

As a joint effort the USACE, FEMA and NOAA work closely with States and have completed approximately 35 hurricane evacuation studies and restudies. From its inception, the hurricane evacuation study program was intended to build State and local capability to plan and carry out rapid and effective evacuations in the face of a hurricane threat. These studies identify areas vulnerable to hurricane storm surge, establish evacuation clearance times, and provide other vital information emergency managers can use to determine the best evacuation routes, shelters and to anticipate sheltering requirements. The relatively modest $17 million federal investment in these studies is more than justified by the lives that have been saved because emergency managers have been better equipped to assess the hurricane threat and to carry out evacuations. It is also worth noting that the amount spent on these studies is only a very small fraction (1/10 of 1%) of the losses experienced in Hurricane Andrew.

There will be an ongoing need to maintain State and local capability through updates and revisions to existing studies. The next generation of evacuation studies need to incorporate advances in technology such as digital storm surge and wind hazard mapping and to integrate these planning tools with State and local Geographic Information Systems(GIS) capabilities. These studies will also need to consider means to insure public safety in the more remote or densely populated coastal areas where evacuation clearance times exceed the National Weather Service's ability to accurately forecast the track of the storm. In such situations, alternatives such as reinforced residential storm shelters for residents located in areas not directly vulnerable to storm surge, may be preferable to evacuation.

During the last 25 years, property losses from hurricanes has dramatically increased, even as loss of life has decreased (Fig. 12.1). These losses are significant and clearly pose a direct threat to the economic and social viability of individuals, their communities and the nation in the wake of these disasters. The Federal commitment for hurricanes, thus far focused on immediate threat to life and public safety must evolve to meet this additional challenge. The Federal government must support and encourage State and local government initiatives to build safer, more hurricane-resistant communities. Stronger, hurricane-resistant construction will not only reduce property damage and economic losses, but will reduce personal injury and loss of life as well.

In 1994, James Lee Witt, then the new Director of FEMA and an experienced state emergency management official, gave his support to promoting an enhanced hurricane program that would include Federal funding for initiatives intended to reduce property damage and other economic and social losses from hurricanes. Aided in this effort by the overwhelming social and economic devastation inflicted by Hurricane Andrew, Director Witt determined to focus national attention on the overall reduction in the threat posed by hurricanes, going beyond their immediate threat to life, to encompass the threat which they pose to the quality of the social, environmental and economic life of the community, the region and the nation.

Toward that end, FEMA, beginning in FY 1994, sought and obtained a three-fold increase, to $2,896,000 in funding for programs aimed at reducing the overall threat from hurricanes. The greatest portion of that increase has been directed each year to State and local governments to assist them in establishing effective hurricane programs that not only save lives, but also support mitigation initiatives that can reduce future property damages and economic losses. For Fiscal Year 97, FEMA has asked for a further increase of $3 million (to $5,896,000) to support and sustain State and local hurricane programs and initiatives. Despite Federal budget restrictions, we should continue to seek a Federal funding level for hurricanes that is more commensurate with risk that these storms pose to life and property.

In the natural hazards arena, as in all others, we are being challenged to do more with less. So, it is imperative that we develop the capability to fully understand, assess and communicate the risks we face not only from hurricanes, but from all natural hazards. With this knowledge, we can pool resources across all loss reduction programs so that we use the resources we do have more wisely and in a way that achieves true, all-hazards risk reduction. Such knowledge will also enable us as emergency managers, planners, scientists and academics to heighten public perception of the risk so that the public will demand safer, more hazard-resistant communities. The federal commitment to mitigate the hurricane hazard has built, thus far, State and local capability to protect against the direct threat to life. The next step is to build State and local capability to reduce the economic and social costs of these storms as well.

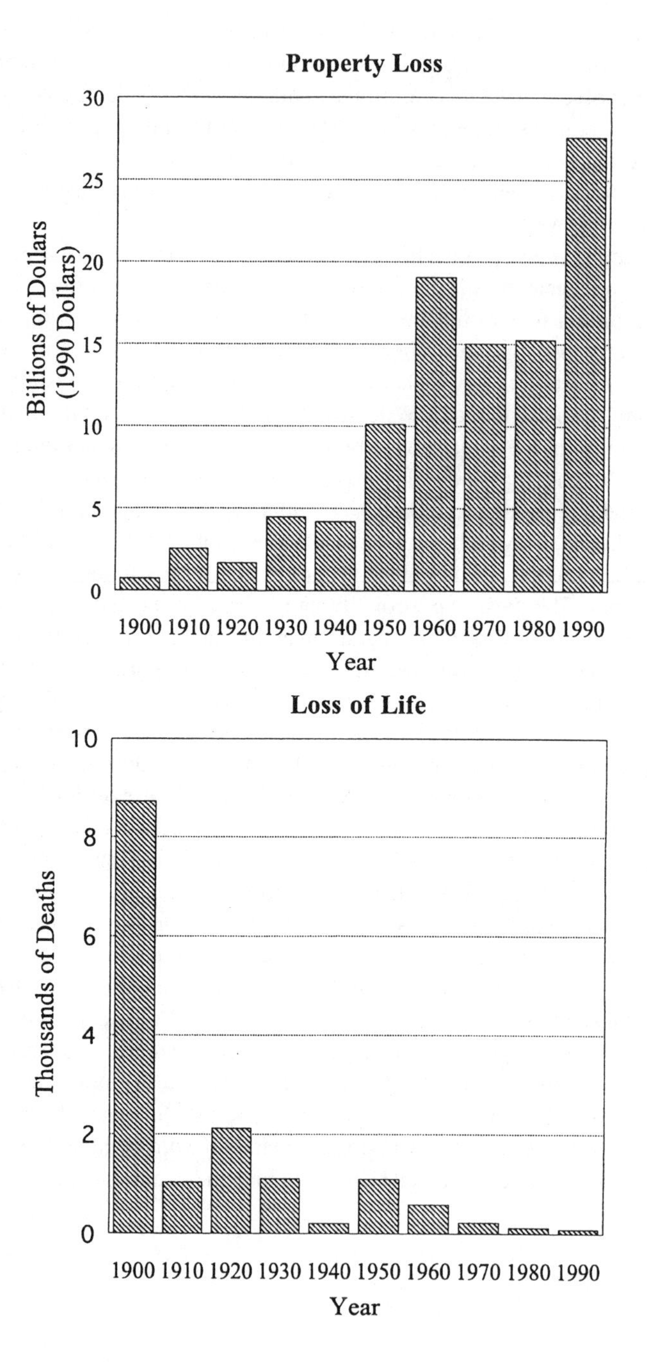

Fig. 12.1 Comparison of Property Losses and Fatalities from U.S. Hurricanes during the 20th Century.

12.2 Building Capability in Risk Assessment

The Federal Emergency Management Agency has embraced an all-hazards approach to emergency management. At the same time, the agency has challenged the emergency management community to take a leading role in working at the Federal, state and local level to implement mitigation initiatives that will achieve a reduction in the number of injuries, loss of life, and the economic and social costs of disasters. In 1995, FEMA published The National Mitigation Strategy which sets a goal to substantially increase public awareness of natural hazard risk and, within 15 years, to significantly reduce the risk of loss of life, injuries, economic costs and disruption of families and communities caused by natural hazards. FEMA recognized that to support this goal we need to establish a consistent means of characterizing risk from the various natural hazards and estimating the losses that could result. Such a standardized loss estimation/risk assessment tool will allow us as emergency managers and mitigators to identify opportunities and underscore the need to employ mitigation measures that reduce those losses.

In furtherance of the above goals, FEMA has been working with the National Institute of Building Sciences (NIBS) to develop a nationally applicable, Geographic Information System (GIS) based, all-hazard loss estimation methodology, known as "HAZUS". Results obtained through the "HAZUS" risk assessment/loss estimation analysis are not limited to direct physical damage to buildings; estimates are made also concerning casualties and shelter requirements, the functionality of lifeline systems and essential facilities, direct and indirect losses and an assessment of the induced hazards of flood, fire and hazardous material releases. Starting in FY'96 and continuing in FY'97, a "mitigation module" will be developed that also will allow "HAZUS" to be used to calculate how the general building stock will be affected with and without mitigation; thereby further demonstrating the cost effectiveness of specific mitigation actions.

Until recently, "HAZUS" development has focused on the earthquake hazard. Funding to expand "HAZUS" by developing loss estimation methodologies for hurricane storm surge and wind hazards and flood hazards has been provided in FY'96. Further expansion is planned to achieve an all-hazards methodology within the next several years. It is FEMA's intention to provide "HAZUS" to States and local governments at no cost to begin the task of building risk assessment capability at the Federal, State and local level that will be used both to leverage and to employ resources for mitigation, disaster planning, response and recovery.

FEMA initiated discussions with the National Emergency Management Association (NEMA) to determine what partnership actions could be taken prior to the release of "HAZUS" to begin to accomplish the risk assessment goals in the National Mitigation Strategy. The development of a sound inventory base is an essential first step in the risk assessment/loss estimation process because the detail and accuracy of that data impacts the credibility and accuracy of the loss estimates obtained from the "HAZUS" analysis. The inventory data set is also the one nsistent feature in risk assessment for all hazards. The diagram below outlines the ısk assessment process. While the hazard information (geologic or atmospheric) may change, the inventory data remains constant.

"HAZUS" contains a considerable amount of inventory information for residential, commercial and industrial buildings. However, additional data as well as more detailed data available at the State and local level can make significant improvements in the results obtained from the analysis. So that users can refine the loss estimates they obtain through a "HAZUS" analysis by incorporating their own building inventory data, FEMA developed a "Building Inventory Tool" as a component of "HAZUS". This tool maps the building inventory data according to a consistent building classification scheme and occupancy uses so that the data can be used for analysis purposes. In anticipation of expanding the methodology to all hazards, this tool can be used also to collect and incorporate information about those features of the general building stock that are unique to assessing the vulnerability to specific other hazards such as storm surge, wind and flood. Figure 12.2, below is a conceptual diagram of the risk assessment system being implemented by FEMA and the States.

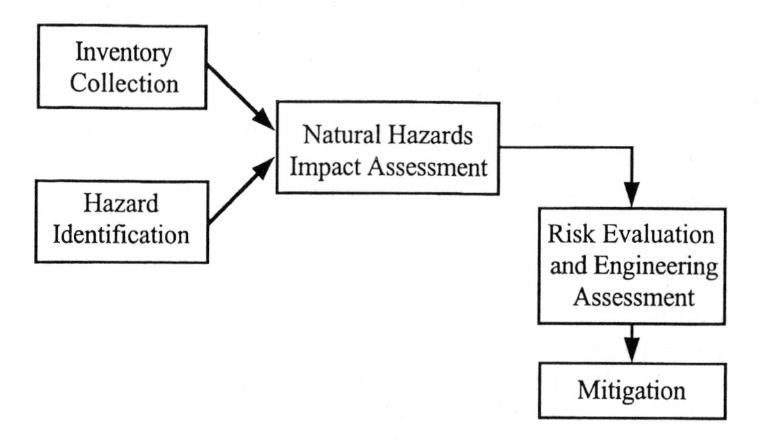

Fig. 12.2 FEMA -State Risk Assessment System

Building State, local and national risk assessment capability is one of the objectives of the agency's Performance Partnership Agreements (PPA) with the States. States can exercise maximum flexibility in determining the projects and activities that will contribute to the PPA goals and objectives and may utilize funding provided through the annual Cooperative Agreement with FEMA to undertake those activities. Beginning with the Cooperative Agreements for FY'97, FEMA is encouraging states to initiate or to expand ongoing activities that will develop and enhance their building inventory database.

12.3 Toward A New Perspective On the Hurricane Hazard

Despite Hurricanes Hugo and Andrew, the public's perception of the hurricane hazard seems to have lessened. This observation is based on statistics which show that between 1980 to 1993, the value of insurable property has increased 179% to $3.15 trillion and over 75 million people now live along a hurricane-prone coastline. In recent years, FEMA has been able to secure increases in its budget for the hurricane program. While this will permit FEMA to sustain current initiatives and promote some new initiatives through State and local emergency managers and floodplain management cadres, the resources available seem to pale in relation to the task.

Similarly, the new risk assessment/loss estimation initiatives we have discussed will contribute greatly to a better public understanding of hurricane storm losses as well as toward developing an understanding of the effectiveness of certain mitigation measures. These initiatives which attempt to influence aspects of societal behavior, development trends and building methods must continue. At the same time, we must reexamine and rethink current plans and practices for dealing with the hurricane threat. For example, should we continue to place total reliance on evacuation for population protection? While prudent evacuation decisions clearly reduce loss of life, we cannot ignore data that suggest that in at least some hurricane-prone areas the existing infrastructure and transportation systems may not support the demands of a quick evacuation decision. With the technology, building science and engineering information available to us today, we must begin to test and evaluate the effectiveness of protecting at least a portion of our population in the buildings and residential structures they occupy.

In many ways, those of us who belong to the partnership that has existed between academia, Federal agencies and State and local government agencies and associations can be proud of our accomplishments under the hurricane program; we have accomplished much in spite of the minimal resource base. Our recent success in securing more resources for this program should not be viewed as an ultimate solution to the hurricane problem, however, as there is no reason to believe this trend will continue given the universal call for budget reductions. Our success in the long term relies on our ability to expand public perception of the hurricane risk and to impel action that reduces that risk to structures and their occupants. All mitigation, whether to save lives or to save property and economic loss, must be implemented and achieved at the local level. As it is stated in the National Mitigation Strategy, our goal must be to substantially increase public awareness of natural hazard risk, so that the public demands safer communities in which to live and work.

Section D

Hurricane Risks and Property-Loss Insurance

13 Insurable Risks, Regulation, and the Changing Insurance Environment

Richard J. Roth, Sr.
26 Timber Lane
Northbrook, IL 60062, U.S.A.

Abstract

The aim of this chapter is to elucidate some facets of the property-casualty business, particularly as it is affected by landfalling hurricanes. Accordingly, a review of insurance principles and a definition of different types of risk are first presented. The impact of hurricane landfall patterns, considered within the context of different forecast lead times, is then discussed. Because the level of risk to life and property is related as much to actions taken by human beings as it is to the level of the underlying natural risk, I underscore the importance of loss mitigation through a proactive strategy that promotes prudent loss-reduction actions, such as improved and better enforced building codes, the implementation of more rational land-use policies, and greater efforts to inform the public of the levels of risks to natural disasters in different regions.

13.1 Introduction

The answer to the question, “Are numerical weather forecasts of hurricanes of interest and of use to the insurance industry?”, is, “It depends.”

The degree of interest and use depends on the time interval involved and the level of skill. Short-range forecasts regarding hurricanes involving time intervals of 24 hours ahead to as much as five days ahead are of much interest, but of limited use by insurers, since the policies they issue cover periods of time, usually for at least one year. Further, cancellation of policies may be done only for failure to pay premium, or gross misrepresentation when obtaining coverage.

What about a seasonal forecast given, let us say, six months ahead of the period within which hurricanes normally occur? Here, the interest and use would vary with the degree of forecasting skill, both actual and perceived.

Would climatological forecasts regarding the variability of hurricane frequency on a decadal or other long-term time scale be of any interest or use to insurance companies? Again, the degree of forecasting skill is a big factor, but interestingly, this knowledge would be of greater use to insuring companies than either seasonal or short-range forecasts. This would give them some flexibility in the tailoring of their marketing plans.

This paper will attempt to shed light on the following questions:

- What use is the insurance industry able to make of short-range hurricane forecasts?
- If seasonal forecasts of hurricane frequency with a significant, measurable degree of skill could be produced, how would this impact the insurance industry?
- What if climatological forecasts of the variability of hurricane frequency on a decadal or other long-term time scale, again with significant skill, were available?
- What would be its effect on insurance companies insuring property against damage from hurricanes?

Some tentative answers to these questions are as follows:

Short-range forecasts, although of much interest to insurers hoping no landfall will occur, have relatively little use except for logistical purposes. They allow the planning and the assembling of catastrophe claims teams to be utilized on as timely a basis as possible. The industry will also, of course, encourage broad dissemination by the media of information regarding measures to mitigate loss of life or injury and to reduce damage to property.

Although some seasonal forecasts are now being made, it is too early to measure their degree of skill with confidence. Forecasts with a measurable degree of skill would be of significant interest although possible strategic actions are limited. In spite of the current uncertain degree of skill in seasonal forecasts, there has developed much interest within the insurance community regarding such predictions.

There is an increasing awareness in the insurance community as to the possibility of gaining knowledge as to expected climatic variability of hurricane frequency on a decadal or long-term scale. For those companies willing to experiment with their strategic planning, the benefit derived would be substantial if a satisfactory level of climatological forecasting skill can be attained.

An understanding of some of the facets of the property-casualty business is necessary to understand why one has arrived at the above answers. Foreknowledge of hurricane occurrence would have a significant effect on the insurance marketplace, but in a manner which is not immediately apparent. Some discussion of insurance basics and the current insurance environment will help to explain why.

13.2 The Property-Casualty Business

13.2.1 Principles of Risk and Insurance

Definition of Risk. Although the term "risk" has been defined in many different ways by economists, insurance scholars, and others, for the purpose of this paper we will accept the concept that defines it as *uncertainty of loss* (Denenberg et al. 1964). There are many type of losses about which uncertainty exists such as the loss of youth, the loss of health, the loss of reputation, etc., but with regard to insurance we are dealing with a particular type of risk: the loss of wealth or profit. This financial risk or uncertainty of financial loss may be classified as either a *speculative risk* or a *pure risk* (Denenberg et al. 1964). Speculative risks present the promise of gain or the chance of loss, while pure risks present the prospect of only loss or no loss. Insurance involves dealing solely with pure risks. Speculative risks may be handled by the technique of hedging, but not by insurance. It is a fundamental principle of insurance that a loss is to be only indemnified with no expectation of profit.

Definition of Insurance. Insurance is the business of transferring pure risk by means of a two-party contract. Only pure risks may be insured as the insurance policy provides only for indemnification of loss. The term "risk" may be confused with the term "peril." The distinction is that "peril" is the source of loss while "risk" is the uncertainty of loss. Similarly, "hazard" is a condition which increases the likelihood of loss such as structures which are at water's edge in hurricane-prone areas.

Property insurance policies insure personal and commercial structures against damage caused by a number of perils. Historically, this line of insurance developed as insurance against peril of fire. Over the years many additional perils have been added to the basic policy, usually by endorsement. Among these additional perils is the peril of winds with excessive speeds which would include hurricanes.

Insurable Perils. Not all perils which create risk of financial loss, and therefore are considered "pure" rather than "speculative" "risks," are insurable. The requirements of an insurable peril are based on both statistical and administrative considerations. These requirements are listed below (Denenberg et al. 1964, pp. 146-148), but it should be stated that there is probably no peril that is currently insured which rigorously meets each and every one of these conditions. Usually, an exception to one or more of the following preconditions is compensated for by practical controls imposed by the insurer through underwriting, policy provisions, and rating techniques.

First, there should be a large number of homogenous exposures to permit the operation of theory of probability. Second, the occurrence of the event should be fortuitous. The timing or the severity of the loss should be out of the control of the insured. Third, the peril must produce a loss definite in time and amount. The insurer must be able to verify the loss in a timely manner and objectively measure its amount. Fourth the insured group of risks must not be exposed to an incalculable catastrophe hazard. It makes it most difficult to insure a peril if there is a significant concentration of values in areas subject to loss from occurrence. And fifth, the premium must be reasonable in relation to the potential financial loss. This is really a practical consideration from the point of view of the insured, rather than a question of the insurability of the peril itself.

13.2.2 Regulation of Insurance Companies

Purpose of Regulation. Since the value of an insurance policy is essentially its "promise-to-pay" in accordance with its provision, it is essential that insurance companies be in excellent financial health. To ensure this, the central and most important purpose of regulation is that of maintaining solvency of the insurance companies being monitored. With regard to property insurance Robert E. Litan, Senior Fellow and Director of the Center for Economic Progress and Employment at the Brookings Institution, in his January 1992 monograph, *Back to Basics: Solvency as the Primary Object of Insurance Industry Regulation*, points out that a number of factors unique to the property/casualty industry makes it particularly important that solvency be the most important objective of regulation in the future. In addition to the usual causes of failure such as mismanagement and fraud, the property/casualty industry is subject to potential catastrophic losses caused by natural disasters such as earthquakes and hurricanes.

Although some believe that solvency should be the only objective, this view has not prevailed. The nature of the insurance enterprise is such that, as the industry developed over the years, additional purposes were felt desirable to be added as regulatory objectives.

Need for Regulation. There are several reasons which can be cited to justify regulation of the insurance industry (Denenberg et al. 1964). The first of these is that without some regulation unbridled competition could lead to inadequate rates and excess commissions which would make production costs unreasonable. It could also lead to very selective underwriting which would severely limit availability of coverage, or to undesirable practices to make the product match the inadequate price. Pricing of insurance products is not an easy matter since it must be determined before costs are known and long before the transaction is completed. This presents a constant temptation on the part of insurers to underestimate costs, and it aggravates the insured's inability to make any sort of rational judgement on the fairness of the price.

Another reason for insurance regulation is the necessity for the insurance contract to be technical and complex to meet possible legal challenges in case of dispute. The result is a document which is complex, legalistic, intangible, and technical. This makes the contract difficult to understand, difficult to evaluate, and usually impossible to bargain over. It is difficult to compare the worth of the premium to the value of a future contingent payoff. A third reason arises from the desirability of sharing loss data with other insurers to ensure accurate pricing. Rarely does one insurer have enough data to be confident that the underlying probabilities in its pricing are accurate to the extent desirable. The pooling of data is to be the benefit of everyone, insured, insurer, and the public at large. This worthwhile endeavor, however, runs directly into problems with the anti-trust laws if no governmental regulation is present.

A final justification for regulation which has been offered in the courts is the great importance of insurance. It has been judicially recognized that there is an almost universal need for insurance, and the significance of insurance is obvious from both a macroeconomics and microeconomics point of view. Thus, regulation is justified, since insurance significantly affects the public interest.

Regulation by the States. Insurance, like banking and public utilities, is a closely regulated industry by the various states (Denenberg et al. 1964). In fact, the states had exclusive jurisdiction up until 1945 as the insurance industry was not considered by the courts to conduct interstate commerce, and, therefore, was not subject to federal jurisdiction. Up until that date the decision of the Supreme Court in the famous case of *Paul v. Virginia* in 1869 had held sway. Its conclusion at that time was that issuing a policy of insurance was not a transaction of commerce. This firmly established the right of states to tax and regulate the business of insurance independent of federal supervision.

In 1944 the Supreme Court overruled *Paul v. Virginia* as a result of their decision in the SEUA case *U.S. v. South-Eastern Underwriters Association* (322 U.S. 533[1944]). It held that the business of insurance was indeed commerce and subject to federal laws, specifically the anti-trust laws. This decision served to cast doubt upon the whole system of state regulation, for it raised serious questions as to the constitutionality of all state legislation. At the same time it created uncertainty as to the application of federal laws, applicable generally to interstate commerce, and left a possible legislative vacuum as there was no specific federal system of insurance regulation.

The potential chaos reigning and the strong entrenchment of state regulation caused Congress to act with little delay. Within less than a year it enacted the McCarran Act, also know as Public Law 15, on March 9, 1945. This preserved the system of state regulation subject to the proviso that the federal anti-trust laws would be applicable to the business of insurance to the extent that such business is not regulated by state law.

Consequently, there was a flurry of state legislation that resulted in rate-regulatory laws and fair trade practice legislation in every state. Both the design of the McCarran Act and the ensuing legislation are generally credited to the National Association of Insurance Commissioners (NAIC), an extralegal council made up of the insurance commissioners of each state. This body, formed in 1871, has been a powerful voice in coordinating and promoting uniformity in the state insurance laws. Even though it has no official or absolute powers on a state or a national basis, its prestige and the publicity given to its actions have encouraged uniformity in insurance regulation throughout the country. States with limited funds and small staffs are particularly inclined to look to the NAIC for guidance in structuring their laws and procedures. Having said this, it should be noted that the various state insurance statutes are notable for their lack of uniformity.

Rate Regulation. A governing department, called the insurance department in most states, is charged with adapting the insurance code. It is usually on a level with other state divisions such as the department of highways. The supervisory authority in most states carries the title of insurance commissioner or commissioner of insurance. One of the insurance commissioner's responsibilities is to regulate the rates that are charged for insurance. In 1992, there were about 6000 insurance companies in the United States. Approximately 3000 of these were property-casualty insurers with another 2100 companies writing life and health insurance (Trieschmann and Gustavson 1995). Both segments of the industry are structurally competitive.

Why, then, is rate regulation necessary? Some of the answers have already been touched on. (Denenberg et al. 1964), points out that in most businesses the cost of production or service is known with some degree of accuracy prior to the time goods are marketed, while, on the other hand, the claimed costs of insurers can vary considerably. This makes judgement a vital factor in rate determinations and can easily lead to improvident rate reductions. Although cutthroat competition is not necessarily undesirable in some other fields, it is a great potential detriment to insureds. Liquidation of insurers may cause financial havoc among it's insured.

Another objective of rate supervision is to ascertain that rates are reasonable, i.e. not too high. Part of this concern arises from the necessity for insurers to pool data to obtain broadest possible base of experience in setting rate levels for types of coverage with low loss frequency. Such cooperation could lead to over-conservatism in setting individual company rates resulting in higher premiums than those justified by the combined loss experience and expenses. Another important objective is to make certain that the rates are not "unfairly discriminatory." There is, however, considerable difficulty in deciding on the difference between "fair" and "unfair" discrimination. It is seldom clear-cut and presents a more difficult administrative problem for regulators than do those of adequacy and reasonableness.

13.2.3 Mitigation

An important aspect of an insurance program is the promotion of steps to mitigate the anticipated losses. This not only benefits the public from immediate problems arising out of property loss, but leads to reductions in premiums as future losses are contained. With regard to property insurance for structures the important areas to address are as follows.

Building Codes. Hurricane Andrew's landfall on August 24, 1992 brought attention to the importance of paying strict attention to building codes. The insured loss from this hurricane was about $17 billion. Estimates have been advanced to the effect that anywhere from 25 to 40% of the insured loses would not have occurred if the properties had been built to the requirements of the then-existing building code. If the lower estimate is taken, society and the insurers would have been spared at least $4 billion in unnecessary losses.

The problem with code enforcement had been recognized by the Insurance Institute for Property Loss Reduction long before Hurricane Andrew hit. The Institute had methodically documented shortcomings and was in the process of working with building code inspectors and the model building code organizations to create what is now called the *Building Code Effectiveness Grading Schedule* (*BCEGS*) (ISO Commercial Risk Services 1995). In June of 1994 the Institute transferred the grading program to another insurance organization, the Insurance Services Organization (ISO), for implementation. This Grading Schedule will score code adoption and enforcement effectiveness of communities. *BCEGS* will function in a manner similar to the *Fire Suppression Grading System* which evaluates the fire-fighting capacities of municipalities and provides a means for granting communities that grade well a financial incentive. *BCEGS* is expected to play a major role in upgrading the nationwide quality of building code enforcement.

Land Use and Public Education. Land-use regulations, although at present not in wide use, will become an increasingly important means of disaster mitigation, particularly in view of the changing demographics and the pressures to continue to build in harm's way. Unfortunately, a pervasive attitude exists in our society which maintains, "It won't happen to me. I will not be affected by a natural hazard event, and if I am, the government will rescue me." To combat this attitude it is essential that the public be educated about the risks they face according to where they live, and that government disaster relief provides only minimal assistance.

13.3 The Changing Insurance Environment

13.3.1 Exposure

Population Change along the U.S. Coast. The demographic changes along both coasts of the United States have been striking. Today, almost one-half of the U.S. population now lives along one of the U.S. coast lines. Between 1970 and 1992 the population in hurricane-prone Florida doubled, compared with an increase of only

25% for the U.S. as a whole (see Pielke and Pielke 1996, Chap. 8. this volume).

The aging of the population is another complicating factor. As the population ages, it is less able to cope with natural disaster. Unfortunately, senior citizens are among those putting themselves directly in the danger zone. Currently, 12.7% of the national population is over 65 years old, but in Florida that figure is 50% higher, or 18.4%. By the year 2030, a full 22% of the national population will be over 65 years old. Natural disasters affect these people more profoundly, which means our social responsibilities to them are greater (see also Rodríguez 1996, Chap. 7, this volume).

To put it simply, there are increasingly more and more people in areas where natural disasters are more frequent. That makes the potential losses significantly greater. The density of the population in coastal counties in 1988 was more than three times the average density of the population in the United States as a whole, and is continuing to increase (Culliton et al. 1990). So, not only are more people in harm's way, but they are packed more tightly together, living and working in millions of structures which may not meet code, and may be subject to catastrophic loss.

The Atlantic coastline is a good example of the foregoing. The residential and commercial insured exposure in the first tier of coastal counties along the Atlantic and Gulf coasts in 1988 was $1.86 trillion (IIPLR and IRC 1995). Now the number had burgeoned to $3.15 trillion, and increase of 75% in just five years. With this concentration in exposures in vulnerable areas, the American coastline has become a disaster waiting to happen.

Catastrophe Severity. Along with this unprecedented increase in exposure, the severity of U.S. insured property losses from hurricane landfalls has been escalating at an alarming rate. This can be seen from the following listing which identifies the five costliest hurricanes in U.S. history making landfall in the continental United States. Their insured losses have been adjusted to 1994 dollars (Bipartisan Task Force 1995).

1	1992	Hurricane Andrew	$15,500 million
2	1989	Hurricane Hugo	$4,195 million
3	1979	Hurricane Frederic	$752 million
4	1965	Hurricane Betsy	$715 million
5	1983	Hurricane Alicia	$676 million

Changing Emphasis on Purposes of Regulation. One of the most important factors which as impacted the insurance industry over the past several years has been the increasing importance being placed on a new concept arising from one of the secondary purposes of insurance- the reasonableness between the company and policyholder. The new concept involves the notion of "availability and affordability," and has been growing in popularity due to an increase in the politicizing of insurance departments. This is particularly true when the state insurance commissioner is elected by the public rather than appointed by the governor.

Availability and Affordability. Availability means everyone's right to purchase insurance without unfair discrimination. Even further, companies are expected to offer insurance for any risk or peril that is perceived essential for the public good. Thus, insurers in California must include the earthquake peril in their policy if the insured desires this coverage, or insureds in Florida have the right to have a policy including wind damage coverage. This would not be too unreasonable if the insurance companies were allowed to charge a rate to cover anticipated losses.

However, affordability is attached to this right to have coverage, which means that the rate must be low enough to meet the insureds' and regulator's view of what a "reasonable" rate is. This, then, acts to the detriment of an insurance company's solvency when, as usually is the case, the rate allowed to be charged is far below that which is needed to cover anticipated losses, or, as in California, the correct rate is impossible to determine.

Residual Markets Mechanisms. The natural consequence of implementing strong "availability and affordability" measures is the withdrawl of insurance companies from a particular market. The next step, then, is to establish a mechanism by legislation that provides that persons who are unable to obtain needed insurance through normal channels may go to an organization, usually jointly sponsored and run by all of the insurance companies doing property insurance business in the state, and obtain the insurance coverage desired. These organizations are known as residual markets mechanisms.

However, this does not solve the solvency problems since the insurance companies which are jointly sponsoring the residual markets organizations are responsible for the losses which arise from them. The net result is that the losses are merely redistributed among the companies depending upon how the assessment procedure is fashioned for the residual markets organization.

Private Insurance Goals vs. Social Objectives. The conflict between primary concern for company solvency and the "availability and affordability" of insurance lies in the confusion of the necessary objectives of private insurance companies and the social objectives perceived necessary by the regulator. There is a critically important difference between the underlying insurance principles of private insurance methods and the underlying principles of insurance methods based on perceived social objectives as illustrated below (Denenberg et al. 1964):

Principle of Equity - Private insurance adopts the principle of equity as a necessity. Each insured pays his fair share of the cost. Socially motivated insurance involves subsidizing as deemed desirable.

Principle of Voluntarism - Private insurance offerings are voluntary. Socially motivated ones are mandatory.

Principle of Reserving and Fund Building - Private insurance creates reserves and accumulates funds to match liabilities. Socially motivated insurance does not concern itself with reserves as it has ultimate authority to collect through levying of taxes.

13.4 Perceptions and Responsibility

A popular concept among the public perceives insurance as a "right," and that it must be made available at an affordable, if need be, subsidized premium. It also has become accepted that people have the right to live where they want, i.e. along the sea coasts or the river's edge, atop a mountain, over a fault line, or in a tornado alley. By exercising the right to choose where they want to live, the following question arises: "Do individuals have the right to pass the risks associated with residing in hazard-prone locales on to the rest of society"?

Traditionally, our society has embraced concepts which anticipate that each individual will be responsible and accountable for his or her own obligations, including "risks" knowingly assumed. In recent years society has moved away from this tradition. This would seem to call for efforts to be made to reinstitute the old standards and expect individual to again assume responsibility for their conscious decisions. this would entail taking the steps necessary to mitigate potential damage and to "pre-fund" their potential financial losses with realistically priced insurance.

13.5 Conclusions

If the increasing exposure, mandatory offering, and inadequate rates continue, insurers will attempt to abandon hurricane-prone areas. This will result in a significant scarcity of insurance against wind damage. Further, the remaining insurers will assume exposure to catastrophic losses. This, then, could result in additional insolvency problems further reducing available insurance, not only against wind damage but for all other lines of property casualty insurance as well.

State regulators must reinstitute insurance company solvency as the primary goal of regulation to avoid the ominous outlook presented above. The regulator must allow rate levels that reflect the risk with appropriate regard for unfair discrimination and reasonableness.

A single 4 or 5 category hurricane making landfall in a large metropolitan area creates enormous exposure being subject to one event. The potential catastrophic losses could threaten the solvency of a significant portion of the insurance industry.

Regulators need to classify risks located in hurricane-prone areas as uninsurable without a federal reinsurance backup. Only the federal government has the financial capacity to weather one of the mega disasters. Private insurers would pay reinsurance premiums to the federal government base on the best available knowledge as to the long-term probabilities.

There is increasing agreement that individuals living in hazard-prone areas should be responsible for risks "knowingly" assumed. However, probably very few people living in these areas now were aware of the risks they assumed when they purchased their property. It is important to develop a policy taking into consideration such ignorance. This policy needs to allow reasonable transition to full assumption of responsibility. To assist in this effort it will be essential that all stakeholders cooperate in mitigation and public education efforts. In addition to the insurance industry, stakeholders include owners; developers; professionals such as architects, engineers, and code officials; and elected officials, professional staff, and enforcement personnel of state and local governments.

If the insurance situation stabilizes through addressing these issues, enabling individual insurance companies to set their rates in a competitive environment, the following results seem likely. As the knowledge base of the atmospheric sciences increases, the accuracy of climatological forecasts for periods of five to ten years should improve. At that point, predicted results will be consistently better than those based on pure chance. Stating the results in probabilistic terms helps clarify the predictions. Forward-looking companies applying these probabilistic forecasts will gain an edge over insurers who are technically unable to use them or which ignore them.

The insurers using these forecasts will increase their marketing efforts and lower rates in expected periods of lower-than-average hurricane frequencies. Conversely, they will increase their rates to reduce their assumption of new business during expected periods higher-than-average frequencies. As a result they can increase their over-all profitability, or at least decrease the amount of losses incurred.

Insurance companies failing to recognize the value of increasingly skillful climatological forecasts will tend to maintain a relatively level rate level. Their book of business will increase during periods of expected higher-than-average frequencies since they will be offering lower rates than their competitors. By contrast, these companies could lose business when predicted losses are low, because their insurance rates will be higher than the forward-looking companies.

Finally, the following are recommendations are proposed to focus for the future and reduce the potential for greater property losses for hurricanes:

First, continued research for the improvement of climatological forecasts. Second, cooperative mitigation efforts by federal, state and local governments including all of the stakeholders in the private sector. Last, vigorous and effective education of the regulators and the public regarding the risks of living in hazard-prone areas.

References

Bipartisan Task Force on Funding Disaster Relief. 1995. United States Senate, Federal Disaster Assistance, U.S. Government Printing Office, Washington, 20 pp.

Culliton, T.J., Warren, M.A., Goodspeed, T.R., Remer, D.G., Blackwell, C.M., and Mcdonough, J.J. III. 1990. *50 Years of Population Change along the Nation's Coasts,* 1960-2010, NOAA, Rockville, MD, 16 pp.

Denenberg, H.S., Eilers, R.D., Hoffman, G., Kline, C.A., Melone J.J., and Snider, H.W., 1964. *Risk and Insurance*, Prentice Hall Inc., Englewood Cliffs, NJ, 615 pp.

Insurance Institute for Property Loss Reduction and Insurance Research Council, 1995. *Coastal Exposure and Community Protection*, Insurance Research Council, Wheaton, IL, 48 pp.

ISO Commercial Risk Services, Inc., 1995. *Building Code Effectiveness Grading Schedule*, Insurance Services Office, New York, 24 pp.

Litan, R.E., 1992. B*ack to Basics: Solvency as the Primary Object of Insurance Industry.* State Farm Insurance Companies, Bloomington, IL, 45 pp. (Copies available at the Insurance Institute for Property Loss Reduction, Boston, MA.)

Pielke, R.A., Jr. and Pielke, R.A., Sr., 1996. Vulnerability to hurricanes along the U.S. Atlantic and Gulf coasts: considerations of the use of long-term forecasts (Chapter 8, this volume).

Rodriguez, H., 1996. A socioeconomic analysis of hurricanes in Puerto Rico: an overview of disaster mitigation and preparedness (Chapter 7, this volume).

Trieschmann, J.S. and Gustavson, S.G., 1995. *Risk Management and Insurance*, South Western College Publishing, Cincinnati, OH, 663 pp.

14 Current and Potential Impact Of Hurricane Variability on the Insurance Industry

Karen M. Clark
Applied Insurance Research
137 Newbury St.
Boston, MA 02116, U.S.A.

Abstract

An analysis of the current and projected U.S. hurricane property loss potential is presented. The estimated loss potential today and the future loss projections are based on a quantitative model developed at Applied Insurance Research for a wide range of uses. It is found that the recent large losses from landfalling hurricanes are due primarily to exponential growth in the valuation of coastal property. The observed variable nature of hurricane landfall frequency in the past century indicate that the loss potential in the future from the onset of a period of more frequent and/or intense hurricanes could increase dramatically from that experienced in recent decades.

14.1 Introduction

Since 1960, hurricane activity, particularly for major storms along the East Coast, has been below average for this century. This has meant that insurers have been relatively free of large hurricane losses. Prior to Hurricane Andrew, the largest insured hurricane loss in the U.S., $3.5 billion, was caused by Hurricane Hugo in 1989. Before Hugo, the worst damage inflicted by a hurricane was just over $1 billion from Hurricane Alicia in 1983. While Hurricane Andrew served as a wake up call to the insurance industry with over $16 billion in losses, its economic toll could have been dramatically worse. Losses could have exceeded $40 billion had Andrew been even more intense or had the storm made landfall on the Miami coast.

One can imagine other scenarios that would result in large losses. A return to a period of high hurricane frequency today, given the current property values along the East Coast, would result in annual hurricane losses much larger than ever before experienced in the past. This paper examines the current U.S. hurricane loss potential based on current insured property values and hurricane frequency and intensity distributions estimated from nearly 100 years of historical meteorological data. It next examines how losses could change with changes in future hurricane activity.

14.2 Current U.S. Hurricane Loss Potential

Based on 1993 insured property values, Table 14.1 gives estimated probabilities of hurricane losses for the entire Gulf and East coasts. The loss estimates are derived using a hurricane computer simulation model and a database of U.S. insured property values, both developed by Applied Insurance Research (AIR). This estimation technique will be discussed in greater detail later in this paper.

Table 14.1 U.S. Hurricane Loss Potential.

Insured Property Loss*	Estimated Probability of Exceeding	Estimated Average Return Interval (years)
7.8	0.100	10
13.2	0.050	20
23.6	0.020	50
30.7	0.010	100
34.5	0.005	200
50.9	0.002	500
51.5	0.001	1000

* In billions.

The loss estimates above show the probabilities of experiencing losses of different magnitudes from a single hurricane in any given year. The probabilities are also expressed in the table as return periods. For example, $7.8 billion, the loss listed in the return period of 10 years represents the 90th percentile of the hurricane loss distribution. This amount is likely to be exceeded 10 percent of the time or one year out of 10, on average. It is estimated that a one in one hundred year insured loss amount is just over $30 billion. It is important to keep in mind that these numbers include only insured damage to building and contents. No time element, i.e. coverage that pays for additional living and business interruption expenses or non-insured losses, are included. No loss to infrastructure is estimated.

For the Southeast U.S. alone, including Florida, the estimates are as shown in Table 14.2. The Southeast region clearly dominates the countrywide hurricane loss potential. For the Gulf Region, including Texas, the estimates are given in Table 14.3. Similar estimates can be derived for other regions. These figures are not static and continue to change (usually increasing) as coastal properties and their values change.

Table 14.2 Southeast U.S. Hurricane Loss Potential.

Insured Property Loss*	Estimated Probability of Exceeding	Estimate Average Return Interval (years)
5.9	0.100	10
12.0	0.050	20
20.7	0.020	50
26.7	0.010	100
34.1	0.005	200
50.9	0.002	500
51.5	0.001	1000

* In billions.

Table 14.3 Gulf of Mexico Hurricane Loss Potential.

Insured Property Loss*	Estimated Probability of Exceeding	Estimate Average Return Interval (years)
3.6	0.100	10
6.2	0.050	20
9.5	0.020	50
11.2	0.010	100
19.2	0.005	200
21.5	0.002	500
49.8	0.001	1000

* In billions.

Standard actuarial techniques that employ historical losses to project future losses are not appropriate for estimating potential hurricane losses for several reasons. Hurricanes, particularly in specific geographic areas, are usually rare events so that there are limited historical loss data. Furthermore, the loss data, especially if it is old, are not very meaningful. Over time, the number of exposed properties, the geographical distribution of these properties and the property values change dramatically. Building materials and designs change, and new structures become more or less vulnerable to hurricanes than the old structures. Changes in building repair costs also affect the dollar damages that will result from hurricanes.

In the past, this lack of historical loss data on hurricanes presented a problem for insurance companies. Standard actuarial techniques rely on past losses to project future losses, but historical hurricane loss information cannot be used to estimate the current or future hurricane loss potential. Applied Insurance Research was the first company to develop an alternative methodology based on Monte Carlo simulation. This approach involves the development of computer programs that describe in detail the meteorological characteristics of hurricanes. Current property values are input into the model to estimate the effects of these storms on exposed properties. All of the important meteorological variables and relationships are included in the model. A high-speed computer then "simulates" the hypothetical hurricanes and estimates the resulting property losses. (For more detail see Clark, 1986.)

Out of this need for better assessment and management of risk for catastrophe loss, AIR was created in the mid 1980s to help the insurance industry better estimate potential losses from catastrophes. Previously, underwriters concentrated on the probable maximum losses (PML), or maximum foreseeable losses, wind premium factor formulas, or even rules of thumb. However, this type of planning did not take into consideration hurricane frequency or true exposures to hurricanes.

To estimate the long-run loss potential, many "years" of hurricane experience are simulated. In this context, "years" actually mean possible scenarios for the current year. This large number of simulations ensures that the resulting probability distribution of losses converges to a stable representative distribution of hurricane losses. The pattern of simulated hurricanes will match closely the pattern of historical hurricanes because meteorological data on the actual events since 1900 were used to estimate the parameters of the hurricane model. The meteorological sources used to develop the AIR hurricane simulation model are the most complete and accurate databases available from various agencies of the U.S. National Weather Service, including the National Hurricane Center.

The five primary characteristics of hurricanes used to simulate each storm are the following: landfall location, central pressure, radius of maximum winds, forward speed, and track direction. The probability distributions of these variables are estimated for coastal segments of equal length from Texas to Maine. Numbers are generated from the probability distributions of these random variables to assign values to the variables for each simulated hurricane. The baseline probability distributions for hurricane frequency and intensity do not incorporate the possible effects of droughts in Africa and El Niño (Gray et al., Chap. 2, this volume). However, all hurricane model probability distributions can be readily modified to capture the possible effects of these climatic influences on hurricane activity. AIR has performed special studies for individual clients to test the impacts that changes in hurricane frequency and/or severity would have on the resulting loss distributions.

After values have been assigned for all of the important meteorological characteristics, the movement of the hurricane is then simulated by the model. Peak wind speeds and wind duration are estimated for each geographical location of interest. Based on peak winds and duration, damages are estimated at each location for different types of structures.

14.3 Key Model Variables

The AIR hurricane model consists of several interconnected pieces that characterize the information required to generate a credible computer-simulated storm. First, detailed historical information on the storm characteristics is required. The hurricane model requires data on the following variables: central pressure, radius of maximum winds, forward speed, storm track, landfall point, and tide height.

Central pressure, or more precisely the difference in central pressure and peripheral pressure, is the primary determinant of wind speed in the hurricane model. Another important storm characteristic is the radius of maximum winds, i.e. the distance from the center of the storm to where the strongest winds are experienced. The radius of maximum winds ranges from 5 to 50 nautical miles. The forward speed indicates the translational speed of the storm. The storm track indicates the direction of the movement after landfall.

Once the specific characteristics of a hurricane are determined, the following information about the sites at which properties are located is incorporated: distance from coast, surface terrain, topography, elevation, building code, and building practices.

Site characteristics are incorporated into the model deterministically, i.e. they are not subject to probabilistic effects. The distance from the coast is a key determinant of how much time each storm would be over land before affecting a particular site and the properties located there. Each simulated hurricane begins to dissipate after landfall. Surface terrain and topography describe local rough features which can decrease or enhance wind speed. Elevation is particularly important for those sites susceptible to storm surge. Finally, local building codes and building practices are incorporated into the modeling process.

The AIR computer simulation model takes into effect two other categories of variables, namely, the structure characteristics and the policy conditions. Structure characteristics include the following: structure replacement value, construction type, height, and architectural detail.

The hurricane model requires full replacement values for all properties even if the insured values are less than the full replacement values. The model also requires identifying codes for the type of construction of the property, for example, wood frame, mobile home, masonry, or steel. Information on height and architectural detail enhances the loss estimation process.

The following policy conditions are finally introduced: insured building value, contents value, time element, and other structures, deductibles, and other conditions. In many instances, buildings are not insured for their full replacement values, but for some lesser amount called the policy limit. The contents of buildings are also assigned values for insurance purposes. Time element coverage pays for additional living and business interruption losses. Most insurance policies include deductibles and other conditions.

For each simulated hurricane, wind speeds are estimated and superimposed on the replacement values of the exposed properties to estimate damages. The hurricane damageability relationships have been derived and refined over a period of many years. They incorporate well-documented engineering studies published by wind engineers and other experts outside of AIR. These damageability relationships also incorporate the results of post-hurricane field surveys performed by AIR and other structural engineers as well as detailed analyses of actual loss data provided by AIR client companies. These relationships are continually refined and validated.

Total industry loss estimates are derived by using AIR's proprietary database of total insured property values by five digit zip code. These property values change over time and must be updated every year along with the hurricane model loss estimates. In the coastal states, coastal property values are growing faster than non-coastal areas.

Table 14.4 shows 1993 insured property values for selected areas. For all of the Gulf and East Coast states, 34 percent of the property value, over $3 trillion, are in the first tier of coastal counties.

Table 14.4 Coastal Property Values.

State	State Total*	Coastal*
Texas	1504	129
Louisiana	351	123
Mississippi	185	25
Alabama	304	37
Florida	1070	872
Total	9232	3147

* In billions.

The hurricane simulation model is able to account for the effects of changing demographics since property values are input into the models. The true value of the model lies in its ability to project potential losses in areas that have not experienced hurricanes in the recent past. For example, the table below shows the model hurricane projections *before* Hurricane Andrew (and based on 1991 property values).

Table 14.5 shows that well before Hurricane Andrew, the AIR hurricane model was projecting hurricane losses much larger than had been experienced in the past. In fact, before Hurricane Andrew, the AIR hurricane model was the *only* source of hurricane loss estimates in excess of $30 billion. This storm illustrated the enormous value of the simulation modeling approach, and indeed, most insurance companies became interested in catastrophe modeling after Hurricane Andrew.

Table 14.5 Pre-Andrew Hurricane Loss Probabilities.

Loss Amount* Insured	Estimated Probability of Exceeding	Estimated Average Return Period (years)
9.5	0.050	20
13.5	0.020	50
18.2	0.010	100
20.8	0.005	200
31.4	0.002	500
39.0	0.001	1000

* In billions.

14.4 Potential Future Impact of Hurricanes

The estimated current loss potential is based on average hurricane frequency this century. Clearly, there have been periods of high frequency and low frequency, particularly with respect to major storms along the East Coast as the following maps show (Fig 14.1).

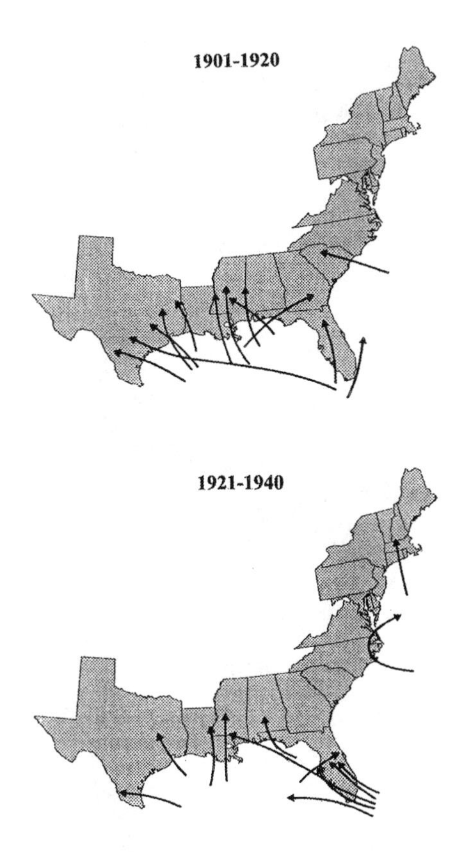

Fig. 14.1. Major U.S. landfalling hurricanes, category 3 or greater. Upper panel, 1901-1920; lower panel, 1921-1940.

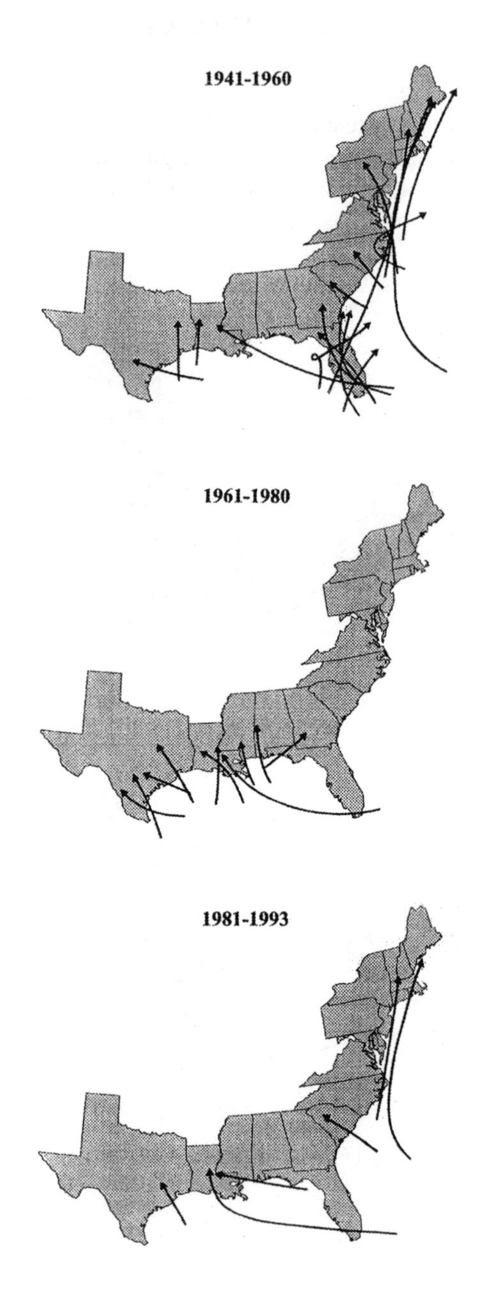

Fig. 14.1. (Cont.) Major U.S. landfalling hurricanes, category 3 or greater. Top panel, 1941-1960; middle panel, 1961-1980; lower panel, 1981-1993.

Table 14.6 summarizes total hurricane activity by 20-year period. The decade of the 1940s was clearly the worst with respect to potential damage producing hurricanes. It is estimated that if the hurricanes that made landfall in this time period were to recur today, the insured losses over a 10-year period would exceed $60 billion. Table 14.7 below shows each of these hurricanes along with their estimated losses based on 1994 property values.

A return to the hurricane activity of the 1950's would also causes significant losses, tallying to over $30 billion over ten years. By comparison, the 1970s and 1980s were characterized by fewer hurricanes and much lower losses. Figure 14.2 summarizes these four decades.

Table 14.6 Number of Significant U.S. Hurricanes by 20 Year Periods

Period	Number of Hurricanes	Annual Frequency
1900-1919	19	0.95
1920-1939	24	1.20
1940-1959	27	1.35
1960-1979	26	1.30
1980-1994	13	0.87
1900-1994	109	1.15

Table 14.7 Estimated Insured Losses from Historical Hurricanes of the 1940s. SS denotes intensity based on the Saffir-Simpson scale.

Year	SS	Landfall	Est. Industry Loss*
1941	3	Texas	0.442
1941	1	Florida	0.024
1942	3	Texas	0.700
1943	2	Texas	0.447
1944	4	North Carolina	4.400
1944	3	Florida	8.070
1945	2	Texas	0.143
1945	3	Florida	8.173
1947	2	Georgia	0.333
1947	4	Florida	21.850
1948	4	Florida	2.117
1948	2	Florida	0.133
1949	3	Florida	4.700
1949	3	Texas	2.020
1950	3	Florida	2.600
1950	3	Florida	4.500
1950	2	Alabama	0.124
	Total		62.776

* In billions.

Variability in hurricane frequency will certainly impact insurance companies. Changes in climate could also lead to changes in hurricane intensity. K. Emanuel (e.g., Emanuel 1987, Chap. 3, this volume) and W. Gray and his associates (e.g.,

Landsea et al. 1994; Gray et al., Chap. 2, this volume) discuss the implications of climatic change on hurricane frequency and intensity elsewhere in this book (see also Clark 1992). It will be noted here that for each one-degree rise in sea surface temperature, we can expect a 1 to 3% drop in the minimum sustainable central pressure, and a 5 to 7% increase in the maximum wind speed of hurricanes (Emanuel 1987).

Table 14.8, taken from Clark (1992) shows the estimated 1990 insured losses for three major hurricanes affecting the United States. It also shows the estimated losses when the maximum wind speed is increased by 5, 10, and 15% corresponding (approximately) to 1, 2, and 3°C sea-surface temperatures rises, respectively. While the precise results vary by storm, the data show that if the maximum wind speeds of these hurricanes were only 15% higher, insured wind losses from these hurricanes would have been more than double! Note that storm surge losses would also have been greater since surge heights also depend on the central pressure.

We can be sure that if the average and/or minimum central pressures of hurricanes do, in fact, fall by even 1 or 2% in response to a rise in sea surface temperature in the regions of hurricane formation, we will be experiencing storms with greatly enhanced destructive power. For major hurricanes, increases in maximum wind speeds of less than 15% could easily result in a *doubling of insured losses.*

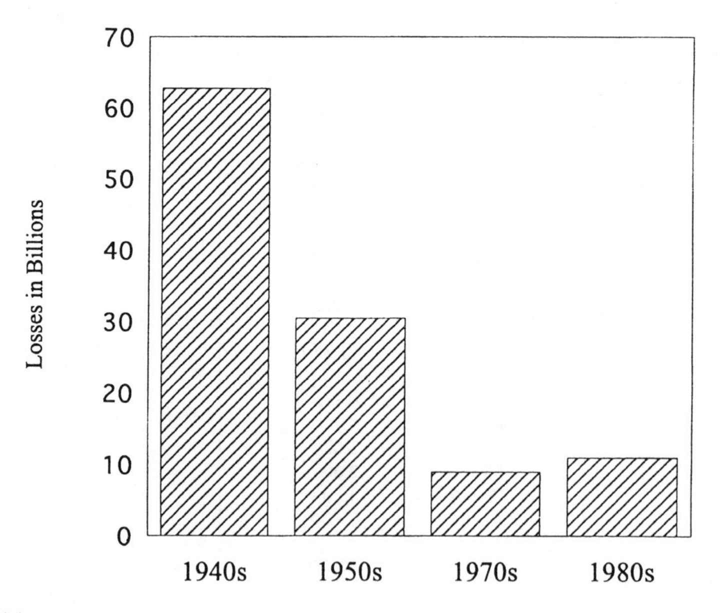

Fig. 14.2. Estimated insured property losses from hurricanes by decade

Table 14.8 Loss Potential in Future Hurricanes.

Storm	Class	Year	Estimated 1990 Insured Losses (000's)	Estimated 1990 Insured Losses (000's) if Maximum Wind Speed Increases		
				5%	10%	15%
Hugo	4	1989	3,658,887	4,902,705	6,514,172	8,542,428
Alicia	3	1983	2,435,589	3,382,775	4,312,884	5,685,853
Camille	5	1969	3,086,201	4,120,733	5,438,332	7,095,008

14.5 Conclusions

Since 1960 the United States has experienced relatively low hurricane activity, while insured coastal values have increased dramatically. Through the use of computer simulation models of hurricanes, we are able to generate a large sample of losses based on *today's* property values to estimate the current hurricane loss potential. The model can also be used to test the impact on insured property losses from changes in hurricane frequency and/or severity. This paper has demonstrated that the natural variability in hurricane frequency dramatically impacts insured property losses from hurricanes. It has also demonstrated that relatively small changes in hurricane intensity, such as have occurred naturally, in previous decades, or from possible future changes in climate can cause dramatic increases in property losses from these storms. For example, Table 14.8 shows that if the wind speeds of Hurricane Hugo, which caused estimated insured losses of $3.7 billion in 1989, had been higher by only 15% the expected loss would have more than doubled to $8.5 billion.

References

Clark, K.M., 1992. Predicting global warming's impact. *Contingencies*, May/June.

Clark, K.M., 1986. A Formal Approach to Catastrophe Risk Assessment and Management, Proceedings of Casualty Actuarial Society, Vol. LXXIII, No. 140, November.

Emanuel, K.A., 1987. The dependence of hurricane intensity on climate. *Nature*, **326**, 483-485.

Landsea, C.W., Gray, W.M., Mielke, P.W. Jr. and Berry, K.J., 1994. Seasonal forecasting of Atlantic hurricane activity. *Weather*, **49**, 273-284.

Closing Remarks

While the impacts on pre-Columbian indigenous societies from landfalling hurricanes are not documented, the effects of these great storms begins to be noted the moment that Europeans arrive in the New World. For centuries thereafter, great loss of life and property is tallied in local histories, and those of the European nations whose vessels were sunk and battered while navigating the waters of the Caribbean, the Gulf of Mexico and the North Atlantic. The chapter by Rappaport and Fernández-Partagás provides a unique compilation of hurricanes and tropical storms that have caused the greatest loss of life in this region since Columbus.

Until quite recently, tropical storms and hurricanes posed a primary threat to human safety, and property losses, while high compared to other natural hazards, was perhaps not as visible an aftermath as is presently the case. With the advent of greatly improved monitoring and prediction systems, it is generally possible to provide enough lead warning time to enable the coastal populations in the path of a hurricane to take the necessary precautions to avoid death and injury. On the other hand, the explosive growth of coastal population and economic development along our shoreline, means that, today, the degree of exposure of most coastal communities to devastation from the effects of a major hurricane is high.

The recognition of such enhanced levels of property-loss exposure to hurricanes in the region is one reason why new public-private partnerships are being developed to address the issue of hurricane-related risks. For example, the Risk Prediction Initiative (RPI), a recently established program that is centered at the Bermuda Biological Station for Research, is supported by the insurance industry, the academic community, and various governmental agencies to help improve the science of hurricane prediction, but, most importantly, to help find ways to implement prudent mitigation strategies aimed at reducing property losses from hurricanes. The chapter by Jamieson and Drury also highlights increased hurricane mitigation efforts at the U.S. Federal Emergency Management Agency (FEMA).

We have tried, in this book, to focus attention to changes in hurricane frequency operating on timescales of a few decades, which result in very different storm tracks and landfalling patterns occurring from one epoch to another. In particular, the period of the 1930s to 1950s saw a high level of hurricane activity in the western Atlantic, with many hurricanes striking the United States. Subsequent decades experienced a slowdown from this level of tropical storm activity, although major storms continued to severely impact the U.S., even during this "low-activity" period (Hurricane Andrew devastated South Florida during one of the least active years of the past few decades).

In recent years, there have been indications that the recent period of relatively low Atlantic tropical cyclone activity may be ending. The 1995 and 1996 hurricane seasons were quite active, with several storms affecting the United States, Mexico, Central America, and the islands of the Caribbean. It is not yet possible to determine whether this increase in tropical storm activity represents a change towards more frequent storm activity, as was prevalent during the earlier decades of this century. Regardless of such changes, the fact remains that the degree of exposure to devastating economic loss from hurricanes is exceedingly high in many areas of the hemisphere. It has been said countless times that Hurricane Andrew was a "wake-up" call to many state and local governments, and particularly to the property insurance industry. This issue is covered in detail in the two chapters on hurricanes and insurance by R. Roth and by K. Clark.

There is an increasing chorus of voices admonishing us "to get serious" about mitigation of hurricane impacts and the root causes of disaster. The chapters by Rodríguez, Pielke and Pielke and by Pulwarty and Riebsame clearly articulate this point. Among the recommendations proposed by these and other authors the following are worth repeating:

Most of the largest losses during the past 50 years in the Caribbean have resulted from freshwater-induced floods, mud slides and landslides. In the U.S. loss of life due to storm surges has steadily decreased primarily due to improvements in the hurricane warning system.

The historical record can provide useful lessons about the impacts of particularly severe events. During the month of October 1870 three hurricanes claimed the lives of over 24,000 people. Four storms in this century have each resulted in fatalities exceeding 8,000 people.

Recent large losses from landfalling hurricanes in the U.S. are due primarily to exponential growth in coastal populations and the valuation of coastal property; furthermore, the relationship between hurricane intensity and damage is nonlinear. Therefore, damages in a particular location can be doubled by relatively small changes in intensity.

Vulnerability to hurricanes may be defined in terms of risk, exposure and capacity to recover; changes in vulnerability can occur without changes in hurricane frequency.

"Voluntary choice" does not provide a sole explanatory model for why people are located in risk-prone areas. There is a need to document the long-term social, economic and political trends that increase vulnerability.

There is a strong need to increase not only vertical but horizontal communication, education, and coordination across sectors, agencies and local groups.

Understanding decadal trends in hurricane activity may be critically dependent on understanding the broader issues of decadal variations of the major ocean circulation. Modeling studies indicate that the upper bounds on hurricane intensity will increase under global warming scenarios but little is known about how the average intensity would change. However, large-scale atmospheric circulation changes resulting in weakening of the Hadley circulation and warming of the upper troposphere (as discussed by Bengtsson, Botzet and Esch) may result in a significant reduction in the number of hurricanes.

Ecosystem dynamics, as discussed by Doyle and Girrod, including structural composition and distribution, are linked to the occurrence of low frequency hurricane events. Natural history may be inextricably tied to decadal-scale variations in hurricane impact.

Crises can provide an opportunity to introduce innovations in mitigation. The success of such programs is inextricably tied to the people, place, values, institutions and environments concerned. Insurability must be coupled to mitigation strategies particularly where private insurance goals and social objectives may be similar.

Recent knowledge about hurricane variability (an area that is comprehensively covered by Gray, Sheaffer and Landsea) indicates that changes in landfall frequency are to be expected, and that the hurricane risk on the U.S. Atlantic seaboard is increasing. Further studies of decadal-scale variability of the conditions favoring hurricane activity can be expected to reduce uncertainty about the physical risk. However, at this time forecasting decadal-scale variability in hurricane climatology is in its infancy (at best). Uncertainty is unavoidable. Thus, the cycle of preparedness, response and recovery must be framed within strategies for longer-term mitigation of the hurricane hazard. FEMA is to be applauded for placing mitigation as a central concern of its disaster policy. A strong need exists for incorporating mechanisms for evaluating programs throughout their planning and implementation stages and in the larger contexts of disaster management. Scientific information and knowledge for understanding and reducing vulnerability exists; however, the mobilization of political action to undertake needed responses remain a central concern.

The Editors

Subject Index

Druck: STRAUSS OFFSETDRUCK, MÖRLENBACH
Verarbeitung: SCHÄFFER, GRÜNSTADT